Advanced Macroergonomics and Sociotechnical Approaches for Optimal Organizational Performance

Arturo Realyvásquez
Instituto Tecnológico de Tijuana, Mexico

Aide Aracely Maldonado-Macías
Autonomous University of Ciudad Juarez, Mexico

Karina Cecilia Arredondo
Universidad Autónoma de Baja California, Mexico

A volume in the Advances in Logistics, Operations, and Management Science (ALOMS) Book Series

IGI Global
DISSEMINATOR OF KNOWLEDGE

Published in the United States of America by
IGI Global
Business Science Reference (an imprint of IGI Global)
701 E. Chocolate Avenue
Hershey PA, USA 17033
Tel: 717-533-8845
Fax: 717-533-8661
E-mail: cust@igi-global.com
Web site: http://www.igi-global.com

Library of Congress Cataloging-in-Publication Data

Names: Realyvasquez, Arturo, 1984- editor. | Maldonado-Macias, Aide Aracely, editor. | Arredondo, Karina Cecilia, 1983- editor.
Title: Advanced macroergonomics and sociotechnical approaches for optimal organizational performance / Arturo Realyvasquez, Aide Aracely Maldonado-Macias, and Karina Cecilia Arredondo, editors.
Description: Hershey, PA : Business Science Reference, [2018]
Identifiers: LCCN 2018020667| ISBN 9781522571926 (hardcover) | ISBN 9781522571933 (ebook)
Subjects: LCSH: Work environment. | Human engineering.
Classification: LCC HD7261 .A278 2018 | DDC 658.3/8--dc23 LC record available at https://lccn.loc.gov/2018020667

This book is published in the IGI Global book series Advances in Logistics, Operations, and Management Science (ALOMS) (ISSN: 2327-350X; eISSN: 2327-3518)

British Cataloguing in Publication Data
A Cataloguing in Publication record for this book is available from the British Library.

For electronic access to this publication, please contact: eresources@igi-global.com.

Advances in Logistics, Operations, and Management Science (ALOMS) Book Series

John Wang
Montclair State University, USA

ISSN:2327-350X
EISSN:2327-3518

Mission

Operations research and management science continue to influence business processes, administration, and management information systems, particularly in covering the application methods for decision-making processes. New case studies and applications on management science, operations management, social sciences, and other behavioral sciences have been incorporated into business and organizations real-world objectives.

The **Advances in Logistics, Operations, and Management Science** (ALOMS) Book Series provides a collection of reference publications on the current trends, applications, theories, and practices in the management science field. Providing relevant and current research, this series and its individual publications would be useful for academics, researchers, scholars, and practitioners interested in improving decision making models and business functions.

Coverage

- Political Science
- Information Management
- Organizational Behavior
- Marketing engineering
- Operations management
- Finance
- Services management
- Production management
- Computing and information technologies
- Decision analysis and decision support

IGI Global is currently accepting manuscripts for publication within this series. To submit a proposal for a volume in this series, please contact our Acquisition Editors at Acquisitions@igi-global.com or visit: http://www.igi-global.com/publish/.

The Advances in Logistics, Operations, and Management Science (ALOMS) Book Series (ISSN 2327-350X) is published by IGI Global, 701 E. Chocolate Avenue, Hershey, PA 17033-1240, USA, www.igi-global.com. This series is composed of titles available for purchase individually; each title is edited to be contextually exclusive from any other title within the series. For pricing and ordering information please visit http://www.igi-global.com/book-series/advances-logistics-operations-management-science/37170. Postmaster: Send all address changes to above address.

Titles in this Series

For a list of additional titles in this series, please visit: www.igi-global.com/book-series

Technological Developments in Industry 4.0 for Business Applications
Luis Ferreira (Polytechnic Institute of Cávado and Ave, Portugal) Nuno Lopes (Polytechnic Institute of Cávado and Ave, Portugal) Joaquim Silva (Polytechnic Institute of Cávado and Ave, Portugal) Goran D. Putnik (University of Minho, Portugal) Maria Manuela Cruz-Cunha (Polytechnic Institute of Cávado and Ave, Portugal) and Paulo Silva Ávila (Polytechnic of Porto, Portugal)
Business Science Reference • copyright 2019 • 451pp • H/C (ISBN: 9781522549369) • US $225.00 (our price)

Advanced Methodologies and Technologies in Business Operations and Management
Mehdi Khosrow-Pour, D.B.A. (Information Resources Management Association, USA)
Business Science Reference • copyright 2019 • 1482pp • H/C (ISBN: 9781522573623) • US $495.00 (our price)

Emerging Applications in Supply Chains for Sustainable Business Development
M. Vijaya Kumar (National Institute of Technology Warangal, India) Goran D. Putnik (University of Minho, Portugal) K. Jayakrishna (Vellore Institute of Technology University, India) V. Madhusudanan Pillai (National Institute of Technology Calicut, India) and Leonilde Varela (University of Minho, Portugal)
Business Science Reference • copyright 2019 • 345pp • H/C (ISBN: 9781522554240) • US $225.00 (our price)

Crowdsourcing and Knowledge Management in Contemporary Business Environments
Regina Lenart-Gansiniec (Jagiellonian University, Poland)
Business Science Reference • copyright 2019 • 304pp • H/C (ISBN: 9781522542001) • US $195.00 (our price)

Burstiness Management for Smart, Sustainable and Inclusive Growth Emerging Research and Opportunities
Andreas Ahrens (Hochschule Wismar, Germany) Ojaras Purvinis (Kaunas University of Technology, Lithuania) Jeļena Zaščerinska (Centre for Education and Innovation Research, Latvia) Diana Micevičienė (Kaunas University of Technology, Lithuania) and Arūnas Tautkus (Kaunas University of Technology, Lithuania)
Business Science Reference • copyright 2019 • 194pp • H/C (ISBN: 9781522554424) • US $185.00 (our price)

Operations Research for Military Organizations
Hakan Tozan (Istanbul Medipol University, Turkey) and Mumtaz Karatas (National Defense University, Turkey)
Information Science Reference • copyright 2019 • 445pp • H/C (ISBN: 9781522555131) • US $195.00 (our price)

Cases on Quality Initiatives for Organizational Longevity
Archana Krishnan (PSG Institute of Management Coimbatore, India)
Business Science Reference • copyright 2018 • 449pp • H/C (ISBN: 9781522552888) • US $195.00 (our price)

IGI Global
DISSEMINATOR OF KNOWLEDGE

701 East Chocolate Avenue, Hershey, PA 17033, USA
Tel: 717-533-8845 x100 • Fax: 717-533-8661
E-Mail: cust@igi-global.com • www.igi-global.com

Table of Contents

Section 3
Macroergonomic Applications

Detailed Table of Contents

workstation. To support the study, information considering the estimated metabolic expenditure in workers was applied to a novel procedure to reduce the metabolic demand of the task according to age and gender. Results of the study indicated that women older than 30 years exceeded the energy limits from moderate to very heavy load activities, and men older than 40 years exceeded the energy limits in heavy and very heavy workloads. The proposal of compensatory equation statistically reduced the energy loads below the recommended limits of energy. The aerobic workload is a sensitive factor for age and gender groups and can be potential risks for developing cardiovascular diseases as well as some musculoskeletal disorders.

Globally, companies are increasingly considering the importance of mental health in workers and their relationship with productivity, which has led to increased research on work stress, which showed that there is a relationship between stress related to work and health disorders, both physical and mental. This chapter addresses the analysis of two of the main consequences that a worker can develop when having work stress: burnout syndrome, measured by the Maslach burnout inventory general survey (MBI-GS) and obesity, through the body mass index (BMI). The study was carried out in 118 people who occupy middle and upper management of the manufacturing industry of Baja California, having as objective to find the relationship that exists between the two variables, using ordinal logistic regression, as well as to characterize the sample using mean difference and hypothesis testing. From this perspective, this chapter can serve as a guide to study the behavior of variables and propose organizational development strategies aimed at reducing and preventing these problems.

Farming provides food, which is the primary need of each and every person, and also provides employment to farmers. Still it is a non-profitable and neglected occupation. Maximum health injuries are observed during field work due to extreme climates and dusty atmosphere. Use of old hand tools and no availability of modernized tools are the cause of 70% of injuries. Although the farming sector plays a vital role on Indian economy, it remains undeveloped and unattractive. The Indian government is also framing many policies for the farmers (i.e., small/nominal farmers), but farming in states like Odisha remains undeveloped. So, in this chapter, a survey is conducted to find the health-related injuries of farmers of Odisha and then a framework is designed by QFD (quality function deployment) to suggest how to avoid injuries and provide occupational safety measures for farmers.

Foreword

The book *Advanced Macroergonomics and Sociotechnical Approaches for Optimal Organizational Performance*, edited by Doctors Arturo Realyvásquez-Vargas, Aide Aracely Maldonado-Macias, and Karina Cecilia Arredondo-Soto offers an important contribution to the emerging field of Macroergonomics, within the science of Ergonomics. The book contains 14 chapters provided by authors from seven countries, and it is divided into three sections named Sociotechnical Systems, Macroergonomic Assessments and Macroergonomic Applications.

The content provided in this book will help promote macroergonomic interventions in more countries and in industrial sectors. Specifically in Mexico, Macroergonomics is a field of Ergonomics that is poorly researched and applied. This topic is not included in most of the educational programs of undergraduate or graduate, reason why Macroergonomics is practically unknown to students. Moreover, few are the people that research and publish about it. All this makes Macroergonomics is unknown for the middle and top management of companies. This book can help fill that gap both in educational institutions and in companies from different sectors of work

All chapters include topics that presently represent challenge in different industrial sectors for enterprises of all sizes that try to optimize their organizational performance. The readers will find new theory, research methods and case studies that he/she could apply in similar situations. All interesting people in Ergonomics will benefit from the content of this book. For all these reasons, this book deserves to be in libraries and bookstores that deal with the topic of Ergonomics, and more specifically with Macroergonomics.

Lilia R. Prado-León
Universidad de Guadalajara, Mexico

Preface

INTRODUCTION

Macroergonomics is the sub-discipline of Ergonomics which is concerned with the analysis, design, and evaluation of work systems. Under this approach, compatibility becomes a key concept, which implies that workers, tools, methods and systems of work can be properly directly and indirectly interwoven (Realyvásquez-Vargas, Maldonado-Macías & García-Alcaraz, 2017). In this matter, macroergonomics focuses in harmonizing all organizational structure of a company, and not only one workstation or one task, as Microergonomics does. According to Hendrick and Kleiner (1999), a work system consists of two or more persons interacting with some form of (1) job design, (2) hardware and/or software, (3) internal environment, (4) external environment, and (5) organizational design (i.e., the work system's structure and processes).

Furthermore, Macroergonomics is a top-down, middle out, and bottom-up approach. In the top-down approach, the overall work system structure may be prescribed to match the organization's socio-technical characteristics. On the other hand, the middle-out approach focusses in the analysis of subsystems and work processes, which can be assessed both up and down the organizational hierarchy from intermediate levels, and also, up and down some changes may be done to ensure the work system design is harmonized. Finally, the bottom-up approach comprises an extensive participation of employees to identification of problems. Currently, Macroergonomics is considered an emergent sub-discipline; then, there is the need to promote current theories and methods, and propose new ones.

Macroergonomic principles are related to human capital factors, organization, tasks, tools, technology as well as environmental conditions in work systems. Macroergonomic elements are: human capital (physical characteristics; motivation and needs; education, skills and knowledge; psychological characteristics), organizational factors (supervision and administration styles; work schedules; organizational culture and worker safety culture; coordination, collaboration and communication; evaluation of performance, rewards and incentives; teamwork and social relations), tools and technology, tasks related to work and environmental conditions (Realyvásquez-Vargas, García-Alcaraz & Blanco-Fernández, 2016).

The general objective of this book is to increase knowledge about Macroergonomics topics and organizational strategies on different work fields (manufacturing, industries, pharmacy, bank, supermarket, educational institutions, etc.). Specific objectives include developing of new theories, applying specific macroergonomic methods, theories and studies for promoting the adoption of Macroergonomics in different work systems around the world in order to achieve enhanced organizational practices in benefit of healthier work environments for human well-being. Readers of this book will find the following aspects for their interest and benefit:

1. It increases the knowledge of researchers, ergonomists, designers, engineers, managers, supervisors, and healthcare professionals around the world about the Macroergonomics field.
2. It provides strategies applied in specific case studies, which will be considered in some other similar problems, which will benefit work systems design.
3. An additional impact of the book is that, for the information that it contains, it serves as a database of competitive strategies that allow increasing organizational performance of companies.

ORGANIZATION OF THE BOOK

This book includes as contributions complete research papers, which complied with the editorial process of IGI Global (norms and guidelines). The call of chapters was distributed among the list of the field for researchers to submit their works to this project. As a result, 29 chapter proposals were received from countries as Albania, Ecuador, Finland, India, Mexico, Poland, Spain, Turkey, United States and Ukraine. Nevertheless, abstracts were subject to a screening process to ensure their quality, authenticity and relevance to this book. Once the revision process concluded, 24 proposals were invited to submit full versions, and three reviewers were assigned to every work to ensure the peer review process. Finally, 15 chapters fulfilled the prestigious reviewers' and editors' requirements and corrections and were accepted accordingly for their publication.

Section 1: Sociotechnical Systems

This section of the book, titled "Sociotechnical Systems" comprises five chapters described as follows:

Chapter 1 named "A Metalearning Approach on Sociotechnical Systems Toward Improving Organizational Effectiveness" is proposed by authors Carlos Raul Navarro Gonzalez, Mildrend Ivett Montoya Reyes, Gabriela Jacobo Galicia, Ismael Mendoza Muñoz from Universidad Autonóma de Baja California (Mexico). This chapter presents a systemic approach in measuring organizational effectiveness, emphasizing differences with short-term and long-term measures. Differences between validating and evaluating any sociotechnical interventions is done, proposing than evaluating could help detecting strongest and weakness in socio technical methodologies and provide a guidance to the organizational improvement. This chapter proposes a tool who can join multiples point of view and help to promote a synergistic action toward technical and social systems; looking to impact organization effectiveness.

Chapter 2 titled "Evaluation of Sociotechnical Systems in Managing Corporate Social Responsibility and Stakeholders' Engagement" is proposed by Toivo Niskanen from Ministry of Social Affairs and Health (Finland). This chapter explore at the Finnish Kemira Corporation, how a corporation manages Corporate Social Responsibility (CSR) and stakeholders' engagement. Greimas' actantial model (1990) and Senge's five disciplines (1994) were applied to evaluate CSR reports with a sociotechnical systems approach in relation to the stakeholders: employees, suppliers, financiers and shareholders, and communities and authorities. It was found that issues of CSR management and stakeholder engagement should be involved in strategic corporate decision making. The business strategy of the corporation emphasizes adherence to CSR guidelines and business standards that reflect the concerns of the stakeholders. CSR strategic procedures indicate the implementation of the corporation's measures to promote and pursue CSR goals that extend beyond their legal responsibilities.

Chapter 3 titled "Socio-Technical Approaches for Optimal Organizational Performance: Air Navigation Systems as Socio-Technical Systems" is proposed by authors Tetiana Shmelova from National Aviation University (Ukraine) and, Yuliya Sikirda from Kirovograd Flight Academy of National Aviation University (Ukraine). In this chapter, the authors made an analysis of the International Civil Aviation Organization documents on risk assessment and the impact of the social environment on the aviation system. The authors obtained the results of the evaluation of non-professional factors: determination of the social-psychological impact on decision making of human-operator by identifying the preferences for organizational performance. The structural analysis of internal and external management environment of aviation enterprise was carried out. And, as follows from the analysis, inhomogeneous factors which influence the aviation activity were classified, formalized and systematically generalized using set-theoretical approach. The influence of factors of internal and external management environment on the aviation enterprise's aviation activity was determined.

Chapter 4 named "A Sociotechnical Systems Approach Applying a Novel Taxonomy to a Survey for the Assessment of Safety Performance" is proposed by Toivo Niskanen from Ministry of Social Affairs and Health (Finland). This chapter approach a novel taxonomy with respect to safety performance. The study applied a combination of qualitative and quantitative methodologies. The following hypotheses were supported: "Activities of the management" had positive impacts on five aggregated variables, namely "Near-accident investigation and instructions" (H1), "Occupational safety and health (OSH) training" (H2), "Operations, technical processes, and the safe use of chemicals" (H3), "Use of personal protective equipment" (H4), and "Measuring, follow-up, and prevention of major accidents" (H5). By undertaking a statistical evaluation and then devising a novel taxonomy, it was possible to gain detailed insights into diverse aspects of a high-risk industry's work with regard to complex sociotechnical systems.

Chapter 5 named "Considerations of the Mental Workload in Socio-Technical Systems in the Manufacturing Industry: A Literature Review" is proposed by the authors Manuel Alejandro Barajas Bustillos, Aide Aracely Maldonado-Macías, Jorge Luis García-Alcaraz, Juan Luis Hernández Arellano and Liliana Avelar Sosa; from Autonomous University of Ciudad Juárez (Mexico). This literature review aims to provide a comprehensive understanding of the use of mental workload in the manufacturing sector. In the search, a combination of keywords was used, classifying each journal according to the mental workload evaluation means, the type of evaluation, and the area of application. Articles not focusing on the manufacturing area were discarded. Regarding the methods used for mental load assessment, the analytic techniques were found to be the most frequently used.

Section 2: Macroergonomic Assessments

This section of the book, titled "Macroergonomics Assessments," comprises five chapters described as follows:

Chapter 6 entitled "Lean Production and Its Impact on Worker Health: Force and Fatigue-Based Evaluation Approaches" is proposed by the authors Murray Gibson from Auburn University (United States) and, Beata Mrugalska from Poznan Univeristy of Technology (Poland). This chapter explores how ergonomics force & fatigue evaluation methods can be applied in a manner to enhance lean initiatives. These methods incorporate ergonomic-related variables of force type, force duration, force frequency and degree of awkward posture; and incorporate the Recommended Cumulative Rest Allowance (RCRA) model as a practical fatigue-based metric.

Chapter 7 named "Exceeding the Recommended Energy Limits Due to Age and Gender in Occupational Aerobic Workloads" is proposed by the authors Cesar Omar Balderrama Armendariz, Jose de Jesus Flores Figueroa and Ludovico Soto Nogueira from Universidad Autonoma de Ciudad Juarez (Mexico) and Judith Lara Reyes from University of Texas at El Paso (United States). This chapter analyzes the physical aerobic work in terms of the metabolic expenditure, and compare it with the recommended boundaries of energy found in literature, proposing an alternative to the potential work overload through a compensatory equation introduced in the standard time of the workstation. The proposal of compensatory equation statistically reduced the energy loads below the recommended limits of energy.

Chapter 8 named "Burnout and Obesity in Middle and Upper Management in the Manufacturing Industry of Baja California" is proposed by the authors Sharon Idali Macías Velázquez, Yolanda Angelica Baez-Lopez, Aide Aracely Maldonado-Macias from Universidad Autónoma de Ciudad Juárez (Mexico), Jorge Limon-Romero and Diego Tlapa from Universidad Autónoma de Baja California (Mexico). This chapter addresses the analysis of two of the main consequences that a worker can develop when having work stress; on the one hand Burnout syndrome, measured by the Maslach Burnout Inventory General Survey (MBI-GS) and obesity, through the Body Mass Index (BMI). This chapter can serve as a guide to study the behavior of variables and propose organizational development strategies aimed at reducing and preventing these problems.

Chapter 9 titled "A Framework Designed for Macro-Ergonomical Analysis of Indian Farmers: Assessment and Analysis of Occupational Injuries of Agricultural Farmers of South Odisha in India" is proposed by the authors Debesh Mishra and Suchismita Sataphaty from KIIT University (India). This chapter analyze Indian community were engaged in farming. Maximum health injuries are observed during field work due to extreme climate, dusty atmosphere. Indian Govt is also framing many policies for the farmers (i.e. small/nominal farmers), still farming states like Odisha remains undeveloped. So a survey is conducted to find the health related injuries of farmers of Odisha and then a framework is designed by QFD (Quality Function Deployment) to suggest how to avoid injuries and provide occupational safety measures for farmers.

Chapter 10 named "The Contribution of Neuroscience and Health Psychology to Macroergonomics: Focusing on Workers as Active Agents" is proposed by the authors Miguel Angel Serrano-Rosa and Francisco Molins from Universidad de Valencia (Spain). The objective of this chapter is to offer a broad and multidisciplinar perspective of prevention risk on the workplace centered in the worker, taking the view of Neuroscience and Health Psychology. For that purpose, stress interventions in the organizations and in the individuals will be outline. This chapter would finish proposing a preventive intervention to improve healthier work environments, taking into account Neurosciences and Health Psychology. This proposal will be based on improving life-style, education for health and self-respect.

Section 3: Macroergonomic Applications

This section of the book, titled "Macroergonomics Applications," comprises five chapters described as follows:

Chapter 11 named "Organizational Development in Improving Operations of a Language Center: Impact on Development of Students" is proposed by the authors Luz Elena Tarango, Manuel Alonso Rodriguez-Morachis, Yolanda Frausto, Eduardo de Jesus Rojas and Marisela Lucero Gaytán from the Instituto Tecnologico de Ciudad Juarez (Mexico). This chapters approach many actions implemented to improve the English communication skills of the students of Instituto Tecnológico de Ciudad Juarez

(ITCJ) in the period of 2013 through 2016. These actions seek to firstly comply with the expectations defined by employers, then to satisfy the English proficiency test for the students to be able to graduate from their career and lastly to meet the requirements established by Tecnologico Nacional de Mexico (TecNM) contained in the 2013-2018 Institutional Innovation and Development Program (IIDP). All these actions were accomplished applying a systemic approach and organizational development.

Chapter 12 titled "Distribution of Food in a Specialized Hospital Using Ambient Intelligence to Improve a Model of Macroergonomics" is proposed by the authors Alberto Ochoa, Juan Luis Hernandez Arellano, Gilberto Rivera and Daniel Azpetia from Universidad Autonoma de Ciudad Juarez (Mexico) and Luis Fernando Maldonado from Universidad Autonoma de Queretaro (Mexico). This chapter propose the Intelligent Food Distribution System as a tool for the distribution of food. It can personalize the way to distribute depending on the medical characteristics of each patient providing foods that contain higher nutrients in the diet set by hospital. A model of decision trees, where based on data from the organization of the United Nations Food and Agriculture is used for decision making. It is simulated three basic food in the diet of Latin American countries, rice, potatoes and lentils from the parameters of fat, energy and protein that contains every type of food, then these three parameters under which food alone would be optimal for receiving the highest amount of nutrients and fats that the body requires.

Chapter 13 named "Trends in Macroergonomics Applications for Improve Work Systems" is proposed by the authors Karina Cecilia Arredondo-Soto from Universidad Autónoma de Baja California, and Arturo Realyvásquez and Guadalupe Hernández-Escobedo from the Instituto Tecnologico de Tijuana (Mexico). This chapter presents the historical contexts of Macroergonomics applications from 1994 to 2013, and the challenges that they posed for Macroergonomics. Next, it examines those works that took place from 2014 to 2018 and attempted to address the raised challenges and the emerged trends in the field. A search in the Scopus database of the word Macroergonomics in the article title, abstract or keywords, and filtering the period from 1994 to 2013 was done, 150 documents resulted from the search and were classified and summarized in 70 conference papers, 64 articles, four book chapters and two editorials. After, the analysis was conveniently limited to only articles and book chapters so that 68 documents were used in this phase. The chapter concludes with a set of predictions and challenges for Macroergonomics in the near future.

Chapter 14 titled "Implementation of an Intelligent Model Based on Machine Learning in the Application of Macro-Ergonomic Methods in a Human Resources Process Based on ISO 12207" is proposed by the authors Edgar Cossio Franco from Universidad Enrique Diaz de Leon (Mexico), Jorge Alberto Delgado Cazarez from Universidad de Guadalajara (Mexico) and Carlos Alberto Ochoa Zezzati from Universidad Autonoma de Ciudad Juarez (Mexico). The objective of this chapter is to implement an intelligent model based on machine learning in the application of macro-ergonomic methods in human resources processes based on the ISO 12207 standard. To achieve the objective, a method of constructing a Java language algorithm is applied to select the best prospect for a given position. Machine learning is done through decision trees and algorithm j48. Among the findings, it is shown that the model is useful in identifying the best profiles for a given position, optimizing the time in the selection process and human resources as well as the reduction of work stress.

Finally, this book aims to highlight the future trends in its consideration to evaluate the organizations from a macroergonomic and systemic approach. This with the aim to determine those factors that have a positive effect on strengthening the performance of enterprises and institutions (Arredondo-Soto, Carrillo-Gutiérrez, Solís-Quinteros & Hernández-Escobedo, 2018).

TARGET AUDIENCE

The target audience of this book is composed of professionals, researchers and students working in the field of Ergonomics that have participated in Macroergonomics projects (proposals of new theories or methods, application of macroergonomic methods). More specifically, the target audience include ergonomists, manufacturing engineers, industrial engineers, industrial designers, researchers, industry practitioners, research scientists and academics involved in work systems design.

As conclusion, this book impacts the emergent field of Macroergonomics since it brings knowledge by means of new theories and methodologies developed by different researchers in different fields of knowledge. Authors have shared their cases of studies and experiences presenting that their theories and methodologies were successful. This fact once again supports the theory that macro-ergonomics provides, in the first instance, well-being and efficiency in the work performance of employees, and secondly, in performance and organizational competitiveness.

Specifically, in Mexico, where most of the authors are from, Ergonomics has begun to take importance both in academic programs and in the labor sector. In this country, new official standards on ergonomics are being developed. Companies, especially manufacturing ones, have begun to implement micro-ergonomics projects to avoid future penalties, so this book provides knowledge to companies develop projects at macroergonomic level. These projects can (and should) be carried out not only in companies of the manufacturing sector, but also in any sector, since Macroergonomics must exist wherever human beings work.

REFERENCES

Arredondo-Soto, K. C., Carrillo-Gutiérrez, T., Solís-Quinteros, M., & Hernández-Escobedo, G. (2018). A Theoretical Framework About the Impact of Human Factors on Manufacturing Process Performance. In J. L. García-Alcaraz, G. Alor-Hernández, & A. A. Maldonado-Macías (Eds.), *New Perspectives on Applied Industrial Tools and Techniques* (pp. 327–352). Cham: Springer. doi:10.1007/978-3-319-56871-3_16

Greimas, A. J. (1990). *Narrative semiotics and cognitive discourses*. Burns & Oates.

Hendrick, H. W., & Kleiner, B. M. (1999). *Macroergonomics: An introduction to work system design*. Santa Monica, CA: Human Factors and Ergonomics Society.

Realyvásquez-Vargas, A., García-Alcaraz, J. L., & Blanco-Fernández, J. (2016). Desarrollo y validación de un cuestionario de compatibilidad macroergonómica. *Contaduría y Administración*, *61*(3), 478–498. doi:10.1016/j.cya.2016.04.002

Realyvásquez-Vargas, A., Maldonado-Macias, A. A., & García-Alcaraz, J. L. (2017). *Macroergonomics for Manufacturing Systems: An Evaluation Approach*. Springer.

Senge, P. (1994). *The Fifth Discipline: The art and practice of the learning organization*. Random House.

Acknowledgment

Editors want to sincerely express their acknowledgement to all the researchers, professors and students who made this book was possible. This book includes 14 chapters about theory, methods and case studies of Macroergonomics and Socio-technical systems. The authors of these chapters are from seven different countries (Finland, India, Mexico, Poland, Spain, Ukraine, United States of America) and this leads to a total of 15 institutions of affiliation.

Institutions to which the editors are affiliated special recognition deserve: Instituto Tecnológico de Tijuana, Universidad Autónoma de Ciudad Juárez and Universidad Autónoma de Baja California. In addition, editors want to highlight the support provided by the National Council of Science and Technology (CONACYT), and PRODEP, organizations that promote research and technological development in Mexico.

Editors also want to acknowledge to the editorial IGI Global Publishing for the opportunity to edit and publish this book; and to the editors' assistants Ms. Marianne Caesar and Josephine Dadeboe, who always guided the editor to complete the book.

Finally, editors would like to recognize the effort of all those people who contribute to the development of this book with their valuable reviews of chapters (being or not being chapters' authors) and to the researcher who contributed with the foreword.

Section 1
Sociotechnical Systems

Chapter 1
A Metalearning Approach on Sociotechnical Systems Toward Improving Organizational Effectiveness

Carlos Raul Navarro Gonzalez
Universidad Autónoma de Baja California, Mexico

Mildrend Ivett Montoya Reyes
Universidad Autónoma de Baja California, Mexico

Gabriela Jacobo Galicia
Universidad Autónoma de Baja California, Mexico

Ismael Mendoza Muñoz
Universidad Autónoma de Baja California, Mexico

ABSTRACT

Sociotechnical systems optimize social and technical systems, but joint optimization should involve autonomy, adaptability, meaningfulness, and feedback as underlying principles. A metalearning approach in the organizational development could affect the process of managing the change inside the organization where innovation, learning, and change produce resistance amount members. A systemic approach in measuring organizational effectiveness is presented emphasizing differences with short-term and long-term measures. Differences between validating and evaluating any sociotechnical interventions is done, proposing that evaluating could help detecting strengths and weaknesses in socio-technical methodologies and provide a guidance to the organizational improvement. This chapter proposes a tool that can join multiple points of view and help to promote a synergistic action toward technical and social systems looking to impact organization effectiveness.

DOI: 10.4018/978-1-5225-7192-6.ch001

INTRODUCTION

Sociotechnical systems are not always designed under principles and guidelines from join optimization, but also organizational interventions should be affect and measure in the organizational effectiveness also promote an inquiry-learning-improving spiral cycle inside company a failure doing this leads to isolated interventions rather than helping the organizational to survive in a complex and changing environment. This chapter discusses join optimization principles and guidelines, how they should be combined with the double cycle learning from organizational learning theory. How organizational effectiveness could be measure in order to support any methodological intervention in the company; but also providing important elements to rather than validating a sociotechnical methodology usage, feedback them to detect specific aspect from organization and the methodology that could impels success or failure from organizational intervention.

Sociotechnical Systems Join Optimization

Sociotechnical systems have two components: Social systems and technical systems; Social systems are related to relations between individuals and with group, include management style and organizational culture. Technical systems include machines and equipment but also work organization (Carayon, et al, 2015). The traditional goal from Sociotechnical systems is use to be focus in the employees looking to make a quality and satisfactory work environment (Mumford, 2006); but according with the systemic concept of wholeness, who states the holistic principle of looking at the whole instead of the parts, not optimizing the separate pieces at risk of sub-optimization the performance of the whole (Pourdehnad, et al, 2011; Dent & Umpleby, 1998).

Thus sociotechnical systems outcomes are both physical and social/psychological whole optimization is not always clear; boundaries optimization between social and technical systems usually leads to isolate sub systems and sub-optimizations. Many authors (Carayon, 2015, Di Maio, 2014, Slomp and Ruël, 2000) states that goal should evolve to optimize joint between both components: Social and technical systems, where synergy and collaboration with booth systems in created. Falling in joint optimization could lead to projects with no or little impact in the organization improvement or sociotechnical initiatives that are not supported by organizational staff rejecting the social component of the whole system. Cherns (1976, 1987) presents nine principles of Socio-Technical Design, that many authors use as starting point for analysis (Hughes, et al, 2017; Di Maio, 2014; Davis, et al, 2014; Ghaffarian, 2011; Slomp & Ruël, 2000), this principle are defined in the Table 1. Thus Di Maio (2014) summarizes those in four underlying principles to achieve joint optimization:

- **Responsible Autonomy:** Individuals/stakeholders take responsibility for the outcome/performance and can make decisions autonomously.
- **Adaptability:** Overall work procedures, tasks and schedules can be adjusted by individual team members to increase optimization.
- **Meaningfulness of Tasks:** Task identity for whole tasks, including skill variety.
- **Feedback Loops:** Recursive interactions and feedback.

Slomp and Ruël (2000) states that a key term in joint optimization of Socio-Technical Systems is self-organized. Self-organization should be development in the organization allowing the production structure and control structure to adapt throws the objectives of organization and its workers.

ORGANIZATIONAL LEARNING SIDE IN SOCIOTECHNICAL SYSTEMS

Organizations have just developed conscience of the greater importance and implications –for their survival, subsistence and success– from to share knowledge and to develop a learning culture (Sepahvand, et al; 2016; Marsanasco & Garcia, 2013; Yeung, 2000; Dixon, 2001). But, how will force and impel the learning within the organization? In this aspect those authors shows a typology where –among others– indicates that the organizations learn through the continuous improvements and emphasizing the work groups participation solving problems; likewise Dixon (2001) declares: "if people begin to share ideas about the subjects that they perceive really important, this creates a learning culture by itself" (Dixon, 2001, p.16).

Many authors (Batras, et al, 2014; Richardson, 2014; Serrat, 2010; Easterby, 2004) examine the double loop learning model that was originally presented by Argyris (2001), where shows that the problem solving is important (single circuit learning); but much more crucial is the double circuit learning, it consists of the critical reflection from behavior and process identifying how people unintentionally

Table 1. Chern's principles of socio-technical systems

Compatibility	The design process should be compatible with its objectives. If the design is intended to foster democracy in work situations, the design processes themselves must be democratic.
Minimal Critical Specification	No more should be specified than is absolutely essential. However, the designers should ascertain what is essential.
The Socio-Technical Criterion	Variances, as deviations from expected standards, must be kept as close to their point of origin as possible. In other words, solution to problems should be devised by the groups that directly experience them, not by supervisory groups.
Multi-Functionality Principle	In order for groups to respond to the changing work environment, they need a variety of skills. These include skills that go beyond what day-to-day production activities require.
Boundary location	Boundaries exist where work activities pass from one group to another and where a new set of skills is required. However, boundaries should facilitate knowledge sharing. All groups should be able to learn from one another despite the existence of the boundaries.
Information	Information must reside where it is principally needed for action. A socio-technical design gives the control authority to the groups whose efficiency is being monitored.
Support Congruence	A social support system must be in place to enjoin the desired social behaviors.
Design and Human Values	High quality work involves: a. jobs to be reasonably demanding; b. opportunity to learn; c. an area of decision-making; d. social support; e. the opportunity to relate work to social life; and f. a job that leads to a desirable future.
Incompletion	Practitioners must recognize the fact that the design is an iterative process. It never stops. The new changes in environment require continual revisions of objectives and structures.

Source: (Ghaffarian, 2011)

contribute with the problems of the organization and then changing their actions accordingly. This model is show in Figure 1.

If double cycle learning is not stated in the organization, defensive routines could be promoted that would inhibit the company's ability to learn. The use of defensive routines inhibits the learning of double circuit, reason why those are key concept who should be continuously minimized inside the organization. These consist of any actions that are taken to avoid a disruption to the participants such as avoiding criticism and blaming anyone; actions who prevent reaching the causes of the trouble (Argyris, 2001; Senge, 1998). Detect and correct an error without questioning or modifying the structure and implicit variables of the system just achieve a single learning cycle (e.g. doing internal rework during production shift). Single cycle learning takes place when adjustments are made; imbalances are corrected by changing the actions. But double cycle learning occurs when beside correcting imbalances a depth examination from regulatory variables from the system was made leading to actions in the system structure (Argyris, 2001) (e.g. questioning the causes of rework and seeking to eliminate them).

Dual-cycle learning is relevant to complex non-programmable aspects and is very important to ensure the long-term survival of the organization; because in this type of learning objectives, values and beliefs that underlie system behavior are re-evaluated; the structural contradictions that lead to the single cycle learning are diminished (Young, 2000). Sterman (1994) affirms that an organization only can subsist and be successful if is able to understand the double circuit learning like a cycle around it must move faster; promoting that the company will be capable to change and affect their structure (ways to see and understand the reality), to create different rules and change the strategy, which results in different policies, guidelines and better decisions. Evolving toward a collective process of solutions search, promoting a institutional dialogue based in trust and reflection (Calton & Kurland, 1995). Were trust is arises by recognizing the interdependence and the understanding that everyone's fortune depends on the combination of individual efforts towards a shared mission, where everyone understands and knows how and why they must cooperate. This could be achieved if the rules and regulations that define the "what to do" are replaced by principles and guidelines that express not only the "what to do" but "the reasons behind the rules" allowing greater flexibility and an intelligent response to the changing circumstances from organization (Heckscher, 1994).

Figure 1. Representation from double learning cycle from Argyris
Source: (Argyris, 2001)

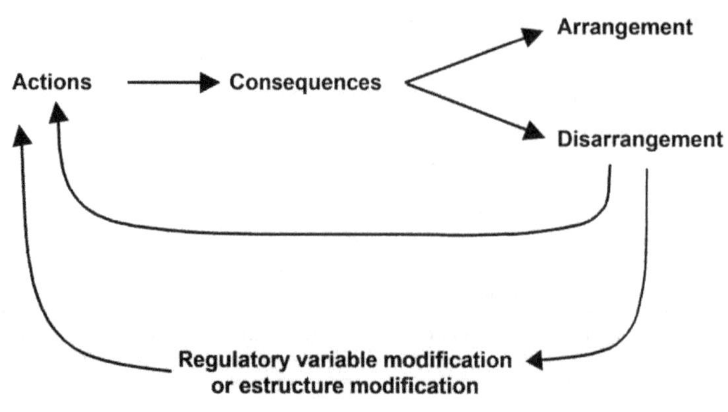

As can be seen from last section immerse in all joint optimization principles proposed by Chern (1976 and 1987) and shows in Table 1, organizational learning is intrinsic; also a important part in the 4 underlying principles presented by Di Maio (2014) to achieve joint optimization in Sociotechnical Systems. Being aware from double cycle learning proposed by Argyris (2001) is essential in adjusting toward the social component from organization, improving the company by the joint optimization in their social and technological systems.

ORGANIZATIONAL EFFECTIVENESS MEASUREMENT ON SOCIOTECHNICAL INITIATIVES

There are many, but traditional model in measuring organizational effectiveness is "Goals model" (Etzioni, 1960, 1987), which assumes that all organization is rationally structured toward achieving goals, thus effectiveness will be measured by the fulfillment from organizational outputs (objectives and goals)- The goals model has the disadvantage that only it could be applied in an environment where goals are clear and consensual; where no uncertainty affects output dates and desired results; It should not be applied in background where goals are not clearly defined, are complex, changing and contradictory (Kweku & Mishra, 2018; Eydi, 2015; Martinez, et al, 2013; Murdaugh, 1998; Cameron, 1980).

Cameron (1986 and 1980) proposes the "ineffectively model" from the organizational effectiveness, who considers than organization his immersed in a confusing context with high complexity, with intricate goals, with changing and contradictory objectives; where the means to achieve results are not clear and the ability from issue solving becomes critical; all this is enclosed in that he calls an "organized anarchy". The ineffectively model conceives the organization as a set of problems and failures, where the factors that inhibit the successful performance from organization should be assessed linking organizational performance with de members ability from making high quality connections who were significant (Ashraf & Abd, 2012; Cheng, 2012; Henri, 2004, Davel & Tremblay, 2003).

The meaningful and valuable connections among employees could be arisen by the elements of the 'organizational culture' and tracked as behaviors, procedures, values, assumptions or beliefs; looking to increase the people ability in coopering within and outside their labor areas. Toward an effective coordination between different organizational areas; strengthen belonging sense from employees to their organization and employees appropriation from organizational purpose and goals. Promoting organizational learning with a genuine dialogue and deliberation around problematic issues; improving company's ability to be flexible and adaptation to the environment. Also some authors support the negative relationship between ineffectiveness with 'organizational culture' (Bhattacharjya & Venable, 2006; Baer and Frese, 2003) or some of the aspects that compose them (Guerra, et al, 2004; Davel & Tremblay, 2003; Stetzer, et al, 1997).

Other models to understand and measure organizational effectiveness exist (Henri, 2004; Murdaught, 1998; Cameron, 1986, 1980), but the "goal model" and "ineffectively model" could exemplify the conflict in investigation related organizational effectiveness, that one model could judge effective but ineffective with according to different models (Cameron, 1980). A complementary posture is recommended, as no one model is appropriate in all circumstances and organizations, they could serve as guidance and complements to evaluate the effectiveness from an organization(Henri 2004), thus introducing nonfinancial indicators in organizational effectiveness measuring is crucial toward addressing strategic and long term plans (Lev & Radnakrishnan, 2005; Amir, et al, 2003; Ittner & Larcker, 2001) even to anticipate financial long-term performance; by this aspect "goal model" is related with short-term performance that could be complemented with long-term survival from "ineffectively model" (Banker, et all, 2005; Henri, 2004).

Falling in validating organizational impact from sociotechnical initiatives leads to isolate interventions not resulting in a spiral cycle from inquiry-learning-improving with reduced and limited benefits. But opposite is true, a good validation and evaluation in organizational impact from sociotechnical initiatives promotes and impels an inquiry-learning-improving spiral cycle who can magnify and expand benefits inside the company. So, it is important evaluate and feedback any soft methodological interventions with the identification and the recognition from own aspects from the organization and team members (Blackman, 2006; Sorensen & Valqui, 2005a, 2005b; Connell, 2001; Barnden & Darke, 2000; Tolvanen, 1998).

Kushner (2002) indicates than an effective measurement from the organizational impact not just shows that something is good for you, but it should include solving the question "How well do I do it?" and providing a guideline about the process. This is the difference between validate an (am I good for you?) and to evaluate them (How well do I do it?). Validating a methodology intervention implies verifying that problem decrease, while evaluation seeks to provide detailed information from methodology intervention to emphasize weaknesses and strengths in the interactive elements that could lead to success or failure from organizational interventions (Checkland, 2000). Thus, over validation a methodological intervention should be evaluated giving a guide for the organizational change (Kushner, 2002) promoting the spiral cycle of inquiry-learning-improving.

METHODOLOGY INTERVENTION

According with Denison (2005), four essential traits for the organizational effectiveness are: Involvement, Consistency, Adaptability and Mission. These four organizational traits were extending throughout 12 dimensions (see Table 2). Denison model for measuring organizational culture has a 60 questions survey where their instrument has an acceptable level from internal consistency (Denison, 1995, 2005).

Table 2. Cultural traits and dimensions from Denison model

Cultural Traits	Cultural Dimensions
Involvement	Empowerment Team Orientation Capability Development
Consistency	Core Values Agreement Coordination and Integration
Adaptability	Creating Change Customer Focus Organizational Learning
Mission	Strategic Direction and Intent Goals and Objectives Vision

An application case from a proposed methodology who was based on searching high leverage points through sharing visions from root problems in production area and grouping them through an affinity diagram; but applying concepts from systems theory, systems dynamics and soft-or. This application case was made in a company from Mexicali, Baja California, Mexico; where this case was made in an electrical industry, where the production area was diagnosed and improved. Denison model for measuring organizational culture was used before any organizational intervention and re-evaluated at the end of the soft intervention.

This implementation process was made during a five-month period; with a four members group team; taking care that at least one of the members was well familiarized with the company and the studied area (thus promote the understanding and clarification of the resulting proposals). Towards the work group followed the underlying rules from "organizational learning", "soft systems", and "system dynamics" then several sessions of training was required, where his "mental models" and "internal routines" was looked for to modify and to sensitize them. This methodology promotes the cycle from inquiry-learning-improving that should be in the organization; where the proposal to actively search of "high leverage points" will be developed in three stages; which are described:

- **Diagnostic Stage:** The first stage consists of using the "modified affinity diagram", for that reason the actual problems were pickup from the diagnostic area then each team collects many complaints, problems or suggestions oriented towards the specific area or department from the authors interest; some were collected within the area, but outside too; this was done asking for the main complaint or problems to 12 employees. Each one of the complaints or problems was rewritten as a simple to understand "phrase problem" that does not shows unclearness with the manifest situations or activities. In case that any pickup problem was expressed like suggestion it was rewritten as a "phrase problem" trying to discover and expose the implicit problem back the pronounced suggestion. Thus the "enriched root problem" was obtained within this stage. Thus if "root problem" is perceived that points for a guilty instead of exposing and showing procedures than are weakness or inadequacy a reframe is required because a system dynamics principle looking to change organizational structure.
- **Analysis Stage:** Starting from the "enriched root problem" obtained of the previous stage, it was reviewed under guidelines from "systemic implications" from system dynamics, from "systemic archetypes" and from "organizational meta-decisions"; that results in the detection of "high leverage points" than should be expressed in a "generic proposal. The conceptualization of "organizational meta-decisions" push us to look for the "metasystem" oriented to the problem resolutions than promotes the "inquiry and learning cycle" and push for better and efficient daily decisions.
- **Stage of Elaboration, Justification, and Understanding of Proposals:** From the "generic proposal", three "specific proposals" from acting has being defined; they should be concise and oriented to attack and to eliminate the "enriched root problem"; later these "specific proposals" are supported, explained and clarified through a "activities system" (such as it was proposed by Checkland (1990, 2001) within the "soft systems methodology"), which come out from writing and connecting the different activities that are in each "specific proposal"; thus the obtained diagram forces to us to clarify the writing of the "specific proposals".

RESULTS

Specific results from soft intervention after collecting expressed problems in the production area and interrelate them a root problem was emerged. The group team looks for patterns and behavior cycles that could be causing problems in the organization. After a general proposal was made, and considering the actual procedures and processes from current way to work, looking for existing policies and guidelines in the organization. Where team defined three "specific proposals" for actions; that were oriented to attack and eliminates the "root problem"; each action proposal was detailed using activities systems. All this was presented to management staff, and an implementation and monitoring stage was started, which lasted two months. Finally, a subsequently measure and evaluation from the organizational impact from intervention was made. A production index was defining from beginning and was monitoring looking to evaluate short-term impact from organizational intervention; that shows improvement (Figure 2).

To evaluate long-term effectively, a survey from "Denison model" of Organizational culture was applied to team members before any organizational intervention and re-evaluated at the end of the soft intervention that last five months (Figure 3 and 4).

DISCUSSION AND FINDINGS

With this organizational intervention starting with Figure 2, It is difficult to discover anything else that "production indicator evolved properly" thus conclude that problem was reduced or solved. Unfortunately, this conclusion was only valid in short-term period; there could not be any guaranty that in long term this problem could not emerge again. So, validity from any sociotechnical intervention using "goal model" is not enough that also could lead to a dangerous misleading about socio technical systems benefits in the organization. Also, not to promote the spiral cycle of inquiry-learning-improving that organization needs to survive.

Figure 2. Reject ratio indicator monitoring

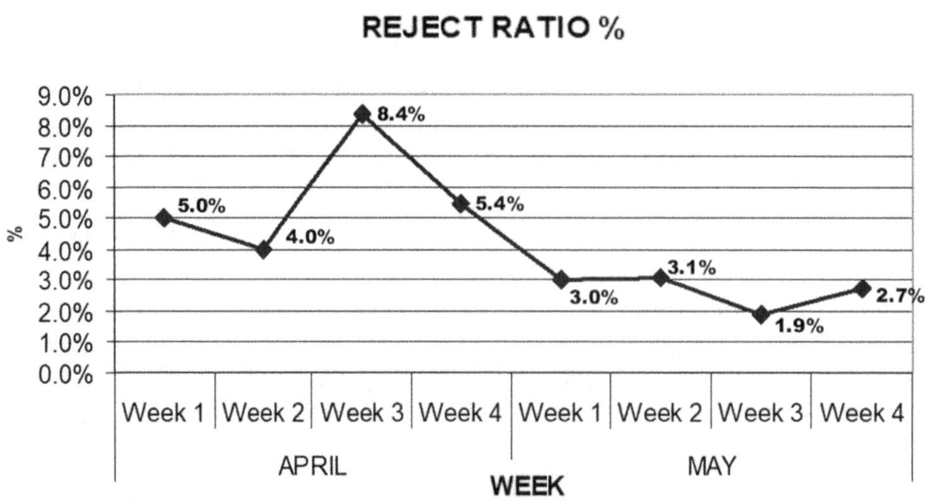

Figure 3. Results from Denison Cultural traits

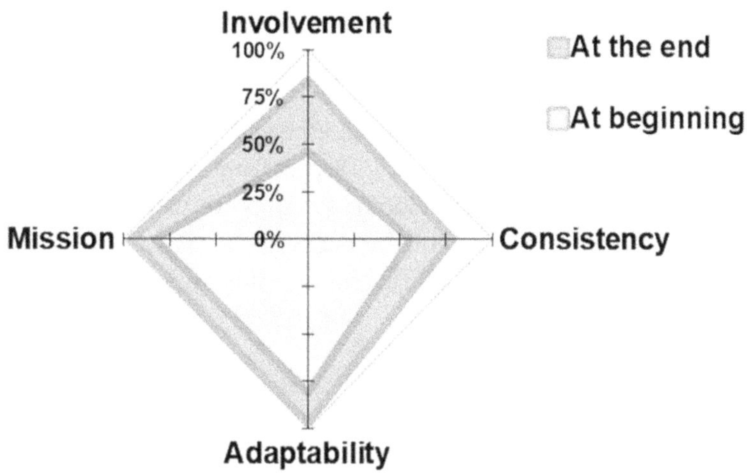

Figure 4. Results from Denison cultural dimensions

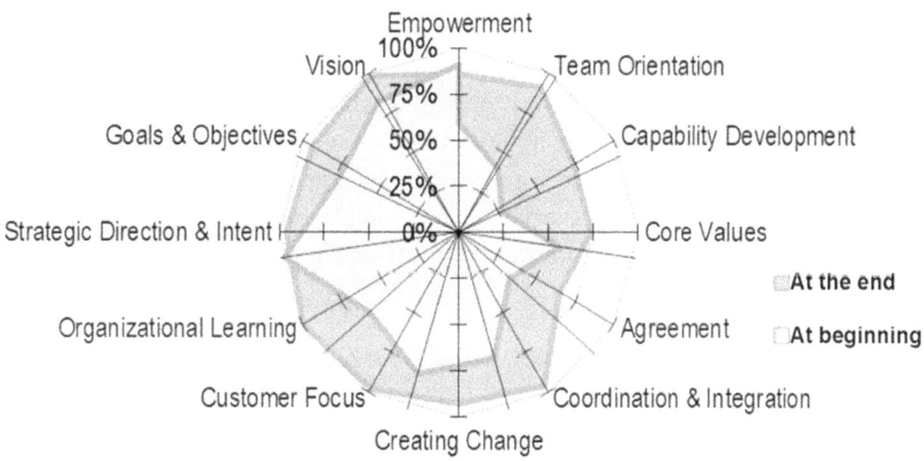

On the opposite side, with Figures 3 and 4, there are plenty of information to not only validate but evaluate the sociotechnical intervention; making ease to validate them in the long-term period and make managers confident about improving validity in long period. So, using "ineffectively model" to support sociotechnical interventions appears to help with support organizational changes validity and evaluation. Also promote the spiral cycle of inquiry-learning-improving that organization needs to survive. Figure 3 supports that we can support than methodology application was useful in long term period in the area culture, this finding is concluded deducted by comparing the people perception before and after the organizational intervention, besides direct impact validation from production rate, so using quantitative and qualitative parameters could help to justify sociotechnical interventions. But Figure 4 provide more information that could help us to evaluate rather than validating the implementation process including qualitative aspects to support the analysis and guidance from methodological intervention; being able to answer the question "How well do I do it?" Kushner (2002). Identification of strengths and weaknesses from sociotechnical methodology, but aspects from the specific organization and work area that could level up or produce failure is important.

In this intervention, looking at Figure 4, it is deduced that methodology applied strengthened the perception that it is better to work as a team than individually; also, the sensibility toward the needs from the "customer focus" was strengthened. But also weak aspects could be finding: "Empowerment" area was high before sociotechnical intervention, thus this aspect from organization could condition results, so in other appliance from sociotechnical methodology if empowerment was not high since beginning the results could not be as good or maybe could be necessary to make a preparation steps before the organizational interventions will be made looking to reinforce this weaknesses; a specific analysis should be made in each individual implementation case.

FUTURE RESEARCH DIRECTIONS

Evaluation and feeding back from sociotechnical systems implementations should be reinforced because this linkage could provide support with management staff from benefits. Also, evaluation should emphasize strong and weak areas in any systemic implementation that could guide through the company to a successful intervention; thus this should be an opportunity area to develop around sociotechnical systems. Rather than just using traditional "goal model" complementing with "ineffectively model" could support sociotechnical implementations, providing with more information that could leads to evaluate and feedback the intervention. The ability from sociotechnical systems to induce a spiral cycle of inquiry-learning-improving could be well received from organizations thus this could provide the softness from their structure adapting to everyday more complex and challenging environment; so this could represent and opportunity area from development and promoting specific interventions from sociotechnical systems.

CONCLUSION

Optimization the intersection from social and technical systems should be the primary objective from sociotechnical interventions, so review from nine Chern's Principles and four Di Maio (2014) underlying guidelines against the sociotechnical methodology before the organizational implementation could lead to detect weakness in methodology and lead to improve them. When implementing socio technical systems in any organization could be helpful consider validating the implementation with the use of quantitative and qualitative indicators, that also are useful not to just validating the implementation rather evaluating them, that could lead to detect strongest and weaknesses from methodology itself. This could be done combining "goal model" within "ineffectively model" in order to measure Organizational effectiveness from intervention. All implementation done from methodologies based in general systems theory, thus sociotechnical systems should promote and impel in the organization a spiral cycle from inquiry-learning-improving that helps the company to subsist rather than an isolated implementation.

REFERENCES

Amir, E., Lev, V., & Sougiannis, T. (2003). Do Financial Analysts Get Intangibles? *European Accounting Review*, *12*(4), 635–659. doi:10.1080/0963818032000141879

Argyris, C. (2001). *Sobre el aprendizaje organizacional.* Oxford University Press.

Ashraf, G., & Abd, S. (2012). A Review on the Models of Organizational Effectiveness: A Look at Cameron's Model in Higher Education. *Educational Studies*, *5*(2), 80–87.

Banker, R. D., Potter, G., & Srinivasan, D. (2005). Association of Nonfinancial Performance Measures with the Financial Performance of a Lodging Chain. *The Cornell Hotel and Restaurant Administration Quarterly*, *46*(6), 394–412. doi:10.1177/0010880405275597

Barnden, A. W., & Darke, P. (2000). A Comparison of SSM with an Organisational Learning Model. *Proceedings of the International Conference on Systems Thinking in Management.*

Batras, D., Duff, C., & Smith, B. J. (2014). Organizational change theory: Implications for health promotion practice. *Health Promotion International*, *31*(1), 231–241. PMID:25398838

Blackman, D. (2006). How measuring learning may limit new knowledge creation. *Journal of Knowledge Management Practice*, *7*(3).

Calton, J., & Kurland, N. (1995). *A theory of stakeholder enabling: Giving óbice to an emerging postmodern praxis of organizational discourse. In Postmodern management and organization theory* (pp. 154–179). Editorial Sage.

Cameron, K. (1980). Critical Questions in Assessing Organizational Effectiveness. *Organizational Dynamics*, *15*(2), 66–80. doi:10.1016/0090-2616(80)90041-8

Cameron, K. (1986). Effectiveness as paradox: Consensus and conflict in conceptions of organizational effectiveness. *Management Science*, *32*(5), 539–553. doi:10.1287/mnsc.32.5.539

Carayon, P., Hancockb, P., Levesonc, N., Noyd, I., Sznelware, L., & Hootegemf, G. V. (2015). Advancing a sociotechnical systems approach to workplace safety – developing the conceptual framework. *Ergonomics*, *58*(4), 548–564. doi:10.1080/00140139.2015.1015623 PMID:25831959

Checkland, P. (2000). Soft Systems Methodology: A thirty year retrospective. *Systems Research and Behavioral Science*, *17*(S1), 11–58. doi:10.1002/1099-1743(200011)17:1+<::AID-SRES374>3.0.CO;2-O

Cheng, Y. C. (2012). *School Effectiveness And School-Based Management: A Mechanism For Development.* Taylor & Francis Group.

Cherns, A. (1976). The Principles of Socio-technical Design. *Human Relations*, *29*(8), 783–792. doi:10.1177/001872677602900806

Cherns, A. (1987). Principles of Socio-technical Design Revisited. *Human Relations*, *40*(3), 153–162. doi:10.1177/001872678704000303

Connell, N. (2001). Evaluating soft OR: Some reflections on an apparently 'unsuccessful' implementation using a Soft Systems Methodology (SSM) based approach. *The Journal of the Operational Research Society*, *52*(2), 150–160. doi:10.1057/palgrave.jors.2601054

Davel, E., & Tremblay, D. G. (2003). Organizational culture and social performance: insights from the experience of family organizations. *Iberoamerican Academy of Management, 3rd International Conference*.

Davis, M. C., Challenger, R., Jayewardene, D. N. W., & Clegg, C. W. (2014). Advancing sociotechnical systems thinking: A call for bravery. *Applied Ergonomics*, *45*(2), 171–180. doi:10.1016/j.apergo.2013.02.009 PMID:23664481

Denison, D. R., Janovics, J., Young, J., & Cho, H. J. (2005). *Diagnosing Organizational Cultures: Validating a Model and Method.* International Institute for Management Development, Working paper 2005-11.

Denison, D. R., & Mishra, A. K. (1995). Toward a Theory of Organizational Culture and Effectiveness. *Organization Science*, *6*(2), 204–223. doi:10.1287/orsc.6.2.204

Dent, E. B., & Umpleby, S. A. (1998). Underlying assumptions of several traditions in systems theory and cybernetics. *Cybernetics and Systems*, 513–518.

Di Maio, P. (2014). Towards a Metamodel to Support the Joint Optimization of Socio Technical Systems. *Systems Journal,* (2), 273-296.

Dixon, N. (2001). *El conocimiento común: Cómo prosperan las compañías que comparten lo que saben.* Oxford University Press.

Easterby, M., Antonacopoulou, E., Simm, D., & Lyles, M. (2004). Constructing Contributions to Organizational Learning: Argyris and the Next Generation. *Management Learning*, *35*(4).

Etzioni, A. (1960). Two approaches to Organizational Analysis: A Critique and a Suggestion. *Administrative Science Quarterly*, *5*(2), 257–278. doi:10.2307/2390780

Etzioni, A. (1987). Normative-Affective factors: Toward a new decision-making model. *Journal of Economic Psychology*, *9*(2), 125–150. doi:10.1016/0167-4870(88)90048-7

Eydi, H. (2015). Organizational Effectiveness Models: Review and Apply in Non-Profit Sporting Organizations. *American Journal of Economics, Finance and Management*, *1*(5), 460–467.

Ghaffarian, V. (2011). The New Stream of Socio-Technical Approach and Main Stream Information Systems Research. *Computer Science*, *3*, 1499–1511.

Guerra, J., Martinez, I., Munduate, L., & Medina, F. (2004). A contingence perspective on the study of the consequences of conflict types: The role of organizational culture. *17th Conference of the International Association for Conflict Management.*

Heckscher, C. (1994). *Defining the post bureaucratic type. In The post-bureaucratic organization: new perspectives on organizational change*. Editorial Sage.

Henri, J. F. (2004). Performance Measurement and Organizational Effectiveness: Bringing the gap. *Managerial Finance*, *30*(6), 93–123. doi:10.1108/03074350410769137

Hughes, H. P. N., Clegg, C. W., Bolton, L. E., & Machon, L. C. (2017). Systems scenarios: A tool for facilitating the socio-technical design of work systems. *Ergonomics*, *60*(10), 1319–1335. doi:10.1080/00140139.2017.1288272 PMID:28277171

Ittner, C. D., & Larcker, D. F. (2001). Determinants of performance measure choices in worker incentive plans. *Journal of Labor Economics*, *20*(2), S58–S90.

Kushner, R. J. (2002). *Action Research Validation of an Inventory of Effectiveness Measures*. Nonprofit Organizational Effectiveness and Performance.

Kweku, F. N., & Mishra, M. (2018). Influence of Human Resource Development (HRD) Practices on Organizational Effectiveness: The Role of Employee Competencies. *International Journal of Management Studies*, *2*(6), 110–124.

Lev, B., & Radhakrishnan, S. (2005). The Valuation of Organization Capital. In C. Corrado, J. Haltiwanger, & D. Sichel (Eds.), *Measuring Capital in a New Economy* (pp. 73–99). National Bureau of Economic Research and University of Chicago Press. doi:10.7208/chicago/9780226116174.003.0004

Marsanasco, A. M., & García, P. S. (2013). The management of knowledge and the learning process in SMES clusters: A study case. *Economía, Sociedad y Territorio*, *13*(41), 1–36.

Martínez, R., Vera, J. G. S., & Vera, M. A. M. (2013). Leadership styles and organizational effectiveness in small construction businesses in Puebla, Mexico. *Global Journal of Business Research*, *7*(5), 47–56.

Mumford, E. (2006). The story of socio-technical design: Reflections on its successes, failures and potential. *Information Systems Journal*, *16*(4), 317–342. doi:10.1111/j.1365-2575.2006.00221.x

Murdaugh J., (1998). *Organizational Effectiveness and Executive Succession: Conclusions About and Implications for Florida's Municipal Police Chiefs.* Senior Leadership Program Publications, SLP-5.

Pourdehnad, J., Wexler, E. R., & Wilson, D. V. (2011). *Systems & Design Thinking: A Conceptual Framework for TeirIntergration.* Organizational Dynamics Working Papers.

Richardson J., (2014). Double Loop Learning: A Powerful Force for Organizational Excellence. *Excerpt from PNSQC 2014 Proceedings.*

Senge, P. (1998). *La quinta disciplina*. Granica.

Sepahvand, R., Aghdam, S. R., Alavi, S. A., & Rezaei, M. (2016). Investigating the Impact of Organizational Learning Capability on Organizational Intelligence in Knowledge Based Organizations. *Proceedings of International Conference on Science, Technology, Humanities and Business Management.*

Serrat, O. (2010). *A Primer on organizational learning*. Asian Development Bank.

Slomp, J., & Ruël, G. C. (2000). *A socio-technical approach for the design of a production control system: towards controllable production units*. University of Groningen.

Sorensen, L., & Valqui, R.V. (2006). Evaluating six soft approaches. *Economic Analysis Working Papers, 7*(9), 1-20.

Sterman, J. D. (1994). Learning in and about complex systems. *System Dynamics Review*, *10*(2-3), 291–330. doi:10.1002dr.4260100214

Tolvanen, J. P. (1998). *Incremental Method Engineering with Modeling Tools.* University of Jyväskylä.

Valqui, R.V. (2005a). *Operational Research: A multidisciplinary discipline.* Informatics and Mathematical Modelling, Technical University of Denmark, DTU.

Valquil, R. V. (2005b). Soft OR approaches. *Engevista*, *7*(1), 4–20.

Yeung, A., Ulrich, D., Nason, S., & VonGlinow, M. (2000). *Las capacidades de aprendizaje en la organización*. Oxford University Press.

Young, S., Tu, Y. M., & Tseng, Y. T. (1999). Organizational Learning as a feedback system: a Conceptual Framework. *17th International Conference of The System Dynamics Society and the 5th Australian & New Zealand Systems Conference.*

Chapter 2
Evaluation of Sociotechnical Systems in Managing Corporate Social Responsibility and Stakeholders' Engagement

Toivo Niskanen
Ministry of Social Affairs and Health, Finland

ABSTRACT

The aim of this chapter is to explore at the Finnish Kemira Corporation how a corporation manages corporate social responsibility (CSR) and stakeholders' engagement. Greimas' actantial model and Senge's five disciplines were applied to evaluate CSR reports with a sociotechnical systems approach in relation to the stakeholders: (1) employees, (2) suppliers, (3) financiers and shareholders, and (4) communities and authorities. It was found that issues of CSR management and stakeholder engagement should be involved in strategic corporate decision making. The business strategy of the corporation emphasizes adherence to CSR guidelines and business standards that reflect the concerns of the stakeholders. CSR strategic procedures indicate the implementation of the corporation's measures to promote and pursue CSR goals that extend beyond their legal responsibilities.

INTRODUCTION AND BACKGROUND

A Sociotechnical Systems Approach in Corporate Social Responsibility

The concept of the sociotechnical system originated with the insights of Tavistock Institute researchers in the early 1950s, specifically with respect to examining the impact of the introduction of novel technical systems (e.g. Trist & Bamforth, 1951). A sociotechnical system is the synergistic combination of humans, machines, environments, work activities and organizational structures and processes that comprise a given enterprise (Mumford, 2006). Carayon et al. (2015) suggest that sociotechnical systems are a type of complex adaptive system, and that analysis from that perspective could significantly enhance our

DOI: 10.4018/978-1-5225-7192-6.ch002

understanding of how sociotechnical systems function and how they might be made to function better. Such systems are frequently described as 'self-organizing' in that adaptations that occur are generally not deliberately or explicitly mandated but instead represent a quasi-organic process of redistributing activity and responsibility across the system (Carayon et al., 2015).

Dobers and Wolff (2000) have argued that the holistic business approach will create sustained success for a company if it is successful in balancing stakeholders' demands and sustainability requirements in sociotechnical systems. Nonetheless, this balancing process is not easy; the management of a business has to be able to deal with value concepts that are conflicting and inherently contain trade-offs (Dobers & Wolff, 2000). To achieve the successful implementation of corporate social responsibility (CSR), managers must build bridge with their stakeholders - through formal and informal dialogues and engagement practices - in the pursuit of common goals and convince them to support the organization's chosen strategic course (Andriof & Waddock, 2002). Business leaders must address the moral complexities that result from the multitude of stakeholder claims and build enduring, mutually beneficial relationships with relevant stakeholders (Maak, 2007). Stakeholder engagement then becomes "CSR in action."

Three elements are key to CSR in systems thinking (Dyllick & Hockerts, 2002, p. 132): 1) working towards a triple bottom line by integrating economic, ecological, and social aspects, and accepting the multiple influences that these aspects have on each other; 2) looking beyond short-term profits driven by shareholders' expectations toward long-term value for all stakeholders; and 3) maintaining not only the capital base of the firm, but also paying attention to the management of natural and social capital. Mounting pressure to implement CSR is applied by stakeholders, e.g. employees, customers, consumers, supply chain partners, competitors, investors, lenders, insurers, nongovernmental organizations (NGOs), media, the government, and society overall (Berns et al., 2009, p. 10). In a regenerative circular economy, the goal is zero waste, renewable energy, recyclable material, and accountability for all materials flowing through the system (Senge et al., 2008, p. 215).

Zink (2014) argues that a sustainable work system must be able to function in the corporation's environment and achieve economic or operational objectives, while development of various human and social resources is also engaged in their sociotechnical systems' operations. Although the concept of sustainability was developed at the macro level rather than at the corporate level, it can also be considered to have a relevant corporate dimension (de Lange, Busch, & Delgado-Ceballos, 2012). The term came into widespread use in 1987, when the World Commission on Environment and Development (United Nations) published a report known as the 'Brundtland Report'; this report stated that 'the "corporate sustainability" seeks to meet the needs of the present without compromising the ability to meet the future generation to meet their own needs' (WCED, 1987, p. 8). Senge et al. (2008, p. 167) suggests that the sustainability innovators creating tomorrow's regenerative economy have all, in their own ways, learned how to see the larger systems in which they live and work. They look beyond events and superficial fixes to see deeper structures and forces at play; they don't allow boundaries (either organizationally or culturally imposed) to limit their thinking; they make strategic choices that consider natural and social limits; and they work to create self-reinforcing cycles of innovation (Senge et al., 2008, p. 167).

Van Luijk (1997) shows that the fundamental task of business ethics is to enhance the ethical quality of decision making and action at all levels of business: at the personal (micro-), organizational (meso-), and systemic (macro-) levels. The corporation is part of a large purposeful system, society. The same is true of CSR, which is a "part" of the corporation. Learning organizations invest in improving the quality of thinking in sociotechnical systems, the capacity for reflection and team learning, and the ability to develop shared visions and shared understandings of complex business issues (Senge, 1994, p. 289).

The five disciplines of what Senge (1994, p. 289) has developed refers to as a "learning organization" as follows: (1) Personal mastery is a discipline of continually clarifying and deepening personal vision, of focusing people's energies, of developing patience, and of seeing reality objectively; (2) Mental models are deeply ingrained assumptions, generalizations, or even pictures of images that influence how people understand the world and how they take action; (3) Building shared vision - a practice of unearthing shared pictures of the future that foster genuine commitment and enrollment rather than compliance; (4) Team learning starts with dialogue, the capacity of members of a team to suspend assumptions and enter into genuine thinking together; and (5) Systems thinking - The Fifth Discipline that integrates the other four."

The growth of sociotechnical systems resources is secured through equal and open interaction among various stakeholders, leading to better mutual understanding and a greater capacity for collaboration (Zink, 2014). Idealized design helps corporations' managers to not only achieve their CSR objectives, but also promotes a better understanding of sociotechnical systems, enhances creativity, simplifies the planning process and accelerates implementation (Ackoff, 2002). Business ethics persistently raises the question of the purpose of business and the economy in economic sciences and business and economic practice and offers well thought-out answers at all levels of action: at the individual, organizational, and sociotechnical systems' levels (Lindgreen & Swaen, 2009). Systems thinking is important in corporations' business strategy because managers are inherently pragmatic and need insights into "current reality", as well as a picture of the future toward which they are moving (Senge, 1994, p. 344). Little research has been carried out to evaluate evidence-based data in CSR. The present study attempts to address this knowledge gap, concentrating on the qualitative analysis of sociotechnical relationships with respect to CSR.

Managing CSR and Stakeholders' Engagement

Carroll and Shabana (2010) state that a business' economic responsibility must be considered as a factor in CSR, which becomes very important in assessing the "business case". Senge et al. (2008, p. 45) conclude that in a world of growing interdependence, it is more important than ever to learn how to expand in CSR performance the scope of normal management attention and concerns in order to see the larger systems in which businesses operate. Several researchers (e.g. Post, Preston, & Sachs, 2002) have stated that the concept of stakeholders is essential to the successful functioning of a commercial enterprise, because they provide resources to the company (e.g. customers, investors, and employees), provide a structure in which the company operates (e.g. supply chain associates and strategic alliances), and make up the sociopolitical arena (e.g. communities and governments). Pirson and Malhotra (2008, pp. 11–12) state that internal stakeholders, such as investors, are most often seeking for evidence of managerial competence: they want to have confidence in the ability of management to effectively control costs and to manage the workforce effectively so that they remain competitive and create financial value. Similarly, Harrison and Wicks (2013) conclude that a firm should serve multiple stakeholders and that the overall performance of a business might be defined as the total value produced by the firm through its systems activities, which is the sum of the utility created for each of a firm's legitimate stakeholders.

Senge et al., (2008, p. 310) discovered that many companies start their sustainability journey with waste reduction or energy efficiency improvements of companies, while a relatively smaller number continually elevate the company's vision, establishing new stakeholders and including possible investors and different sources of capital. Inevitably, this new systems approach starts with engaging the company in a new conversation with stakeholders (Senge et al., 2008, p. 310). While financial performance may be important to many of a firm's stakeholders, it may not be the only aspect of value (Harrison & Wicks, 2013; Harrison, Freeman, & Abreu, 2015). Lee (2008) moves from a discussion of macrosocial effects to an organizational-level analysis of CSR and its impact on organizational processes and performance. In addition, researchers (e.g. O'Riordan & Fairbrass, 2014) have also shifted their focus away from explicitly normative and ethics-oriented arguments towards implicitly normative and performance-oriented managerial studies.

The gaps in knowledge with respect to CSR highlight the need to examine more precisely how decision makers in a commercial enterprise manage their day-to-day CSR strategies and stakeholders' engagement. Although the research described above has resulted in a deeper understanding of how to manage CSR and stakeholder engagement, a critical gap still exists in the literature. Systems-thinking knowledge in this area needs to be developed further to consider how stakeholders can be involved in order to develop and promote positive support, and so that the corporation can understand and adapt to their concerns, e.g. develop its CSR market-related and ethics initiatives.

THE AIM OF THE RESEARCH

The aim of this study is to explore at the Finnish Kemira Corporation how a corporation manages corporate social responsibility (CSR) and stakeholders' engagement. In this respect, the aim of the present study is to explore different kinds of CSR practices presented as "discourses" in the sustainability reports. The study also considers how the sustainability reports of Kemira to implement CSR practices in relation to employees, suppliers, financiers and shareholders, and communities and authorities. CSR holds that organization exists within networks of stakeholders, face the potentially conflicting demands of these stakeholders, and translate the demands into CSR objectives and policies. It also investigates how the stakeholder framework can be analyzed via the contents and impacts of Kemira Corporation's CSR reports. Rather than a single, comprehensive activity, CSR comprises many different activities from which an organization can choose. Accordingly, relevant questions include the following: How should the level of an organization's CSR activity be evaluated? What are the different criteria and indicators that may assess the level of CSR effectively?

The research questions with respect to managing CSR and stakeholder engagement in the Kemira Corporation are as follows:

1. How can sociotechnical systems consist of interactive, mutually engaged and responsive business interactions?
2. How can sociotechnical systems represent not only the core aspects of modern CSR, but also change, challenge and advance the promotion of transparency and accountability?
3. How can sociotechnical systems move from a focus on stakeholders being managed by corporations to a focus on CSR interaction practices that companies implement with their stakeholders based on a process-oriented perspective?
4. How can sociotechnical systems manage and develop strategic corporate decisions and engage stakeholder approval of CSR strategy?

RESEARCH MATERIALS AND METHODS

Research Materials

An online questionnaire was carried out by the Finnish Kemira Corporation (2016c) globally considered with "High Performance and Engagement Model". A sampling method was applied and a target group of 4150 employees was used. The response rate to the questionnaire was 85%. The respondents were as follows: blue-collar workers (n = 1358), white-collar workers (n = 1223), managers and team leaders (n = 627), senior managers (n = 166), and directors (n = 143). The qualitative research data came from evaluations and reviews of the Kemira Corporation's CSR strategy in its Responsibility Report (Kemira Corporation, 2016a) and Annual Report (Kemira Corporation, 2016b). The Kemira Corporation, headquartered in Helsinki, has 4,685 employees and a global manufacturing network of 64 factories.

Quantitative Methods and Analysis

The questionnaire was specifically developed by the Kemira Corporation company to explore CSR. Analysis of the questionnaire with respect to the 'High Performance & Engagement Model' (Kemira Corporation, 2016c) drew on data obtained from CSR respondents (n = 3520), using the Likert scale ranging from 1 to 5 as follows: 1 = strongly disagree, 2 = disagree, 3 = neither disagree nor agree, 4 = agree, and 5 = strongly agree.

Qualitative Methods and Analysis

A Case Study Approach

Yin (1994, p. 13) defines a case study as, "an empirical inquiry that investigates a contemporary phenomenon within its real-life context, especially when the boundaries between phenomenon and context are not clearly evident." The case study is often used to develop knowledge of individual, group, organizational, social, political, and related phenomena (Yin, 1994) and is suitable for exploratory, descriptive and explanatory reasons (Yin, 1981). The studies under discussion can individually and collectively be said to embody all three. Case studies have been criticized for being less rigorous than other research methods and for exhibiting various biases (Yin, 1994). Case studies try to reduce or offset such bias by obtaining deep insights into a particular situation and how the people involved interpret it.

The focus of case studies is on processes and interactions, and their purpose is to consider entities rather than individual phenomena. Another concern about case studies is their tendency to induce generalized conclusions (Yin, 1994). While organizations are socially constructed systems, they are also parts of larger sociotechnical systems, so it can be difficult to draw generalized conclusions unless the investigated part-system is analyzed as a part of a larger entity.

The Constructionism Approach

Constructionism starts with the assumption that access to shared dynamic, changing and individually constructed reality only occurs through social constructions such as language and shared meanings in everyday interactions (Lord & Dinh, 2012). Social reality is not separate for different individuals, but it is intimately interwoven and shaped by everyday interactions (Uhl-Bien, Maslyn, & Ospina, 2012) and focuses on relationality (Cunliffe, 2008). In this light, CSR relationships can be considered as emergent from and co-constructed as an interactive dynamic of CSR (applied from Fairhurst & Grant, 2010). The study is iterative in design as the process and structure of the boundary analysis is iterative (applied from Noda and Bower, 1996).

The Framework of Senge's Five Disciplines (1994)

Senge (1994) presents a sociotechnical systems framework that describes the five disciplines of organizational learning as: personal mastery, mental models, shared vision, team learning, and systems thinking. A systems thinking approach is needed to enable corporations to manage "the interrelationships rather than linear cause-effect chains and [to see] processes of change rather than snapshots" (Senge, 1994). Senge et al. (2008, p. 125) show that companies that perform well in four characteristics ("Innovations and repositioning", "Growth path and trajectory", "Cost and risk reduction", and "Reputation and legitimacy") maximize shareholder value over time by thinking more comprehensively about their business, which also enables them to manage CSR and stakeholders' engagement more effectively.

Senge (1994) presents the five disciplines of organizational learning as follows:

1. **Personal Mastery:** This aspirational discipline involves formulating a coherent picture of the results that corporations most sincerely wish to achieve, alongside a realistic assessment of their current situation. Learning to cultivate the tension between vision and reality can expand a corporation's capacity to make better choices and to achieve more of the goals they set for themselves.
2. **Mental Models:** This discipline of reflection and inquiry focuses on developing awareness of the attitudes and perceptions that influence thought and interaction. By continually reflecting upon, talking about and reconsidering internal perspectives of their environment, corporations can improve their ability to govern their actions and apply their making process more effectively.
3. **Shared Vision:** This collective discipline establishes a focus on mutual purpose. Individuals in corporations learn to nourish a sense of commitment to the group or organization by developing shared images of the future that they seek to create, and the principles and guiding practices which they hope will lead them to that future.
4. **Team Learning:** This discipline of organizational interaction involves different techniques, such as dialogue and skillful discussion, which encourage teams to transform their collective thinking and learn to mobilize their energies and abilities in a synergistic manner. The overall outcome of the team is greater than the sum of its individual talents.
5. **Systems Thinking:** In this discipline, individuals in organizations learn to better understand interdependency and change, and in that way they start to deal more effectively with the forces that shape the consequences of their actions. Systems thinking is based upon a growing theory regarding the behavior of feedback and complexity—the innate tendencies of a system that lead to growth or stability over time.

The Greimas Actantial Model (1990)

The application of Greimas' actantial model (1990) makes it possible to analyze how a company practically implements CSR principles, to identify factors that affect this implementation, and to evaluate how seriously a company incorporates CSR measures into its operations.

The actantial model was applied in discourse analysis to provide a summary of CSR outcomes. The six actants were divided into three oppositions, each of which forms an axis of the actantial description (Hébert, 2014):

- The axis of goal: subject/object.
- The axis of power: helper/opponent.
- The axis of transmission and knowledge: sender/receiver.

RESEARCH RESULTS

Questionnaire Survey Results at the Kemira Corporation

High Performance and Engagement Model

Results of the questionnaire with respect to the 'High Performance & Engagement Model' (Kemira Corporation, 2016c) for most favorable answers of the Likert scale ranging (5 = "Strongly Agree" and 4 = "Agree") are presented in Table 1. The statistical difference of the results (Likert scales 1 + 2 + 3 vs. 4 + 5) for comparing the years 2013 and 2015 indicated with the chi-squared test the following p-values of the sub-categories: (1) $p < 0.001$; (2) $p < 0.001$; (3) $p < 0.001$; (4) $p < 0.001$; (5) not significant; (6) $p < 0.01$; (7) $p < 0.01$; (8) $p < 0.001$; and (9) $p < 0.001$. Likert scale ranging from 1 to 5 is as follows: 1 = strongly disagree, 2 = disagree, 3 = neither disagree nor agree 4 = agree, 5 = strongly agree.

Table 1. Percentage of survey respondents (n = 3520) who selected the Likert scale's most favorable answers (5 = "Strongly Agree" and 4 = "Agree")

High Performance and Engagement in CSR at the Kemira Corporation	2015%	2013%
(1) Kemira Corporation shows a commitment to ethical business decisions and conduct.	81	71
(2) I feel that Kemira Corporation is a socially responsible company.	79	69
(3) At Kemira Corporation we live the value "We care for people and the environment."	77	66
(4) At Kemira Corporation we live the value "We are dedicated to customer success."	75	70
(5) At Kemira Corporation we regularly use customer feedback to improve our work processes.	63	63
(6) At Kemira Corporation senior management is committed to providing high quality products and services to external customers.	76	73
(7) At Kemira Corporation we regularly use customer feedback to improve our work processes.	63	59
(8) At Kemira Corporation we live the value "We succeed together."	55	49
(9) At Kemira Corporation we live the value "We drive performance and innovation."	61	49

Source: The author

Systems Approach to Stakeholder Engagement at the Kemira Corporation

Kemira Corporation's approach to stakeholder engagement is based on active stakeholder relationship management (Kemira Corporation, 2016a). The most recent review of stakeholders' expectations towards Kemira Corporation was conducted in spring 2015 and it includes activities ranging from information sharing to active dialogue and collaboration on issues of mutual interest, as follows:

- **Employees**
  - **Stakeholder Engagement:** Employee surveys; performance management process; town hall meetings; compliance and ethics hotline.
  - **Key Topics:** Performance management and rewards; visible role modelling of Kemira Corporation's values.
  - **Kemira Corporation's Applications:** Performance and development discussions; leadership development, skills development and training programs; documented action plans produced in response to the findings of employee surveys.
- **Suppliers**
  - **Stakeholder Engagement:** Working closely with core suppliers to help them meet performance expectations and take corrective actions if needed; supplier performance management.
  - **Key Topics:** Customers expect responsibility throughout the supply chain; business ethics and compliance.
  - **Kemira Corporation's Applications:** Code of conduct for suppliers, distributors and agents; supplier sustainability assessments.
- **Customers**
  - **Stakeholder Engagement:** Direct customer contact by Kemira Corporation's sales organization; exhibitions and trade shows; product testing and plant trials; customers' requests regarding CSR performance.
  - **Key Topics:** Product safety; transportation safety; reduction of environmental impacts in manufacturing and products.
  - **Kemira Corporation's Applications:** Product stewardship; transportation safety programs; sustainability checks in new product development.
- **Shareholders**
  - **Stakeholder Engagement:** Regular events like Capital Markets Day, roadshows, conference calls and one-to-one meetings;
  - **Key Topics:** Alignment of CSR with Corporation's strategy and business activities; potential business risks and opportunities related to CSR and transparency;
  - **Corporation's Response:** Kemira Corporation's approach to CSR management; transparent reporting and disclosure (e.g. CDP ("carbon disclosure project") aims to reduce companies' greenhouse gas emissions and mitigate climate change risk.
- **Local Communities**
  - **Stakeholder Engagement:** Collaboration with local communities at major sites to understand and address their concerns; collaboration with schools and universities.
  - **Key Topics:** Safety and environmental risks; employment opportunities.
  - **Kemira Corporation's Applications:** Transparency, regular and open dialogue with local communities; site-specific activities, e.g. open-door days; regular industrial risk assessments.

- **Regulatory Bodies**
  - **Stakeholder Engagement:** Participation in the activities of industrial trade associations; subject-specific dialogue with regulatory bodies at national and European Union) (EU) level; managed relationships with selected non-profit organizations.
  - **Key Topics:** Resource efficiency; safety of shale oil and gas and enhanced oil recovery; chemicals safety.
  - **Kemira Corporation's Applications:** Circular economy position paper; active participation in Responsible Care working groups; membership of Global Cleantech and climate councils.

Business Strategy at the Kemira Corporation: A Five Disciplines Classification

The major theme raised in the qualitative discourse is how the Finnish Kemira Corporation shall apply "corporate sustainability" to their business strategy. A five disciplines classification is presented in Table 2, Table 3, Table 4, and Table 5 (Appendix) with respect to different stakeholders. In this qualitative systems analysis, the author of this article, as the investigator, has selected the most important individual sub-categories in the iterative study process of Senge's (1994) sociotechnical systems approach.

The major themes raised in the five disciplines classification (Senge, 1994) as they apply to the employees of the corporation are presented in Table 2 (Appendix). Regard the major themes raised in the five disciplines classification (Senge, 1994) as they apply to customers, suppliers and consumers are presented in Table 3 (Appendix). Similarly, the major themes raised in the five disciplines classification (Senge, 1994) in relation to financiers and shareholders are presented in Table 4 (Appendix). Finally, the major themes raised in the five disciplines classification (Senge, 1994) as they apply to the community and authorities are presented in Table 5 (Appendix).

Qualitative Discourse Analysis of CSR Report at the Kemira Corporation With the Greimas Model

Figures 1-4 outline different dimensions of qualitative stakeholder discourse and corresponding levels of engagement. The Greimas (1990) model serves as a frame of reference in the discourse analysis of sociotechnical systems. In this qualitative systems analysis, the author of this article, as the investigator, has selected the most important individual sub-categories in Greimas' (1990) systems approach and makes possible an evaluation of the extent to which the company's stakeholder's discourses are either participatory and inclusive or hierarchal and exclusive. In order for discourse to be participatory, the stakeholders and organization must be open-minded and willing to listen to alternative and critical voices that may bring new ideas and insights to bear on the issues that the company is trying to resolve. Interpreting the stakeholders' roles does not necessarily mean that the company is committed to taking action. The response filter pertains to the difference between decisions resulting from the stakeholder discourse and the actual implementation of initiatives and their related impacts. While decisions following from stakeholder discourse may be fairly representative of the stakeholders' interests, it may still prove difficult to translate these decisions into action.

Overall, the Greimas framework describes how CSR activity is perceived by the individual stakeholders, produces benefits (axis of power: helper) and challenges (axis of power: opponent) in relation to individual stakeholders, influences some kind of proactive relationship (transmission and knowledge: sender), and results in behavioral outcomes (transmission and knowledge: receiver). The model provides three main insights. First, it shows that stakeholders respond to CSR initiatives based on the degree to which the individual stakeholder derives benefits as a result of the company engaging in that activity. Second, the model shows that the nature of the stakeholder-company relationship is determined by the type of benefits derived from the CSR activity. Finally, the model emphasizes the importance of distinguishing between different measures of CSR investment and the perceptions that stakeholders hold about the company's initiatives.

Discourse in Relation to Employees at the Kemira Corporation

Figure 1 presents the semiotic actantial model developed by Greimas (1990), connecting CSR theory and discourse data as it applies to employees.

Figure 1. Actantial model of the CSR applications in relation to employees at the Kemira Corporation
Source: The author

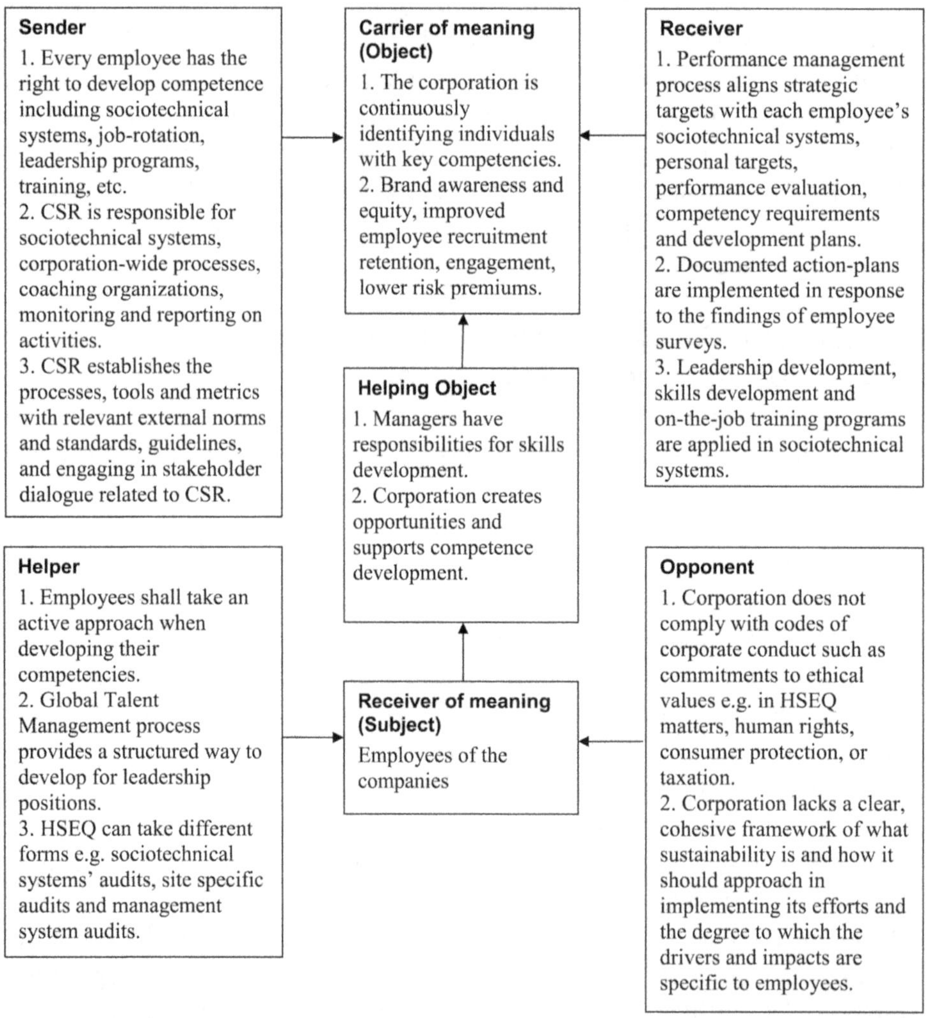

Discourse in Relation to Customers and Suppliers at the Kemira Corporation

Figure 2 presents the semiotic actantial model developed by Greimas (1990), connecting CSR theory and discourse data as it applies to customers and suppliers.

Discourse in Relation to Financiers and Shareholders at the Kemira Corporation

Figure 3 presents the semiotic actantial model developed by Greimas (1990), connecting CSR theory and discourse data as it applies to financiers and shareholders.

Figure 2. Actantial model of the CSR applications in relation to customers and suppliers at the Kemira Corporation
Source: The author

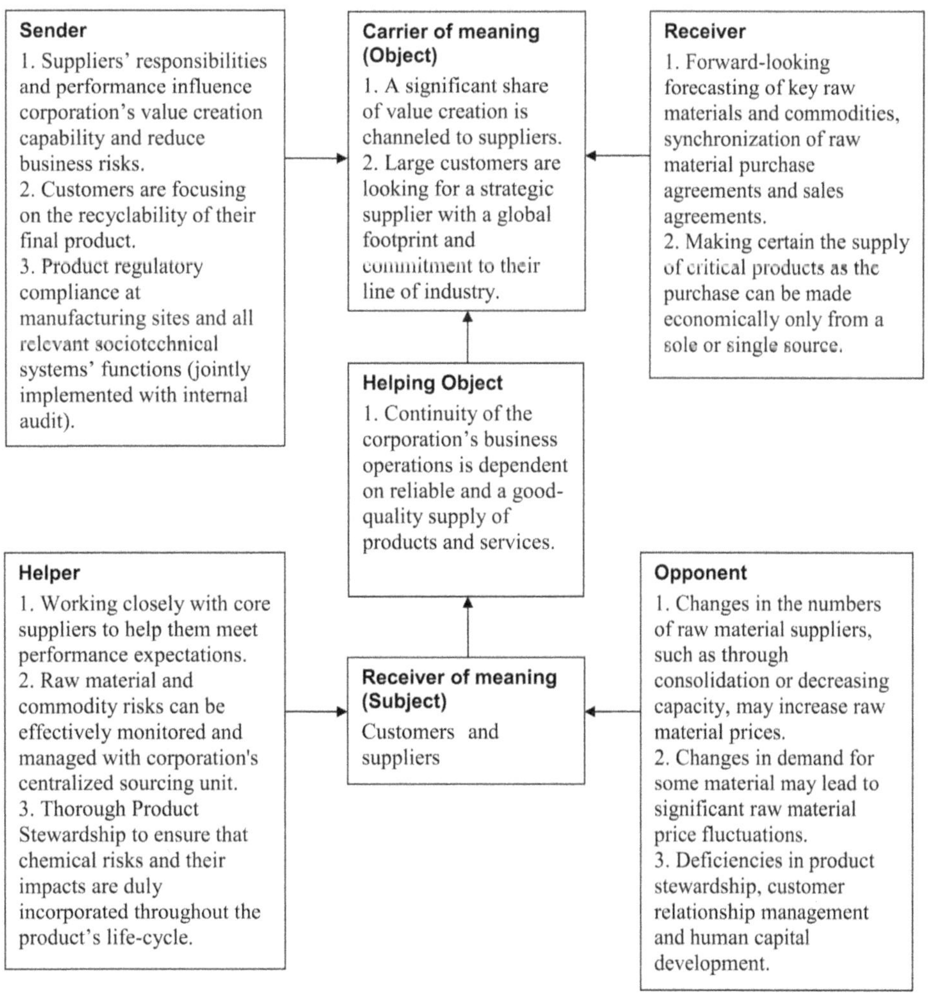

Figure 3. Actantial model of the CSR applications in relation to financiers and shareholders at the Kemira Corporation
Source: The author

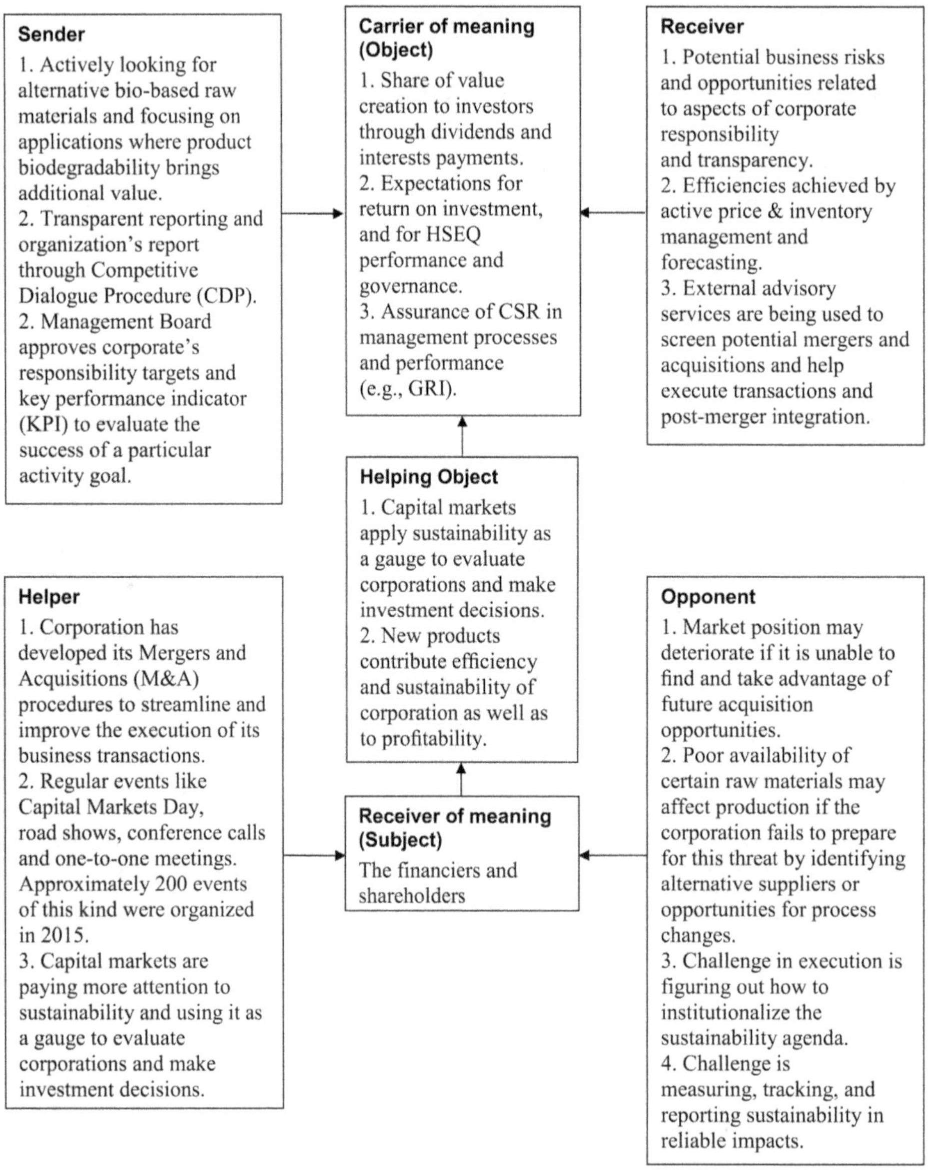

Discourse in Relation to Communities and Authorities at the Kemira Corporation

Figure 4 presents the semiotic actantial model developed by Greimas (1990), connecting CSR theory and discourse data as it applies to the community and authorities.

Figure 4. Actantial model of the CSR applications in relation to community and authorities at the Kemira Corporation
Source: The author

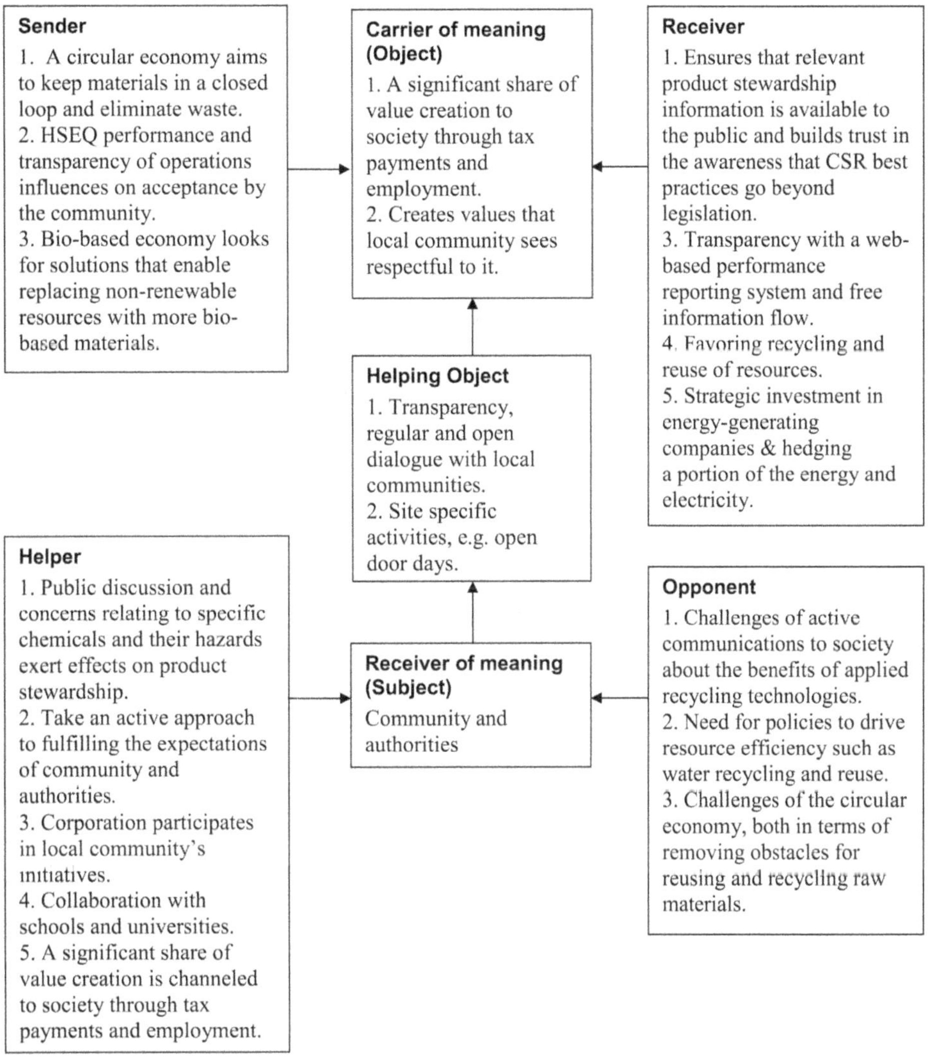

DISCUSSION, SOLUTIONS AND RECOMMENDATIONS

Theoretical Implications

This study presents at the Kemira Corporation a sociotechnical systems approach to the operationalization and empirical assessment of CSR structuration, i.e. how different operations and their degree of CSR structuration within sociotechnical systems can be conceptualized within the corporation's strategies. CSR marketing opportunities within visionary strategies can be looked in systems thinking with a resource-based view, where CSR development is deep seated in the normative level of the company (Baumgartner & Ebner, 2010, p. 78). This study shows that theories of sociotechnical systems may well benefit the management at the Kemira Corporation within corporate social responsibility and stakeholders' engagement. A sociotechnical approach integrates CSR management and stakeholders' engagement, which in turn can enhance strategy, decision-making processes, participation, and collaboration. The present study supports the theoretical framework presented by Freeman (1994), in which CSR and stakeholder engagement should be involved in corporate planning and business policies that focus on developing strategic corporate decisions.

Practical Implications

This study supports the finding of Jackson (2003) that with increasing complexity, change and diversity, managers at the Kemira Corporation have inevitably sought the receive help of the following: scenario planning; benchmarking; value chain analysis; continuous improvement; total quality management; learning organizations; process re-engineering; and knowledge management. The present results support Senge (1994, p. 128) conclusion that a sociotechnical systems approach is of greatest benefit to distinguish high- from low-leverage changes in highly complex situations. In effect, the art of systems thinking lies in seeing through complexity to the underlying structures that generate change (Senge, 1994, p. 128). The present results support the conclusions of Dare (2016) that a good corporation operates as follows: (1) it assesses its sustainability strategies as it would any investment, systematically evaluating each value-creation level; (2) it adopts a broad, systems-thinking approach to its business; (3) it improves scenario-planning capabilities that allow it to effectively build resilience to unpredictable future environments and external shocks; (4) it develops tracking, measuring, and reporting capabilities, particularly as the bar for transparency continues to rise; (5) it retools research and development (R&D), product development, and sales and marketing to reimagine how products are designed, made, used, and recycled; and (6) it enhances capabilities in innovating organizational models and management practices.

The present results support the recommendation of OECD (2011, p. 25) (Organization for Economic Cooperation and Development) that when dealing with consumers, enterprises should act in accordance with fair business, marketing and advertising practices and should take all reasonable steps to ensure the safety and quality of the goods or services they provide. This study supports the results of Senge et al. (2008, p. 340) that the development of new processes, products, and services will be a key area for rethinking and change and that R&D and innovation work (including the redesign of core processes) is crucial and must be designed into systems. This systems thinking typically means that significant changes must be made to how mainstream business is carried out. It is often project-oriented but has clear links to mainstream functions, e.g. using a full life-cycle review to bring more discipline to the front end of how to design, engineer and undertake. The new, proactive approach forces people to reflect in advance before making major design decisions (Senge et al., 2008, p. 340).

In practice, the Kemira Corporation is compliant with international systems initiatives (e.g. GRI (2014), ISO 9001 (2008), ISO 14001 (2004), BS OHSAS 18001 (2007), AA1000 (2008) and OECD (2011)). Senge et al. (2008, p. 341) argue that the department of health, safety and environment should help the whole system of a company (top management, line management, human resources (HRs), communications, finance and legal teams) to increase performance sustainability but should not be held accountable for achieving it, as ownership remains with the core business units.

The present results support the conclusions of other investigators (e.g. Dare, 2016; DesJardins, 2016; Greenwood & van Buren, 2010; Harrison & Wicks, 2013;). Harrison and Wicks (2013) show that rather than focusing on economic measures of performance, a stakeholder-based performance measure challenges managers to examine the CSR values that their firms are creating from the perspective of the stakeholders involved in creating it. DesJardins (2016) found that good management practices must understand the organizational-level implications that follow from their business activities according to the macro-level model of global economic development. Dare (2016) argues that commercial firms are pressured by both internal and external stakeholders to engage in CSR to address a myriad of ever-changing expectations about the role of business in society. Greenwood and van Buren (2010) showed that the trust is a fundamental aspect of the moral treatment of stakeholders within the organization-stakeholder relationship and furthermore stakeholders trust the organization to return benefit or protections from harm commensurate with their contributions or stakes.

The present results support the findings of other investigators (e.g., Pedersen, 2006; Berns et al., 2009; Harrison & Wicks, 2013; Orlitzky, Schmidt, & Rynes, 2003). Pedersen (2006) states that the degree of transparency is an important element in stakeholder CSR reporting, because neither the involved parties nor outsiders are able to hold the company (or the stakeholders) accountable without access to information about CSR processes and outcomes. Berns et al. (2009) argue that transparency is of great importance in financial and business transactions, rewarding ethical behavior and making commitments to all stakeholders. Harrison and Wicks (2013) found that that a stakeholder-based perspective of value is important from a managerial viewpoint, because managers tend to focus their attention on aspects that lead to higher performance based on what can actually be measured. Orlitzky et al. (2003) showed that the company's executives and top-leadership must be attentive to the perceptions of third parties, regardless of whether they are market analysts, public interest groups or the media.

Generalization of the Present Results

The current study has several limitations that should be noted. Qualitative analysis was based on classifications chosen by the author of this article as an investigator. Results may thus be subject to inaccuracies related to the inability of the investigator to correctly recall information. In the qualitative analysis the discourse is a dual concept: first, it refers to the interactive process in which implications are produced; and second, it refers to the final outcome of this process (van Dijk, 1995). Qualitative research is iterative rather than linear, so a good qualitative researcher moves back and forth between design and implementation to ensure congruence between question formulation, literature, recruitment, data collection strategies, and analysis (Morse, Barrett, Mayan, Olson, & Spiers, 2002). The findings of the present qualitative study are judged to be reasonably reliable and valid, but the question remains whether the results are of local interest or are transferable to other subjects and situations.

The limitations related to the applicability of Greimas' model (1990) are twofold. First, as an analytical tool, although Greimas' model may appear clear and simple, it may prove to be rather complicated. The construction of the relationships between the actantial units of the text is always relative and depends on the researchers' definitions. However, despite the mentioned limitations, the present results support the proposal that Greimas' model (1990) can be used as a theoretical framework and as a practical semiotic tool to describe and interrogate CSR.

FUTURE RESEARCH DIRECTIONS

Future studies may benefit from the adoption of research designs that permit an examination with longitudinal intervention designs of causality in stronger stakeholders' engagement with CSR. In addition, an investigation could be conducted to explore how stakeholders can practically effect CSR. In this context, stakeholder theory encourages managers to clarify to their stakeholders that they have a shared sense of value and to emphasize the values that bind its core stakeholders together (Freeman, Wicks, & Parmar, 2004). This study supports the findings of Enderle (2016) that further explorations are needed to develop best practices about how business ethics (e.g. CSR) take inspiration from the sociotechnical system's framework, including natural, economic, human, and social capital, persistently raising up responses at individual (micro-), organizational (meso-), and systemic (macro-) levels of action.

CONCLUSION

This study shows that in order to achieve effective stakeholder participation and commitment, a sociotechnical systems approach can be utilized in management processes. The present systems thinking approach includes the identification of stakeholder roles (e.g., employees, owners, communities, suppliers and customers) at the Kemira Corporation. The study discovered that a systems thinking approach allows the modeling of system-wide effects of CSR initiatives over the long term to form more effective alliances on the macro levels of global economics. In CSR reporting, the business strategy of the corporation revealed practical and concrete results in quantitative and qualitative measurements, CSR guidelines, business standards, norms, and expectations that reflect concerns for the stakeholders. Yet the way in which organizations choose to manage their stakeholder relationships in practice varies considerably, such that stakeholder engagement can represent different features and various theoretical perspectives. Inconsistencies, equivocalness, and practical implications related to the role of stakeholder engagement in CSR therefore still need synthesis and resolution. Therefore, from a narrow view, the business case justifies CSR initiatives only when they produce direct and clear links to firm financial performance. In contrast, a broad view would note that CSR initiatives produce direct and indirect links to firm performance that enables the corporation to benefit from CSR opportunities. Effective CSR requires developing appropriate CSR strategies, and effective CSR activities are those directed at improving both stakeholder relations and social welfare.

In CSR reporting the present and future business implications of CSR issues were explained in the context of CSR strategy and key value drivers, which were established in the context of the corporation's wider business strategy. The author described which resources were allocated for the achievement of the corporation's CSR objectives and how CSR issues are managed as a part of corporation's business strategy. CSR reporting presented future CSR targets in order to develop CSR-enhancing products and services. Kemira Corporation's CSR strategy helps to achieve CSR efficiency and effectiveness for the benefits of the Kemira and its stakeholders. Corporation's values were factored into the strategy and into its responses to the external environment and stakeholder expectations. CSR issues were described as advantages of corporation's internal capability within the limits of available resources.

ACKNOWLEDGMENT

The author is grateful for help of the EHS (Environment, Health and Safety) Unit of the Kemira Corporation regarding the questionnaire survey results. The author is very grateful to the two anonymous reviewers for their important comments, which improved substantially the quality of the chapter.

REFERENCES

AA1000APS. (2008). *AccountAbility Principles Standard.* Retrieved March 22, 2018, from http://www.accountability.org/standards/aa1000aps.html

Ackoff, R. (2002). The corporation as a community, not as a corpus. *Reflections: The SoL Journal*, *4*(1), 14–21. doi:10.1162/152417302320467517

Andriof, J., & Waddock, S. (2002). Unfolding stakeholder engagement. In J. Andriof, S. Waddock, B. Husted, & S. S. Rahman (Eds.), *Unfolding Stakeholder Thinking: Theory, Responsibility and Engagement* (pp. 19–42). Sheffield, UK: Greenleaf; doi:10.1162/152417302320467517

Baumgartner, R. J., & Ebner, D. (2010). Corporate sustainability strategies: Sustainability profiles and maturity levels. *Sustainable Development*, *18*(2), 76–89. doi:10.1002d.447

Berns, M., Townend, A., Khayat, Z., Balagopal, B., Reeves, M., Hopkins, M., & Kruschwitz, N. (2009). *The business of sustainability: Findings and insights from the first annual business of sustainability survey and the global thought leaders' research project*. Massachusetts Institute of Technology. MIT Sloan Management Review. Retrieved March 22, 2018, from http://www.ideiasustentavel.com.br/pdf/mitsloan_sr2009-dl.pdf

BS OHSAS. (2007). *Occupational health and safety management systems*. Retrieved March 22, 2018, from http://www.isoqsltd.com/iso-certification/bs-ohsas-18001

Carayon, P., Hancock, P., Leveson, N., Noy, I., Sznelwar, L., & van Hootegem, G. (2015). Advancing a sociotechnical systems approach to workplace safety - developing the conceptual framework. *Ergonomics*, *58*(4), 548–564. doi:10.1080/00140139.2015.1015623

Carroll, A. B., & Shabana, K. M. (2010). The Business Case for Corporate Social Responsibility: A Review of Concepts, Research and Practice. *International Journal of Management Reviews*, *12*(1), 85–105. doi:10.1111/j.1468-2370.2009.00275.x

Cunliffe, A. (2008). Orientations to social constructionism: Relationally responsive social constructionism and its implications for knowledge and learning. *Management Learning*, *39*(2), 123–139. doi:10.1177/1350507607087578

Dare, J. (2016). Will the Truth Set Us Free? An Exploration of CSR Motive and Commitment. *Business and Society Review*, *121*(1), 85–122. doi:10.1111/basr.12082

de Lange, D. E., Busch, T., & Delgado-Ceballos, J. (2012). Sustaining sustainability in organizations. *Journal of Business Ethics*, *110*(2), 157–172. doi:10.100710551-012-1425-0

DesJardins, J. (2016). Is it Time to Jump off the Sustainability Bandwagon? *Business Ethics Quarterly*, *26*(1), 117–135. doi:10.1017/beq.2016.12

Dobers, P., & Wolff, R. (2000). Competing with 'soft' issues - from managing the environment to sustainable business strategies. *Business Strategy and the Environment*, *9*(1), 143–150. doi:10.1002/(SICI)1099-0836(200005/06)9:3<143::AID-BSE239>3.0.CO;2-C

Dyllick, T., & Hockerts, K. (2002). Beyond the business case for corporate sustainability. *Business Strategy and the Environment*, *11*(2), 130–141. doi:10.1002/bse.323

Enderle, G. (2016). How Can Business Ethics Strengthen the Social Cohesion of a Society? *Journal of Business Ethics*. doi:10.100710551-016-3196-5

Fairhurst, G., & Grant, D. (2010). The social construction of leadership: A sailing guide. *Management Communication Quarterly*, *24*(2), 171–210. doi:10.1177/0893318909359697

Freeman, R. E. (1994). The politics of stakeholder theory: Some future directions. *Business Ethics Quarterly*, *4*(4), 409–421. doi:10.2307/3857340

Freeman, R. E., Wicks, A., & Parmar, B. (2004). Stakeholder Theory and 'The Corporate Objective Revisited'. *Organization Science*, *15*(3), 364–369. doi:10.1287/orsc.1040.0066

Greenwood, M., & van Buren, H. J. III. (2010). Trust and stakeholder theory: Trustworthiness in the organisation–stakeholder relationship. *Journal of Business Ethics*, *95*(3), 425–438. doi:10.100710551-010-0414-4

Greimas, A. J. (1990). *The social sciences: A semiotic view*. Minneapolis, MN: University of Minnesota Press.

GRI. (2014). *Global Reporting Initiatives: Sustainability Reporting Guidelines*. Retrieved March 22, 2018, from https://www.globalreporting.org/reporting/g4/Pages/default.aspx

Harrison, J. S., Freeman, R. E., & Abreu, M. C. S. (2015). Stakeholder theory as an ethical approach to effective management: Applying the theory to multiple contexts. *Review of Business Management*, *17*(55), 858–869. doi:10.7819/rbgn.v17i55.2647

Harrison, J. S., & Wicks, A. C. (2013). Stakeholder Theory, Value, and Firm Performance. *Business Ethics Quarterly*, *23*(1), 97–124. doi:10.5840/beq20132314

Hébert, L. (2014). *The Actantial Model.* Retrieved March 22, 2018, from http://www.signosemio.com/greimas/actantial-model.asp

ISO 14001. (2004). *Environmental management systems*. International Organization for Standardization. Retrieved March 22, 2018, from http://www.isoqsltd.com/iso-certification/iso-14001

ISO 9001. (2008). *Quality management systems*. International Organization for Standardization. Retrieved March 22, 2018, from http://www.isoqsltd.com/iso-certification/iso-9001

Jackson, M. C. (2003). *Systems thinking: creative holism for managers*. John Wiley & Sons Ltd. Retrieved July 18, 2018, from http://webcourses.ir/dl/Systems%20Thinking.pdf

Kemira Corporation. (2016a). *Responsibility Report 2015*. Retrieved March 22, 2018, from http //www.kemira.com/en/newsroom/publications/sustainability-report/pages/default.aspx

Kemira Corporation. (2016b). *Annual Report 2015*. 220 pages. Retrieved March 22, 2018, from http://www.kemira.com/en/newsroom/publications/annual-report/pages/default.aspx

Kemira Corporation. (2016c). *High performance engagement model applied in Kemira corporation: A survey in collaboration between Kemira and IBM*. Helsinki: Kemira.

Lee, M. D. (2008). A review of the theories of corporate social responsibility: Its evolutionary path and the road ahead. *International Journal of Management Reviews*, *10*(1), 53–73. doi:10.1111/j.1468-2370.2007.00226.x

Lindgreen, A., & Swaen, V. (2009). Corporate social responsibility. *International Journal of Management Reviews*, *12*(1), 1–7. doi:10.1111/j.1468-2370.2009.00277.x

Lord, R. G., & Dinh, J. E. (2012). Aggregation Processes and Levels of Analysis as Organizing Structures for Leadership Theory. In D. V. Day & J. Antonakis (Eds.), *The Nature of Leadership* (pp. 29–65). Thousand Oaks, CA: Sage Publications.

Maak, T. (2007). Responsible leadership, stakeholder engagement, and the emergence of social capital. *Journal of Business Ethics*, *74*(4), 329–343. doi:10.100710551-007-9510-5

Morse, J. M., Barrett, M., Mayan, M., Olson, K., & Spiers, J. (2002). Verification Strategies for Establishing Reliability and Validity in Qualitative Research. *International Journal of Qualitative Methods*, *1*(2), 13–22. doi:10.1177/160940690200100202

Mumford, E. (2006). The Story of Socio-technical Design: Reflections on Its Successes, Failures and Potential. *Information Systems Journal*, *16*(4), 317–342. doi:10.1111/j.1365-2575.2006.00221.x

Noda, T., & Bower, J. L. (1996). Strategy making as iterated processes of resource allocation. *Strategic Management Journal*, *17*(S1), 159–192. doi:10.1002mj.4250171011

O'Riordan, L., & Fairbrass, J. (2014). Managing CSR Stakeholder Engagement: A New Conceptual Framework. *Journal of Business Ethics*, *125*(1), 121–145. doi:10.100710551-013-1913-x

OECD. (2011). *OECD Guidelines for Multinational Enterprises*. OECD Publishing. Retrieved March 22, 2018, from http://www.oecd-ilibrary.org/governance/oecd-guidelines-for-multinational-enterprises_9789264115415-en

Orlitzky, M., Schmidt, F. L., & Rynes, S. L. (2003). Corporate social and financial performance: A metaanalysis. *Organization Studies*, *24*(3), 403–441. doi:10.1177/0170840603024003910

Pedersen, E. B. (2006). Making Corporate Social Responsibility (CSR) Operable: How Companies Translate Stakeholder Dialogue into Practice. *Business and Society Review*, *111*(2), 137–163. doi:10.1111/j.1467-8594.2006.00265.x

Pirson, M., & Malhotra, D. (2008). Unconventional insights for managing stakeholder trust. *MIT Sloan Management Review*. Retrieved July 18, 2018, from http://www.hbs.edu/research/pdf/08-057.pdf

Post, J., Preston, L., & Sachs, S. (2002). *Redefining the corporation: Stakeholder management and organizational wealth*. Stanford, CA: Stanford University Press.

Senge, P. M. (1994). *The Fifth Discipline: the Art & Practice of the Learning Organization* (2nd ed.). New York: Bantam Doubleday/Currency Publishing Group.

Senge, S., Smith, B., Nina Kruschwitz, N., Laur, J., & Schley, S. (2008). *The necessary revolution: How individuals and organizations are working together to create a sustainable world*. London: Nicholas Brealey Publishing.

Trist, E., & Bamforth, K. (1951). Some Social and Psychological Consequences of the Longwall Method of Coal Getting. *Human Relations*, *4*(1), 3–38. doi:10.1177/001872675100400101

Uhl-Bien, M., Maslyn, J., & Ospina, S. (2012). The nature of relational leadership: A multitheoretical lens on leadership relationships and processes. In D. V. Day & J. Antonakis (Eds.), *The Nature of Leadership* (pp. 289–330). Thousand Oaks, CA: Sage Publications.

van Dijk, T. A. (1995). Discourse Semantics and Ideology. *Discourse & Society*, *6*(2), 243–289. doi:10.1177/0957926595006002006

van Luijk, H. J. L. (1997). Business ethics in Western and Northern Europe: A search for effective alliances. *Journal of Business Ethics*, *16*(14), 1579–1587. doi:10.1023/A:1005819216111

WCED. (1987). *Our common future. World Commission on Environment and Development*. Oxford, UK: Oxford University Press.

Yin, R. K. (1981). The Case study Crisis: Some Answers. *Administrative Science Quarterly*, *26*(1), 58–65. doi:10.2307/2392599

Yin, R. K. (1994). *Case Study Research: Design and Methods* (2nd ed.). Thousand Oaks, CA: Sage Publications.

Zink, K. J. (2014). Designing sustainable work systems: The need for a systems approach. *Applied Ergonomics*, *45*(1), 126–132. doi:10.1016/j.apergo.2013.03.023 PMID:23608710

APPENDIX: MAJOR THEMES BEING RAISED IN CSR DISCOURSES AT THE KEMIRA CORPORATION'S CSR REPORTS WITH A FIVE DISCIPLINES (SENGE, 1994) CLASSIFICATIONS

See Tables 2-5.

Table 2. Discourses in relation to employees at the Kemira Corporation: A Five Disciplines (Senge, 1994) classifications

	Discourses in Relation to Employees at the Kemira Corporation
Personal Mastery	• "With respect to competitiveness and growth, as well as to improve operative efficiency, it is essential to attract and retain personnel with the right skills and competences. By systematical development and improvement of compensation schemes, learning programs, and career development programs, corporation aims to ensure the continuity of skilled personnel also in the future." • "With respect to diversity in recruitment decisions, the Code of Conduct obligates us to respect the principles of equal opportunity and treatment without regard to age, race, birth, gender, creed, political persuasion, social status or origin."
Mental Models	• "Equal opportunity principles are also highlighted in our Recruitment Policy. In 2015, the share of women in the total workforce was 26%, in executive positions 22%, in Management Board 22%, and in Board of Directors 33%." • "Skilled leaders and employees are key to the successful execution of the Corporation's strategy. Performance management process consists of Performance and Development Discussions (PDD) and performance evaluation, supported by online management tools."
Shared Vision	• "In recruitment decisions, the Corporation does not apply gender and/or minority quotas as such but aim at finding the best professional knowhow for each position". • "The corporation has screened its current portfolio for priority substances that are subject to future regulatory restrictions or associated with particular concerns of safety, stakeholder relations and reputation. Management plans are prepared for these substances."
Team Learning	• "The corporation conducted an eLearning program on human rights in 2015 and the training was completed by 80% of the target group of 2,850 persons. Every employee receives regular training on our Code of Conduct, which is available in 21 languages and distributed to all employees." • "Our Code of Conduct, which is aligned with values, sets the minimum standards of expected behavior for employees and business partners. Internal policies and procedures provide more detailed guidance to steer daily work and decision making."
Systems Thinking	• "EHSQ management system is based on ISO 14001, ISO 9001 and OHSAS 18001 standards. Inputs (raw materials) and outputs (products) are mainly chemicals, including some classified as harmful or hazardous substances. EHSQ Policy requires that all manufacturing sites are certified in accordance with OHSAS 18001 management system standards for OHS. By the end of 2015, 86% of manufacturing sites were covered by OHSAS 18001 certification." • "The Corporation is committed to internationally recognized principles, such as the OECD Guidelines for Multinational Enterprises, the United Nations Global Compact and the chemical industry's Responsible Care."

Source: The author

Table 3. Discourses in relation to customers and suppliers at the Kemira Corporation: A Five Disciplines (Senge, 1994) classifications

	Discourses in Relation to Customers and Suppliers at the Kemira Corporation
Personal Mastery	• "Through Product Stewardship, the Kemira Corporation strives to ensure that products can be safely used by stakeholders, and that chemical risks and their impacts are duly incorporated into decision making and operations throughout product life cycle management. Customers have their own sustainability targets and follow several voluntary certification schemes, including eco-labeling schemes, which set further expectations on product offerings."; • "The suppliers invited to participate in EHSQ assessments have mainly been direct material suppliers, but some service and logistics providers have also been assessed. The Kemira Corporation has implemented improvement plans together with certain suppliers who were rated in the high risk category. In addition, suppliers with ongoing improvement plans are reassessed."
Mental Models	• "Supplier Performance Management Program, comprising of supplier Performance Evaluations and Supplier Sustainability Assessments, helps in identifying any possible challenges relating to suppliers' performance, including sustainability." • "Continuous improvements of sustainability performance is of strategic importance, and gives a competitive advantage through products that enable cost and resource savings in customer processes (e.g., water, energy, fibers)."
Shared Vision	• "The Carbon Index monitors CO_2 performance from both consolidated and individual manufacturing perspectives. The target is to reduce the value of the Carbon Index to or below 80 by the end of 2020, in comparison to the Carbon Index value of 100 in the baseline year 2012."; • "The Kemira Corporation is an innovation partner for all customers and many of customers are sustainability leaders in their business sectors and only work with partners with similar high standards and commitments. The Corporation has screened the current portfolio for priority substances that may be subject to future regulatory restrictions or associated with particular concerns. Management plans are prepared for these substances.
Team Learning	• "Supplier Sustainability Assessments are conducted by an external third-party corporation specialized in standardized supplier evaluation and auditing, based on the principles of the UN Global Compact, the Responsible Care program and Performance Management Program, comprising of supplier Performance Evaluations and Supplier Sustainability Assessments." • "The Kemira Corporation focuses on the sustainability of its business and is further improving the coordination and cooperation between Business Development, R&D and Sales Units in order to better understand the future needs and expectations of its customers."
Systems Thinking	• "The Kemira Corporation's business is supply chain intensive and has over 650 direct material suppliers, with 60 core suppliers, who deliver 80% of direct material spend. The Corporation drives responsibility in the supply chain through the Code of Conduct for Suppliers. By the end of 2015, 93% of suppliers had signed the Code. Some 27% of all raw materials used by Corporation in 2015 were recycled or industrial by-products from external partners." • "Growth and expansion to new applications can occur as follows: (1) Continued faster, industry leading innovations with key customers; (2) Manufacturing expansion investments with high return on capital employed; (3) Targeted bolt-on acquisitions to enable acquisition of equipment and provision of services."

Source: The author

Table 4. Discourses in relation to financiers and shareholders at the Kemira Corporation: A Five Disciplines (Senge, 1994) classifications

	Discourses in Relation to Financiers and Shareholders at the Kemira Corporation
Personal Mastery	• "Kemira corporate responsibility work is tied closely with strategy in helping create long-term sustainable value to the corporation and its stakeholders, improving operational efficiency, ensuring compliance and managing risks. The innovation capability is measured through an innovation sales target. In this respect, 10% of sales revenue by the end of 2016, will be from new chemistries, product upgrades and tailored chemistries marketed into new applications, developed and launched within the last 5 years." • "Steady growth and cash flow generation with strategic objective e.g. as follows: (1) Further strengthening the base business in raw and waste water treatment; (2) Ensuring efficiency & cost and product leadership through operational and commercial excellence and innovation; (3) Building on top of our strong customer base with innovation driven Advanced Water Treatment applications."
Mental Models	• "The Kemira Corporation's future market position and profitability depend on its ability to understand and meet current and future customer needs and predict market trends and its ability to innovate new differentiated products and applications." • "The Kemira Corporation holds over 1,300 patents and in 2015, innovation sales were EUR 189 million (8%). In 2015, the Kemira Corporation launched a program for operational excellence, addressing the efficiency and optimization of material and information flows from suppliers to customers in order to drive forward the fulfillment of its growth and profitability targets."
Shared Vision	• "Innovation and R&D related risks are being managed through an efficient management of the R&D portfolio by close collaboration between R&D and business segments." • "The following group policies are ready: Investment Policy (2013), Cash Management Policy (2012), Credit Management Policy (updated 2014), Acquisition and Divestment Policy (2014), Inventory Policy (2012), Risk Management Policy (updated 2014), Tax Policy (updated 2014), Treasury Policy (updated 2014), Transfer Pricing Policy (updated 2014)."
Team Learning	• "The Kemira Corporation monitors systematically leading indicators and early warning indicators that focus on market development. Growth through acquisitions also involves risks, such as the ability to integrate acquired operations and personnel successfully." • "Sourcing activities cover the identification and selection of suppliers, the consequent negotiations and contract management, and the management of supplier relationships."
Systems Thinking	• "The management consists of formal Performance and Development Discussions (PDD)" and Performance Evaluations, which are also linked with the Talent Management process and the Annual Salary Review Process. An on-line tool is used to document and follow-up Performance Management Process globally. Product quality, property or product yield optimization is 50%; Process or energy efficiency is 20%." • "Management system approach is supported by a LEAN manufacturing culture, which aims to achieve continuous improvements and create more from less, while providing increasing value to the Kemira Corporation. Its measures include the optimization of value chains and flows of information, and improvements in production efficiency."

Source: The author

Table 5. Discourses in relation to communities and authorities at the Kemira Corporation: A Five Disciplines (Senge, 1994) classifications

	Discourses in Relation to Communities and Authorities at the Kemira Corporation
Personal Mastery	• "The Kemira Corporation regularly reviews stakeholders' expectations and concerns to help us update sustainability priorities. Collaboration includes activities ranging from information sharing to active dialogue and collaboration on issues of mutual interest. The corporation employs approximately 4,685 professionals around the world." • "Stakeholder engagement is based on active stakeholder relationship management. Stakeholders continue to highlight the importance of topics such as sustainable products, safety, employee development, business ethics and compliance, responsibility along the supply chain, and reductions in environmental impacts both in its own operations and throughout the value chain."
Mental Models	• "Stakeholder groups include the local communities where corporation operates, regulatory bodies, trade associations, decision makers and opinion leaders. The Kemira Corporation regularly reviews stakeholders' expectations and concerns to help us update sustainability priorities. The employees are encouraged to raise any compliance issues via multiple channels." • "The Kemira Corporation is actively looking for alternative bio-based raw materials and focuses on applications where product biodegradability brings additional value, e.g. in wastewater treatment and packaging and board."
Shared Vision	• "The management scope of ethics and compliance is based on the topics included in the Code of Conduct. Management is based on activities designed to promote the principles of ethical business behavior. The Code of Conduct obligates the company to respect the principle of equal opportunity and treatment without regard to age, race, birth, gender, creed, political persuasion, social status or origin. The same equal opportunity principle is also highlighted in our Recruitment Policy." • "Impacts on communities mainly relate to environmental impact through resource efficiency in its own operations and throughout the value chain while the most important social impacts relate to safety, employee engagement and ethical business conduct."
Team Learning	• "Product stewardship goes beyond regulatory compliance, which in itself sets tight controls on the manufacture and sale of chemicals. The target is to reduce the value of the Kemira Corporation Carbon Index to or below 80 by the end of 2020, in comparison to the Carbon Index value of 100 in the baseline year 2012." • "Continuous improvements of sustainability performance is of strategic importance and by the end of 2015, the Corporation had certified 86% of major site locations according to ISO 14001, ISO 9001 or OHSAS 18001 standards."
Systems Thinking	• "The community's well-being derives in different ways e.g. from direct and indirect employment and paying taxes. Product stewardship goes beyond regulatory compliance, which in itself sets tight controls on the manufacture and sale of chemicals." • "The approach to reducing greenhouse gas emissions is based on improving energy efficiency at manufacturing sites and on purchasing energy with lower emission levels. Key principles of product stewardship for chemicals focus on efficient risk assessment: identifying the intrinsic properties of the substance, the use conditions and the potential exposure to that chemical."

Source: The author

Chapter 3
Socio-Technical Approaches for Optimal Organizational Performance:
Air Navigation Systems as Sociotechnical Systems

Tetiana Shmelova
National Aviation University, Ukraine

Yuliya Sikirda
Flight Academy of National Aviation University, Ukraine

ABSTRACT

In this chapter, the authors present a socio-technical system for optimal organizational performance at aviation enterprises such as air navigation system as socio-technical system. The authors made an analysis of the International Civil Aviation Organization documents on risk assessment and the impact of the social environment on the aviation system. The authors obtained the results of the evaluation of non-professional factors: determination of the social-psychological impact on decision making of human-operator by identifying the preferences for organizational performance. The structural analysis of internal and external management environment of aviation enterprise was carried out. And, as follows from the analysis, inhomogeneous factors that influence the aviation activity were classified, formalized, and systematically generalized using set-theoretical approach. The influence of factors of internal and external management environment on the aviation enterprise's activity was determined.

DOI: 10.4018/978-1-5225-7192-6.ch003

INTRODUCTION

Air Navigation System (ANS) in conformity to the principles of functioning may be referred to Socio-technical System (STS) within which close co-operation between human and technological components occur. The distinguishing feature of the STS is an availability of the hazardous kinds of activity as well as usage of the high-level technologies in production. Since operations in STS generally involve high-risk / high-hazard activities, the consequences of safety breakdowns are often catastrophic in terms of loss of life and property (International Civil Aviation Organization [ICAO], 2004). The more a human-operator (H-O) is trying to control a production process being aided by high-level technologies, especially in case of distant operation, the more non-transparent becomes the result of the operation of a system, which is accompanied by a high degree risk of the catastrophe causing (ICAO, 2002). Most researches were conducted with a view to the provision of safety in nuclear power production (Keating, 2001; Bertsch, 2007; Flueler, 2006). The provision of flight safety in the ANS by means of high-level technological processes depends primarily on the reliability of H-O as well as his timely professional decisions. Currently, one of the main strategic problems of mankind on the path to sustainable development is the safety and stability of technogeneous production (ICAO, 2013a). As noted technogeneous production is a complex system that contains interrelated technical, economic and social objects. It has a multilevel hierarchical structure and a high level of risk. Recent results show that there are frequent and common emergency such as disaster, accidents, crashes in hydraulic engineering, chemical and military industries, gas and oil pipelines, nuclear power plants and transport (Clegg, 2000; Keating, 2001; Flueler, 2006; Carayon, 2006; Bertsch, 2007; Baxter, 2011).

BACKGROUND

Sociotechnical systems theory is a theory about the social aspects of people and society and technical aspects of machines and technology (Kuchar, & Yang, 2000). Sociotechnical refers to the interrelatedness of "social" and "technical" aspects of an organization. The sociotechnical theory, therefore, is about "joint optimization", with a shared emphasis on achievement of both excellence in technical performance and quality in people's work lives. The sociotechnical theory, as distinct from socio-technical systems, proposes a number of different ways of achieving joint optimization. They are usually based on designing different kinds of organization, ones in which the relationships between socio and technical elements lead to the emergence of productivity and wellbeing (Kuchar, & Yang, 2000).

Statistical data show that human errors account for up to 80% of all causes of aviation accidents. A significant part of aviation accidents (33%) is in violation by crew members of the laws, rules, and regulations, as well as violations of preparation for training (42%) (Leychenko, Malishevskiy, & Mikhalic, 2006). The existing approaches to checking separate aspects (psychophysiological, behavioral, ergonomic, professional, etc.) do not consider the functional state of H-O in the conditions of the dynamic change of external and internal factors (Makarov, Nidziy, & Shishkin, 2000). Representation of the ANS in the form of STS first makes possible to take into account the influence of the social, cultural environment of people who decision making (DM). Culture surrounds people and affects their values, convictions, and behavior, which they share along with other members of different social groups. Culture serves to bind us together as members of groups and to provide clues as to how to behave in both normal and unusual situations. The psychologist Hofstede suggests that culture is a "collective programming of the mind"

(ICAO, 2004). Thus, fatal mistakes can be committed by normal, healthy, highly motivated and well-equipped personnel. Scientists have lately used the term "conscience deviation" when they analyzed the causes of aviation events conditioned by the insufficient development of the appropriate cultural values in a person that makes decisions (Leychenko, Malishevskiy, & Mikhalic, 2006).

The authors classified the factors that affect the H-O, who DM: professional factors (knowledge, habits, skills, experience); non-professional factors (individual-psychological, psychophysiological and socio-psychological). Aviation systems with its complex interrelation between a man and technologies have been evolved as complex STS (Figure 1). The distinction between "professional" and "non-professional" factors, features of factors are indicated in manuscripts and articles of authors (Kharchenko, Shmelova, & Sikirda, 2011a, 2011b, 2012a, 2012b, 2016, 2018; Shmelova et al., 2012, 2013, 2015, 2016, 2017).

The authors propose to get the evaluation of non-professional factors and determine the stability of ANS in the performance of professional active:

- Determination of the social-psychological impact on DM of H-O by identifying the preferences;
- Diagnosing individual-psychological qualities of H-O ANS in the development of flight situation;
- Monitoring of the psychophysiological factors as emotional state H-O ANS for timely diagnosis transition to a potentially dangerous mental activity.

The process of DM of operators in STS in Figure 2.

The ambient conditions determine the reaction of H-O and this reaction changes the environmental conditions accordingly. One of the possible approaches to the solution of these problems is formalization and mathematical presentation of the ANS operators' activities in the form of a complex STS on the base of the systemic analysis. It is necessary to timely diagnose the properties of the operator for all factors and forecasting of situations. Taking into account in the act of DM by a H-O within ANS, besides the separate professional factors (knowledge, habits, skills, experience) also the factors of non-professional nature (individual psychological, psychophysiological and socio-psychological) enables to predict the H-O's actions on the basis of modelling the "large-scale" outcomes of individual action with the aid

Figure 1. Factors impact in ANS as STS

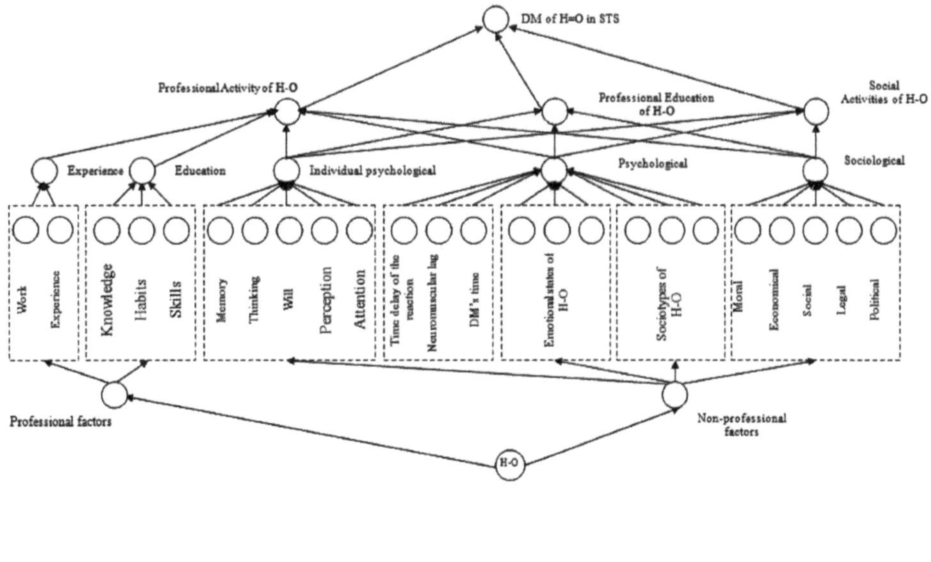

Figure 2. Graf of DM by H-O

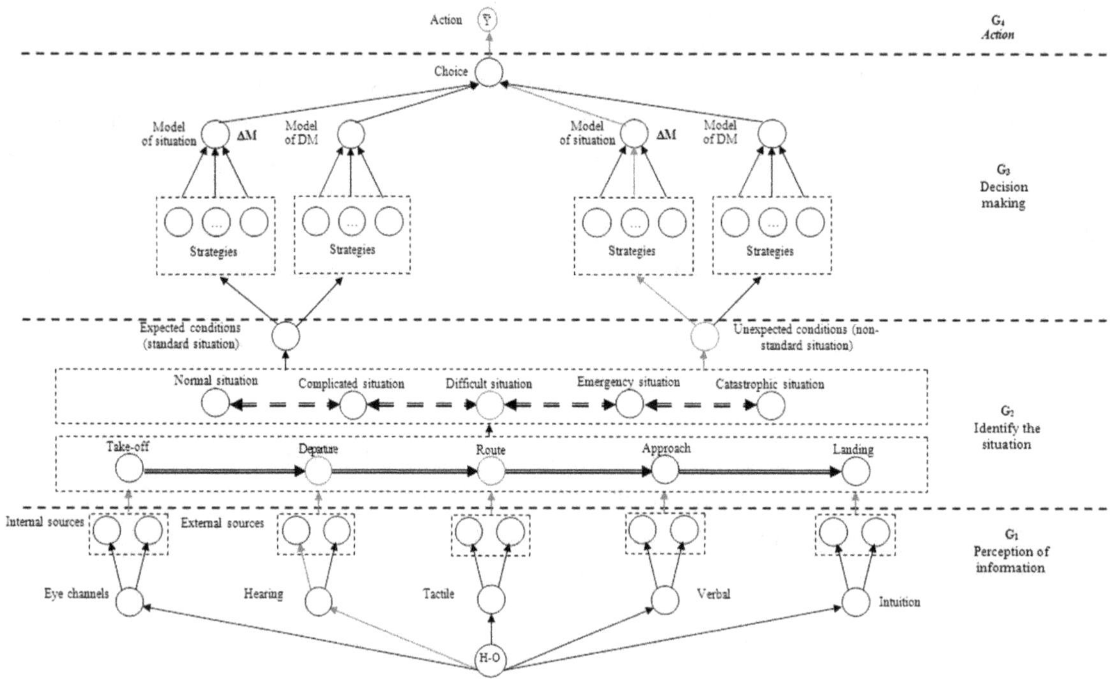

of the reflexive theory (Lefebvre, 2001, 2008; Lefebvre, & Adams-Webber, 2002). For example, modeling of DM of H-O in the unexpected conditions of flight (difficult flight situation, the hearing channel of identification informational by H-O, route stage of flight) functional $\overline{Y}$ (Figure 3) (1):

$$\overline{Y} = F\left(\overline{G}_{y}^{I}, \overline{G}_{p3}^{II}, \overline{G}_{s3}^{II}, \overline{G}_{c2}^{II}, \overline{G}_{Mdm}^{III}, \overline{G}_{i}^{IV}\right)\left(\overline{F_{inp}}\left\langle f_{ipt}, f_{ipa}, f_{ipp}, f_{ipth}, f_{ipi}, f_{ipn}, f_{ipw}, f_{iph}, f_{\exp}\right\rangle, F_{ip}\right) \tag{1}$$

On the basis of the methods for analysis of DM by the H-O ANS using the graph, stochastic, GERT, Markov's networks, the reflexive theory the methodology for analysis of flight situation development under influence DM by H-O ANS in-flight emergency has been developed (Kharchenko, Shmelova, & Sikirda, 2012, 2016). The proposed models will allow timely diagnosing and predicting the possible actions of H-O in the expected and unexpected conditions of operation of the aircraft. The practical value of the research carried out is to develop a method for conducting a prolonged socio-psychological correction of the H-O ANS in the process of training and professional activity, as well as the application for investigators of the approach to the evaluation of the activity of H-O during the investigation of the air crash.

Figure 3. Decomposition of the process DM by H-O ANS and the systemic analysis of factors which affect the DM by H-O in ANS

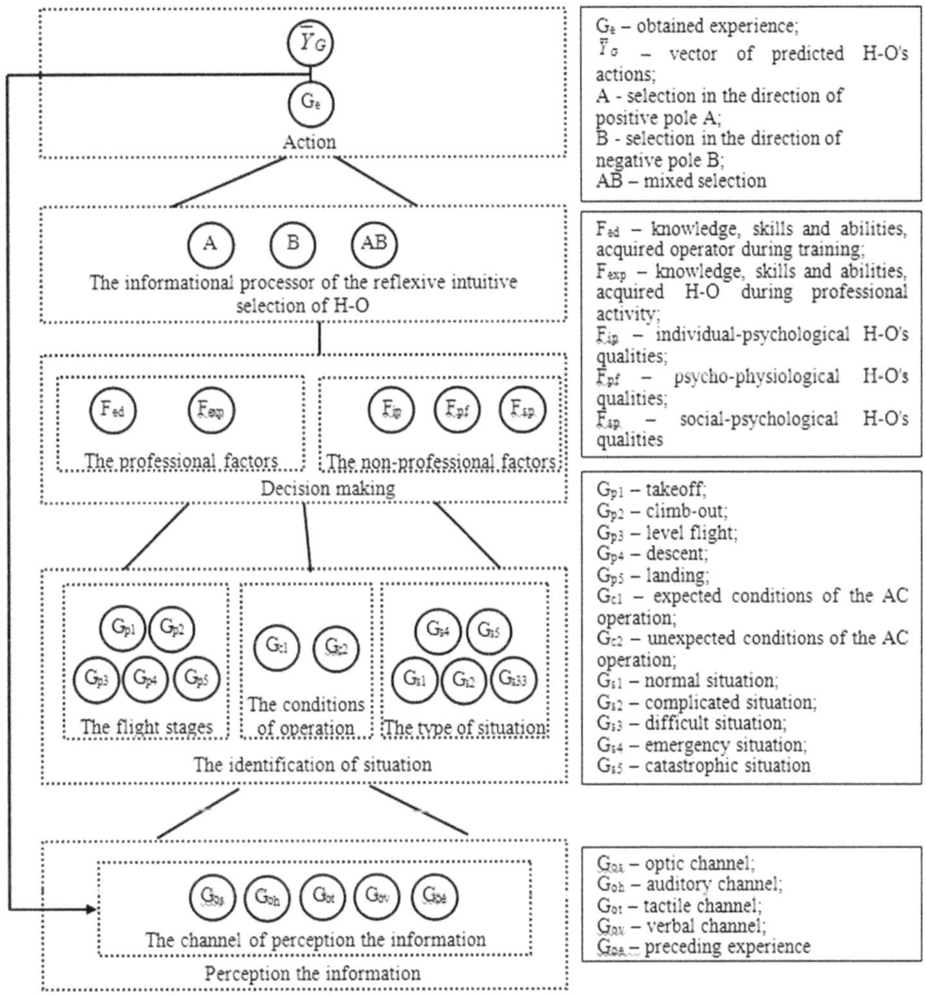

Reflexive Model of Bipolar Choice of Human Operator of the Air Navigation System as Sociotechnical System

Analysis of the influence of the factors of professional and non-professional activities on the DM in Air Navigation Socio-technical system (ANSTS) was made (Kharchenko, Shmelova, & Sikirda, 2016). The result of the evaluation of non-professional factors is determination the social-psychological impact on DM of H-O by identifying the preferences, diagnosing individual-psychological qualities of H-O ANS in the development of flight situation, monitoring of the psychophysiological factors (emotional state) H-O ANS for timely diagnosis transition to a potentially dangerous mental activity and determine the stability of ANS in the performance of professional actives was presented (Figure 4). In order to take into account, the complexities of the factors that influence on H-O of the ANSTS in the expected and unexpected conditions of operation of an aircraft, a reflexive model of bipolar choice of H-O has been worked out. In the "Informational processor of the reflexive intuitive choice" by H-O is selected in the directions of positive pole A, negative pole B; mixed selection AB according to reflexive theory (Lefebvre, 2001). The choice of H-O ANS is described by the function (2):

$$X = f\left(x_1, x_2, x_3\right) \tag{2}$$

where X – is a probability, that H-O is ready to choose a positive pole A in the reality; x_1 – is a pressure of the environment on H-O toward the positive alternative at the moment of the choice, $x_1 \in [0, 1]$; x_2 – is a pressure of the previous experience of H-O toward the positive alternative at the moment of the choice, $x_2 \in [0, 1]$; x_3 – is a pressure of the intention of H-O toward the positive alternative in moment of the choice, $x_3 \in [0, 1]$.

Verified the reflexive theory, i.e. unexpected emergency situations H-O performs automatic selection influenced by previous experience (x_1) and environment (x_2). In expected emergency situations H-O has influenced the choice of the intention (x_3) of the H-O. The alternative solution B - is the choice of H-O, which is determined by the H-O preferences system under which any form of arrangement of F-set is understood, i.e., removing the uncertainty of choice of some element $f^* \in F$ on the basis of selection rule K. A selection of a rule K shows the concept of a rational behavior of individual γ and his preferences system ρ in a particular situation of choice $\left\{\gamma, \rho\right\} \to K$. The H-O ANS preferences systems are influenced by professional $\overline{F}_p$ and non-professional $\overline{F}_{np}$ factors as it is expressed by Equation (3) and Equation (4), respectively.

$$\overline{F}_p = \left\{\overline{F}_{ed}, \overline{F}_{\exp}\right\} \tag{3}$$

$$\overline{F}_{np} = \left\{\overline{F}_{ip}, \overline{F}_{pf}, \overline{F}_{sp}\right\} \tag{4}$$

where $\overline{F}_{ed}$ – are the knowledge, skills and abilities, acquired H-O during training; $\overline{F}_{\exp}$ – are the knowledge, skills and abilities, acquired H-O during professional activity; $\overline{F}_{ip} = \left\{f_{ipt}, f_{ipa}, f_{ipp}, f_{ipth}, f_{ipi}, f_{ipn}, f_{ipw}, f_{iph}, f_{\exp}\right\}$ – is a set of H-O individual-psychological factors (temperament, attention, perception, thinking, imagination, nature, intention, health, experience); $\overline{F}_{pf}$ – is a set of H-O psychophysiological factors (features of the nervous system, emotional types, sociotypes); $\overline{F}_{sp} = \left\{f_{spm}, f_{spe}, f_{sps}, f_{spp}, f_{spl}\right\}$ – is a set of H-O socio-psychological factors (moral, economic, social, political, legal factors).

The priority of economic and social factors in the result of the influence of *socio-psychological* factors on the professional activities by military pilots and navigators were obtained with using Expert Judgment Method (EJM) (see Figure 4 and Table 1).

The preferences models and the priorities of socio-psychological factors $\overline{F}_{sp}$ (f_{sps} – social factors; f_{spe} – economic factors; f_{spl} – legal factors; f_{spp} – political factors; f_{spm} – moral factors) for military pilots and navigators also were obtained by means of Equation (5):

Figure 4. The influence of socio-psychological factors on the professional activities of military pilots and navigators

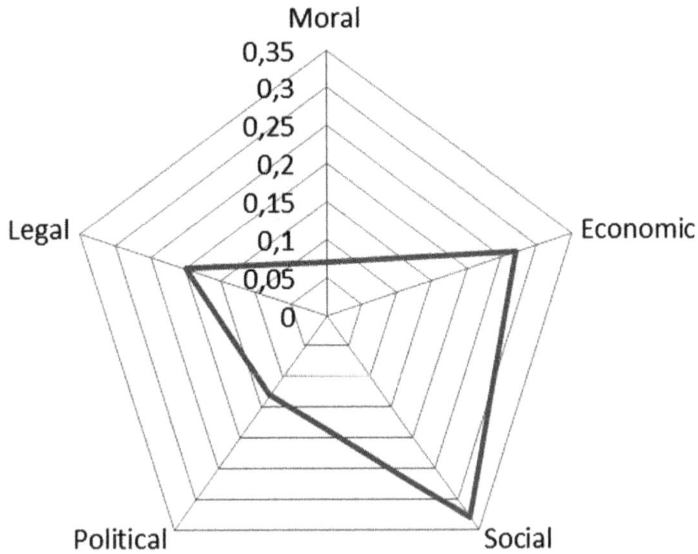

$$f_{sps} \succ f_{spe} \succ f_{spl} \succ f_{spp} \succ f_{spm} \quad (5)$$

The previous studies the factors that affect the DM by an H-O ANS have been determined, namely: level of knowledge, skills, abilities, preceding experience as well as the factors of non-professional nature. The analysis of social-physiological factors conducted by the authors allowed making a conclusion that the activities of pilots are influenced by the own image, the image of the corporation as well as by interests of a family (Kharchenko, Shmelova, & Sikirda, 2012). At the same time, respondents – Air Traffic Control's Operators (ATCO) pay special attention to interests of their families, their own economic status, and professional promotion.

We researched the influence of socio-psychological factors on professional activities H-O (civil pilots and controllers) too. Graphs of the preferences models for civil pilots and ATCO also the priorities of socio-psychological factors were obtained and showed (Figure 5).

Table 1. The preferences system of military pilots and navigators

Factors	Pilot			Navigator		
Socio-Psychological	**Average of Factor**	**Weight of Factor**	**Rank of Factor**	**Average of Factor**	**Weight of Factor**	**Rank of Factor**
Moral factors	4.67	0.07	5	4.75	0.07	5
Economic factors	2.00	0.27	2	2.20	0.27	2
Social factors	1.00	0.33	1	1.60	0.33	1
Political factors	4.33	0.13	4	4.2	0.13	4
Legal factors	3.00	0.20	3	2.30	0.20	3

Figure 5. The influence of socio-psychological factors on the professional activities of pilots and ATCO

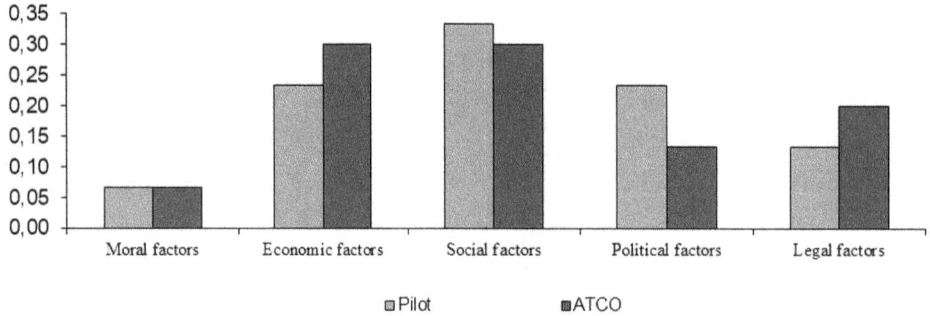

In next case, using EJM obtained the preferences system of the pilot on the set of individual-psychological factors $\overline{F}_{ip}$, which reflect the objective characteristic of DM and thinking psychology of H-O is guided by an action, in cases of normal and catastrophic situations by Equation (6) and Equation (7), respectively.

$$\left(f_{iph}, f_{\exp}\right) \succ f_{ipa} \succ f_{ipw} \succ f_{ipt} \succ f_{ipi} \succ f_{ipp} \succ f_{ipth} \succ f_{ipn} \tag{6}$$

$$\left(f_{iph}, f_{\exp}\right) \succ (f_{ipt}, f_{ipp}) \succ f_{ipa} \succ f_{ipw} \succ f_{ipth} \succ f_{ipi} \succ f_{ipn} \tag{7}$$

where fi_{ph} – is health; f_{ipexp} – is the experience; f_{ipa} – is attention; f_{ipw} – is intention; f_{ipt} – is temperament; f_{ipi} – is imagination; f_{ipp} – is perception; f_{ipth} – is thinking; f_{ipn} – is nature.

With the bipolar reflexive behavioral model of H-O in extreme situations received *W*-functions of positive (*A*) and negative (*B*) choice. The model represents the subject (H-O), who is before the bipolar choice of one of the alternatives: *A* (positive pole) and *B* (negative pole). The alternative solution is the choice of H-O, which is determined by the H-O decision-making system in risk (stochastic uncertainty). The optimal solution is finding by applying the Equation (8) of the criterion of expected value with the principle of risk-minimizing as (8):

$$A_{opt} = \min\left\{R_{ij}\right\} \tag{8}$$

where R_{ij} – is an expected risk for solution A_{ij}, which is determined by Equation (9):

$$R_{ij} = \sum_{j=1}^{m} p_{ij} u_{ij}, \ i = \overline{1,n}, \ j = \overline{1,m} \tag{9}$$

where p_{ij} – is a probability of *j*-factor influence during *i*-alternative solution choice, and $\sum_{j=1}^{m} p_j = 1$; u_{ij} – is a loss, associated with choosing *i*-alternative solution during *j*-factor influence.

For the formalization of the behavior of ANS H-O in-flight situations, the graphics models' relationships between a cause and an impact – graphs, trees, events and functional networks of stochastic structures – might be useful. To study the impact of DM by H-O during the flight situations development applied the stochastic network type GERT (Graphical Evaluation and Review Technique). GERT allows modeling the increase of flight situations complication as well as its decrease and/or simplification. GERT is an alternative probabilistic method of network planning, applicable in the case when these actions can only start after completion of a prior action including cycles and loops. In stochastic networks of the flight situation development of GERT type, the tops are represented by stages of the situation (normal, complicated, difficult, emergency or catastrophic), and the arcs are represented by a process of transition between stages of the situation.

Let's consider the stochastic network model of the flight situation development GERT $G = (N; A)$ with a set of tops N and a set of arcs A. The time t_{ij} of transition from *i*-flight situation to *j*-flight situation is a random variable. Transition (i, j) can be executed only if *i*-top has been done. For calculation of transition time t_{ij} from *i*-flight situation to *j*-flight situation, it is necessary to know conditional probability (in discrete case) or the density of distribution (in continuous case) of random variable Y_{ij}. This allows researching the performance of the whole network $G = (N; A)$ and to identify the moments of time distribution t_{ij} of network G, calculate mathematical expectation μ_{jE} and variance of the execution time δ^2 of network G in case of complicated, complex, catastrophic or emergency situation. Let f_{ij} be the conditional probability (density of distribution) of time to make the transition from flight situation G_i to flight situation G_j. The conditional producing function of moments of random variable Y_{ij} is defined by Equation (10):

$$M_{ij}(s) = E\left[e^{Y_{ij}}\right] \tag{10}$$

In continuous and discrete cases, the random variables Equation (10) is being transformed into Equation (11) and Equation (12), accordingly:

$$M_{ij}(s) = \int e^{sy_{ij}} f\left(y_{ij}\right) dy_{ij} \tag{11}$$

$$M_{ij}(s) = \sum e^{sy_{ij}} f\left(y_{ij}\right) \tag{12}$$

If $y_{ij}=a=const$, then $M_{ij}(s) = E\left[e^{sa}\right] = e^{sa}$.

W- is the function for random variable Y_{ij} as transmission coefficient of GERT-network is introduced in Equation (13):

$$W_{ij}(s) = p_{ij} M_{ij}(s) \tag{13}$$

where p_{ij} – is a probability, that *j*-flight situation will come and transition *(i; j)* has been made; $M_{ij}(s)$ – is a conditional producing function of moments of random variable Y_{ij}.

The Algorithm of the Stochastic Network Analysis: An Example of GERT-Network

1. For obtaining close stochastic network *G* enter in the open stochastic network $W_E(s)$ additional dummy arc with *W*-function $W_A(s)$, which connects the drainage of the open network with a source *s*.
2. For modified network *G* to determine all *k*-loops, $k = \overline{1, n}$.
3. The equivalent transmission coefficient for all *k*-loops of *G*-network, $k = \overline{1, n}$ is being calculated by Equation (14):

$$T(L_n) = \prod_{k=1}^{n} T_k = \prod_{k=1}^{n} \left[\prod_{(i,j)\in L_{k_1}} t_{ij} \right] \tag{14}$$

where $T_k = \prod_{(i,j)\in L_{k_1}} t_{ij}$ – is an equivalent transmission coefficient of *1*-loop L_{k1}; t_{ij} – is a time of transition from *i*-flight situation to *j*-flight situation.

4. To apply Mason's rule for topological equation close stochastic network *G*, Equation (15) is applied:

$$H = 1 - \sum T(L_1) + \sum T(L_2) - \sum T(L_3) + ... + (-1)^k \sum T(L_k) + ... = 0, \tag{15}$$

where $\sum T(L_k)$ – is a sum of equivalent transmission coefficients for all possible *k*-loops.

5. From the topological equation of close stochastic network *G* transmission coefficient of an open network, $W_E(s)$ were determined.
6. The first and the second moments of random variable Y_{ij} are determined by means of Equation (16):

$$\mu_{jE} = \frac{\partial^j}{\partial s^j}\left[M_E(s)\right], \tag{16}$$

where μ_{1E} – is a mathematical expectation of the execution time of network *G*; μ_{2E} – is a standard deviation of the execution time of network *G*.

Thus according to results of stochastic network analysis of the flight situation development from normal to catastrophic the following values have been obtained (Shmelova, Sikirda, & Jafarzade, 2013;

Shmelova, & Sikirda, 2016): mathematical expectation of flight situation's development time μ_{jE}; variance of flight situation's development time δ^2; probability of flight situation's development p_{ij}.

For example, let's analyze catastrophic situation development under hazardous weather conditions using the decision tree and stochastic network GERT (Kharchenko, Shmelova, & Sikirda, 2012, 2016). According to data of the National Transportation Safety Board (NTSB) (Aviation Accident Statistics, 2017), during the last 10 years, 21,3% aviation accidents happened due to weather conditions, of which 39,1% – in bad weather conditions. The major cause of aviation accidents in bad weather conditions (68%) considered improper and untimely DM by the crew of the aircraft. Based on the *W*-functions of the positive and negative of H-O choice the Markov's network of flight situations' development from normal to catastrophic was constructed (Figure 6). Markov's process with discrete states W_{ij} is called the process of death and life.

Expected risks R_A, R_B of H-O obtained in DM during the approach performed in bad weather conditions under the influence of external environment x_1, previous experience of H-O x_2 and intention of H-O x_3. Expected risk of H-O DM is given by Equation (17):

$$R_{DM} = \begin{cases} R_A = \min\left\{R_{ij}\right\} \\ R_B = \left\{\gamma, \rho\right\} \\ R_{AB} = \left\{X(x_1, x_2, x_3), \gamma, \rho\right\} \end{cases}, \quad (17)$$

where R_A is an expected risk of H-O in DM taking into account the criterion of minimizing of expected value; R_B is an expected risk of H-O DM taking into account his preferences model; R_{ij} is an expected risk of A_{ij}-decision; γ is a concept of rational behavior of individual; ρ is a preferences system of individual in a particular situation of choice; R_{AB} is a mixed choice of H-O.

Figure 6. Stochastic models: GERT-network, decision tree, and Markov's network: W_{ij}, $W_E(s)$, $W_A(s)$ – are the transmission coefficients of (i,j)-arc, of the open network and of dummy arc; G_1, G_2, G_3, G_4, G_5 – are the normal, complicated, difficult, emergency, catastrophic situations; A, B – are the positive or negative choice

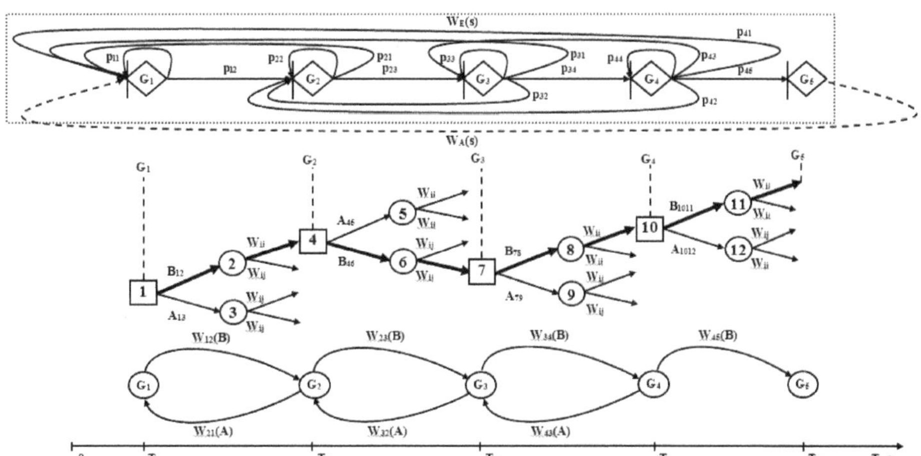

So, based on the reflexive theory of bipolar choice, the expected risks of DM of the ANS's operator have been studied and the influence of the external environment, previous experience and intention of the H-O have been identified.

Socio-Technical Approaches for Optimal Organizational Performance of the Air Navigation System as a Socio-Technical System

It is considered that aviation is the most fail-safe type of transfer. In as little as century, aviation, in the sphere of flight safety, rose through the ranks from an unstable system to the first «ultra-safe» system in the history of transport, it means system in which the number of catastrophic failures, in the sphere of safety, make up less than one per one million of production cycles. According to ICAO's data (ICAO, 2013b), compared with 2011, in 2012 the number of incidents in the world decreased on 21%, in 2013 – on 13%; the number of fatalities in 2012 decreased on 10%, in 2013 – on 55%. Consequently, the least number of deaths was fixed in 2012-2013, starting from 2004. As a result of the decreased number of incidents and increased number of departures, the frequency of incidents in the world in 2012 was reduced to 3,2 events per million departures and in 2013 – to 2,8 events per million departures (ICAO, 2014). This is the lowest value since the ICAO started to monitor the frequency of incidents in the world.

Aviation systems cannot be wholly free from dangerous factors and connected with the risks, while, the elimination of aviation events and serious incidents continues to be the final goal of human activity in the sphere of aviation safety. Neither human activity nor systems created by it guarantee a total absence of operating errors and their consequence (ICAO, 2013a). In such a way, safety is a dynamic characteristic of aviation with the help of which risk factors for flight safety should steadily decrease. It is important to note that adoption of effective indices of ensuring flight safety is frequently influenced by internal and international standards and also by cultural features (ICAO, 2004). While risk factors for flight safety and operating errors are under control, such opened and dynamic system as civil aviation may be controlled providing the necessary balance between flight performance and safety requirements for passengers and their baggage. ICAO constantly develops and improves proactive, based on the risks evaluation, methods, directed on the farther decrease in the number of aviation events in the world. Also, ICAO encourages aviation communities to recognize the importance of adherence to the single global approach for safety improvement and monitoring. A modern approach, founded on the characteristics (performance-based approach – PBA), based on the next three principles: the main accent on desired/necessary results; decision making, oriented on desired/necessary results; using facts and data while decision making. Herein the principle "using facts and data while decision making" admits that tasks shall comply with the widely known in Western management criteria SMART, that correspond to an abbreviation of five English words: specific, measurable, achievable, relevant and timebound (ICAO, 2009). Such level of accuracy of tasks determination may be achieved only using the way of consistent and structural description of inhomogeneous components of aviation – aviation enterprises, aviation personnel, aviation infrastructure, techniques, rules and information that is used for creating of conditions and applying of aerospace by aircraft users.

In aviation document, the safety balance model is presented (Figure 7). Efficiency addresses the operational and economic cost-effectiveness of gate-to-gate flight operations from a single-flight perspective. In all phases of flight, airspace users want to depart and arrive at the times they select and fly the trajectory they determine to be optimum.

Figure 7. Civil aviation in context: Safety balance model
Source: ICAO, 2005

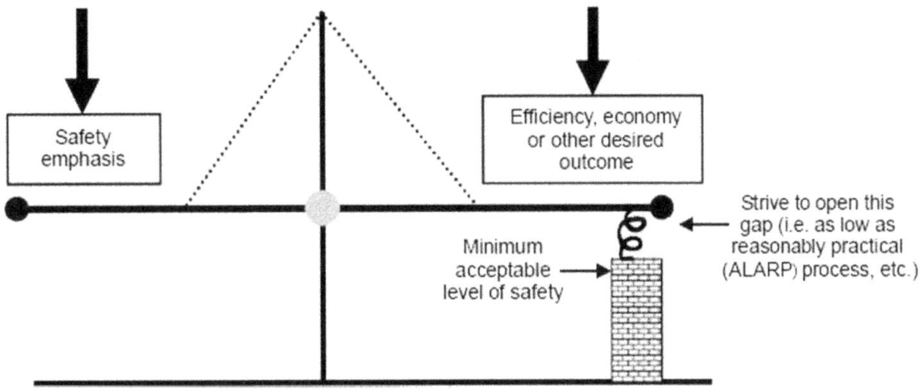

The aviation in the whole is a complex system that requires research of the human contribution to safety and an understanding of how human performance may be affected by its multiple and interrelated components such as technical, political, physical, social, economic, culture, etc. (ICAO, 2005). The human factor remains a major cause of aviation accidents. That is why the current management system needs to improve. The human factor as a term requires a precise definition because when it is used in everyday life, it often encompasses all aspects of human activity (ICAO, 1998, 2002). People are the most flexible, adaptable and important element in the aviation system and the most vulnerable in terms of opportunities effect on its activity. At the initial stage of development of aviation, many problems have been associated with exposure to human noise, vibration, heat, cold and acceleration forces. But optimization of the human role in complex systems is related to all aspects of human activity, such as DM processes and knowledge; design configuration displays, controls and equipment of cockpit; maintaining communications, and software; preparation of plans and maps.

An approach to forecasting the efficiency of the airline's operating activity is presented using combinatorial and regression analysis methods. The developed regression model allows finding the optimal aircrafts' composition to guarantee the minimum cost of air transportations provided that the demand for transportation is ensured maximally. In modern conditions, when opening a new airline or changing demand for air transportation routes, the airline faces problems in determining the optimal composition of the air fleet and the distribution of existing aircrafts on selected flights in order to maximize the satisfaction of existing demand for air transportation, provided that the air transport company's costs are minimized. The mathematical apparatus that is expedient to use for calculating the optimal composition of the airline fleet park is combinatory and regression analysis, for distribution of aircraft in the established directions of air transportation – transport model (Shmelova, & Sikirda, 2017).

An Algorithm for Analyzing the Efficiency of an Enterprise as STS and Optimal Organizational Performance

1. A system analysis and formalize factors affecting the STS's operators. Analysis of the influence of the factors of professional and non-professional activities on the DM by H-O:

   a. Influence of the factors of non-professional activities (individual-psychological, psychophysiological and socio-psychological) on the DM by H-O.
   b. Influence of the factors of professional activities (knowledge, habits, skills, experience, etc.) on the DM by H-O. Socionic and sociometric diagnosing of ANS's operator.
   c. The methods of diagnosis of H-O qualities (influents of professional and non-professional factors on activities).
   d. Diagnosis and evaluation of a psycho-emotional state of the STS's operator in emergencies.
   e. Identification and analysis of organizational factors for optimal organizational performance.
2. Models of DM in STS and forecasting the development of the situation for optimal organizational performance:
   a. Deterministic models of DM in STS.
   b. Stochastic models of DM in STS by H-O in extreme situations (under conditions of stochastic reflexive bipolar choice).
   c. The Neural network, Markov's, GERT-models of DM in STS by H-O in extreme situations.
   d. Models of diagnosis of the emotional state of H-O in extreme situations.
3. Decision support systems and Expert systems for H-O of enterprises.
4. Level of influence of factors of internal and external management environment of aviation enterprise on organizational activity. Creation of "Passport of the enterprise safety" (in aviation, for example).
5. Forecasting the efficiency of the enterprises operating activity.

Example of Identification of the Organizational Factors that Affect Flight Safety in Air Traffic Control System

In order to identify organizational factors that affect flight safety in air traffic control (ATC), an expert questioning was conducted on 30 area ATCO at Lviv Regional Branch of UkSATSE (Sikirda, Shmelova, & Tkachenko, 2017). The questionnaire was based on "Swiss-Cheese" model by Professor James Reason (ICAO, 2013a) and the variety of structural elements for the development of "organizational accidents" were outlined. It consisted of eight selected main groups of organizational factors: operational environment, procedures and manuals, engineering procedures and maintenance, cooperation between ATC sectors, ATC systems and equipment, infrastructure, airspace structure, company management, and structure (Table 2).

Each expert filed the matrix of individual preferences. With the help of the pairwise comparison method and ranking, the significance rank of each group of factors according to individual expert`s priorities has determined. Next step was to form the group preferences matrix and to obtain the average index of the group of experts concerning each group of organizational factors R_{gri} and rank of each group R'_{gri}. The competences of experts have considered being equal. Two stages of questioning have performed in order to achieve the agreement among professionals on the level of influence on safety by each group. Level of significance of each group of organizational factors has described with weight coefficients ω_i. The weight coefficient is determined by the Equation (18):

$$\omega_i = \frac{C_i}{\sum_{i=1}^{n} C_i}, \tag{18}$$

Table 2. The groups of organizational factors that influence flight safety in ATC

Group Number	Group Name	Group Description
1	Operational environment	Factors, connected with the physical environment (temperature, air circulation, illumination, level and timing of noise, atmospheric pressure, etc.).
2	Procedures and manuals	Factors, associated with adequacy/inadequacy of procedures/manuals, failure to comply or possibility/impossibility of compliance with them.
3	Engineering procedures and maintenance	Factors, connected with the work of engineers, including routing technical checks and equipment maintenance after the failure. In addition, procedures of design, installation, and implementation of new equipment.
4	Cooperation between ATC sectors	Factors, relating to the technical aspects of system operations between ATC control sectors as well as between adjacent ATC systems (compatibility of coordination procedures, Letters of Agreement, acceptance of information from other sources).
5	ATC systems and equipment	Factors, associated with work of the hardware, software and its compatibility.
6	Infrastructure	Factors, connected with aerodrome (physical parameters, the configuration of maneuvering areas, restriction zones) and environmental layout.
7	Airspace structure	Factors, relating to classification of airspace structure, route network, capacity, the configuration of sectors.
8	Company management and structure	Factors, connected with the style of company management at all levels, company ethics.

where $C_i = 1 - \frac{R'_{gri} - 1}{n}$ – are the intermediate assessments; R'_{gri} – are the ranks of i-group of organizational factors.

The obtained weight coefficients of groups of organizational factors that affect the safety of flights in ATC are presented in Table 3.

As a result of performed expert survey, the following order of groups of organizational factors depending on the level of their influence on flight safety has achieved (from the most significant to the least significant one):

Table 3. Calculation of weight coefficients of organizational factors' groups

Group Number	Group Rank, R'_{gri}	Intermediate Assessment, C_i	Weight Coefficient, ω_i
1	7	0.250	0.06
2	3	0.750	0.17
3	6	0.375	0.08
4	2	0.875	0.19
5	1	1.000	0.22
6	5	0.500	0.11
7	4	0.625	0.14
8	8	0.125	0.03
Σ		4.5	1

1. ATC systems and equipment ($R_{gr5} = 2{,}63$; $R'_{gr5} = 1$; $\omega_5 = 0{,}22$).
2. Cooperation between ATC sectors ($R_{gr4} = 3{,}12$; $R'_{gr4} = 2$; $\omega_4 = 0{,}19$).
3. Procedures and manuals ($R_{gr2} = 3.5$; $R'_{gr2} = 3$; $\omega_2 = 0.17$).
4. Airspace structure ($R_{gr7} = 3.98$; $R'_{gr7} = 4$; $\omega_7 = 0.14$).
5. Infrastructure ($R_{gr6} = 4.38$; $R'_{gr6} = 5$; $\omega_6 = 0.11$).
6. Engineering procedures and maintenance ($R_{gr3} = 5.27$; $R'_{gr3} = 6$; $\omega_3 = 0.08$).
7. Operational environment ($R_{gr1} = 6.37$; $R'_{gr1} = 7$; $\omega_1 = 0.06$).
8. Company management and structure ($R_{gr8} = 6.75$; $R'_{gr8} = 8$; $\omega_8 = 0.03$).

The results of the expert questioning have presented as a system of advantages as indicated by Equation (19):

$$R'_{gr5} \succ R'_{gr4} \succ R'_{gr2} \succ R'_{gr7} \succ R'_{gr6} \succ R'_{gr3} \succ R'_{gr1} \succ R'_{gr8}, \tag{19}$$

where R'_{gri} is the rank of *i*-group of organizational risk factors.

It is clear that out of all organizational groups of factors "ATC systems and equipment" group has the most significant impact on safety in ATC, and "Company management and structure" – the least one. The graphics interpretation of the results of an expert questioning in the form of a histogram is shown in Figure 8.

The results of flight safety assessment at Lviv Regional Branch of UkSATSE on the basis of analysis of organizational factors are presented in Table 4. In accordance with the matrix of the risk index (ICAO, 2013a), which takes into account the probability and severity of possible consequences, the scale of acceptability (tolerance) of risk factors is determined on the basis of the theory of fuzzy sets with the use of linguistic variables: extreme risk (100 points), high risk (80 points) moderate risk (60 points), low risk (35 points) and scarce risk (20 points). To ensure a sufficient level of flight safety, the risk indicators must be no more than 60 points taken at the maximum permissible level of danger. The actual significance of the level of danger to the groups of organizational factors is determined by questioning the ATCO at Lviv Regional Branch of UkSATSE and statistical processing of the results, which

Figure 8. Graphics interpretation of expert opinions
Source: The authors

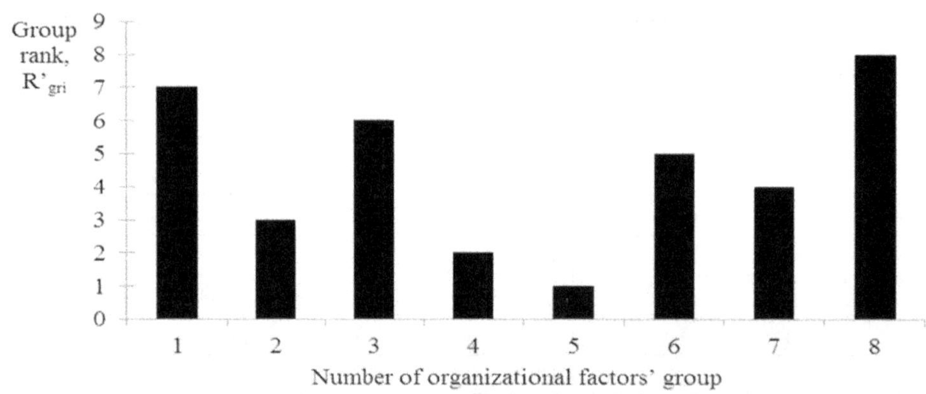

confirmed the consistency of expert opinions. From the multi-parametric indicator of the flight safety state in ATC on the basis of the analysis of organizational factors to the scalar indicator was carried out in a multiplicative way (Sikirda, Shmelova, & Tkachenko, 2017) by means of Equation (20):

$$W = \prod_{i=1}^{n} L_i^{w_i} = \prod_{i=1}^{n} P_i, \quad (20)$$

where L_i is a level of hazard of i-group of organizational factors; ω_i is a weight coefficient (degree of influence) of i-group of organizational factors; and P_i is a parameter of the hazardous level of i-group of organizational factors.

The results of the expert questioning are presented graphically using a spider diagram (Figure 9).

With the help of data from Table 4 the authors were received maximum allowable W_{all} and actual W_{act} value of the multiplicative function for assessing the level of flight safety in ATC at Lviv Regional Branch of UkSATSE on the basis of organizational factors:

$$W_{all} = 60^{0,06} \cdot 60^{0,17} \cdot 60^{0,08} \cdot 60^{0,19} \cdot 60^{0,22} \cdot 60^{0,11} \cdot 60^{0,14} \cdot 60^{0,03} = 60,99;$$

$$W_{act} = 20^{0,06} \cdot 35^{0,17} \cdot 25^{0,08} \cdot 35^{0,19} \cdot 20^{0,22} \cdot 25^{0,11} \cdot 30^{0,14} \cdot 35^{0,03} = 27,33.$$

Table 4. Results of the flight safety assessments in ATC at Lviv Regional Branch of UkSATSE

Group of Organizational Factors	Operational Environment	Procedures and Manuals	Engineering Procedures and Maintenance	Cooperation Between ATC Sectors	ATC Systems and Equipment	Infrastructure	Airspace Structure	Company Management and Structure
Weight coefficient (degree of influence), w_i	0.06	0.17	0.08	0.19	0.22	0.11	0.14	0.03
The maximum allowable level of hazard, L_{alli}	60	60	60	60	60	60	60	60
The actual level of hazard, L_{acti}	20	35	25	35	20	25	30	35
The parameter of the maximum allowable hazardous level, P_{alli}	1.28	2.01	1.39	2.18	2.46	1.57	1.77	1.13
The parameter of expertise of the actual hazardous level, P_{acti}	1.20	1.83	1.29	1.97	1.93	1.42	1.61	1.11
The difference between maximum allowable and actual parameters of the hazardous level, $\Delta P = P_{alli} - P_{acti}$	0.08	0.18	0.10	0.21	0.53	0.15	0.16	0.02

Figure 9. Graphics interpretation of the expert questioning results for the flight safety in ATC at Lviv Regional Branch of UKSATSE on a basis of organizational factors

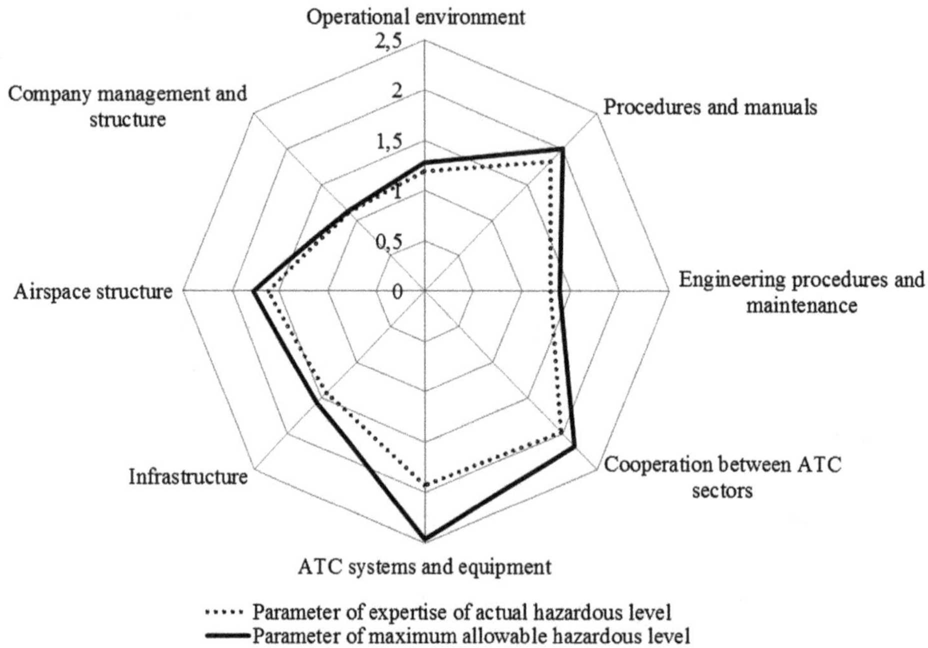

Comparison of flight safety results in Lviv Regional Branch of UkSATSE was performed on the basis of the maximum allowable level of danger and actual danger level expertise of organizational groups of factors $\Delta W = W_{all} - W_{act} = 60{,}99 - 27{,}33 = 33{,}66$. An example of the expertise results, presented in Figure 9, shows the correspondence of the values of all groups of organizational factors to the maximum permissible level of the hazard, which indicates a high flight safety index of ATC at this Regional Branch. The presented researches were performed within the NETCENG TEMPUS Project "New Model of the Third Cycle Engineering Education Due to the Bologna Process" framework which is funded with the support of the European Commission. The core of the flight safety management practical mechanism in the ATC system is the purposeful search for risk factors that lead to aviation accidents in order to protect from their influence. Identification of risk factors is implemented in the form of regular monitoring, collection, processing, and accumulation of information about factors that have caused aviation accidents. The model of "Swiss-Cheese" by Professor James Reason shows that the organizational risk factors play a significant role in the causality of aviation accidents and require a comprehensive research. Organizational factors that influence the safety of flights in ATC were systematized in eight groups: operational environment, procedures, and manuals, engineering procedures and maintenance, cooperation between air traffic control sectors, ATC systems and equipment, infrastructure, airspace structure, company management, and structure. It was revealed that the most significant influence on ATC had "ATC systems and equipment" group of organizational factors and the least – "company management and structure". The multiplicative function for assessing the safety of flights in ATC was obtained, which allows checking the conformity of the values of organizational factors to the maximum permissible level of the hazard. The proposed methodology for assessing the impact of organizational factors on flight safety in ATC will allow developing the safety passports that can be applied by the aviation authorities during the certification inspections of air navigation service providers to compare the normative and actual indicators of their activities (Sikirda, Shmelova, & Tkachenko, 2017).

The Structure of Internal and External Management Environment of Aviation Enterprise

Nowadays a new approach for flight safety provision is forming in global practice. The ICAO's term "Safety Management System" (SMS) (ICAO, 2013a) integrates operations and technical systems with the management of financial and human resources to ensure aviation safety or the safety of the public. Given statement implies binding into one, in a single system different objects and subjects of aviation activity. The subjects of aviation activity are designers and manufacturers of aviation equipment, airline operators, organizations engaged in aircraft maintenance and repair, and also handling companies. The air traffic services providers, aviation educational institutions, research and design aviation organizations, aviation authorities are included to them. Each subject of aviation activity is used to provide manufacturing processes and functioning of civil aviation the appropriate objects: aircrafts, their components and equipment, industrial equipment, ground vehicles, aerodrome and on-route equipment (radio- and electrotechnical facilities), engineering technical buildings, other movable and immovable assets, etc. And it is a very complex task to join the named elements of aviation.

Since the period of "organizational era" (the 1990s – present), flight safety has been considered systematically, taking into account organizational, human and technical factors. Also, at that date, a notion "aviation incident because of organizational reasons" appeared in aviation, taking into account the impact of organizational culture and politics on the control systems efficiency for flight safety risk factors. The "Swiss Cheese Model" by James T. Reason shows that significant role in the incidents causation belongs to organizational and management factors (ICAO, 2013a).

Building an effective SMS of aviation activity requires an integrated research of the environment in which aviation enterprises operate. The structure of the enterprise environment is considered in depth in the works of such prominent scholars in the field of management process (Mescon, Albert, & Khedouri, 2008). Many types of research related to the determination of the influence of factors of internal and external management environment of aviation enterprise on the efficiency of its business. Estimating the level of influence of factors of aviation enterprise's management environment on the efficiency of SMS is a perspective task. Safety management of aviation enterprise is directed on the formation and provision the achieving its goals rationally using present resources (labor, material, financial, informational, etc.). Aviation enterprise is under continuous pressure from different branches of the internal and external environment. Aviation enterprise's management environment is a set of circumstances and factors inside and around that affect DM process (Mescon, Albert, & Khedouri, 2008) (Figure 10).

The internal environment of aviation enterprise is a complex of components connected with each other by means of certain structures within it. The main variables of the internal environment of aviation enterprise management:

- **Enterprise Goals:** Specific final state or desired result, which must be achieved (safety, regularity and economic efficiency of aviation activity; while in the context of aviation, achievement of safety means to achieve state in which the possibility of harm to persons or of property damage is reduced to, and maintained at or below, an acceptable level (ICAO, 2013a);

Figure 10. The environment of the aviation enterprise's management

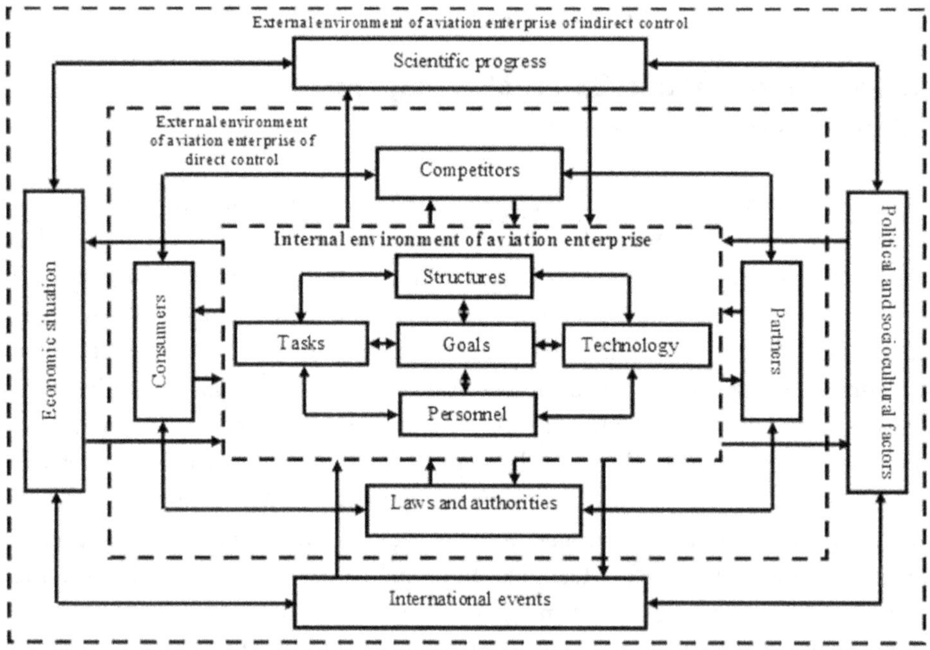

- **Enterprise Tasks:** Predictable work, series of works or part of it, which must be done with the help of previously defined method at the predetermined time (detection of negative factors that influence on safety, regularity and economic efficiency of aviation activity; management of risk factors);
- **Aviation Enterprise Structures:** Logical relationships, the interaction of organizational forms of technical processes, inputs processing, constructed in such a way that it is possible to achieve the goals of aviation enterprise in the most effective way (functional interaction diagrams);
- **Technologies:** Conversion process in the system, which consists of programs testing and implementation of operations over resources in order to transform them into the desired product (established sequence of operational development, production, and sale of air, air navigation, airport, maintenance, handling, etc. services);
- **People of Aviation Enterprise (Central Factor):** Aviation personnel with their abilities, aptitudes, needs, expectations, perceptions, attitude and values (crew members, personnel of technical maintenance, etc.).

Internal variables, as usual, are called sociotechnical subsystems because they have the social component (people) and technical components (other internal variables).

The external environment of aviation enterprise management is a complex of elements which are not part of aviation enterprise but have some influence on it. The main characteristics of the external environment of aviation enterprise are the relation of its factors, complexity, mobility, and uncertainty. The ***external environment*** of aviation enterprise management is divided into two types:

1. ***The External Environment of Direct Influence*:** It is an environment that includes elements that directly affect the operations of aviation enterprise and feel the direct impact of its operations. They are:
   a. Consumers of air (passengers, corporate clients, VIP-clients, etc.) and other services in aviation - natural or legal persons who use services for personal needs or for needs of their organization;
   b. Competitors - producers of similar air, air navigation, airport, maintenance, handling, etc. services on the market;
   c. Partners - legal or natural persons who are co-owners of the aviation enterprise or temporary partners according to any transaction, agreement (air navigation service providers, airports, handling and cleaning companies, catering firms, hotels, etc.);
   d. Laws and authorities – legal and normative acts of bodies of higher legislative and executive powers or their special bodies that oversee the compliance with state requirements (international and local regulations, applicable regulations of the former USSR Ministry of Civil Aviation);
2. ***The External Environment of Indirect Influence*:** It is is a complex of factors that do not influence directly, don't an immediate impact on the operations of aviation enterprise, but will eventually be shown to them. They are:
   e. Economic situation – the tendency and level of inflation, interest rate, level of unemployment;
   f. Progress in science and technology – correspondence between techniques and present requirements/standards;
   g. Political factors – a tendency in the sphere of legislation, court cases, etc.;
   h. Sociocultural factors – a set of attitudes, values, norms, beliefs, behavior, etc.;
   i. International events – the processes which take place outside the country and influence on the central rate, business struggle, etc.

Any aviation enterprise is operating in the environment and functioning in the case when the environment gives it that opportunity. There should be an idea as of internal so of the external environment in order to determine the level of safety of certain aviation enterprise, its potential, and development trend and also its place in the environment. The study of the internal and external environment allows aviation enterprise to reduce the negative impact of the elements on the level of safety, and increase opportunities to improve safety. Decomposition of factors which influence on safety was done in order to provide the structural analysis of management environment of aviation enterprise (Shmelova, Sikirda, Assaul, & Stasiuk, 2015; Shmelova, Sikirda, & Assaul, 2015) (Figure 11).

Content and formal description of parameters are given in Tables 5–7. The structural analysis of management environment of aviation enterprise was carried out in order of significance decreasing of its factors: firstly, the factors of internal management environment were analyzed, secondly – factors of external management environment of direct influence and, ultimately, factors of external management environment of indirect influence. In such a way, from the point of the systematic approach, the factors of internal and external management environment of aviation enterprise which influence on its aviation activity were determined.

Generalization of inhomogeneous factors of internal and external management environment of aviation enterprise was carried out using set-theoretical approach (Shmelova, Sikirda, & Sunduchkov, 2013).

Figure 11. Decomposition of factors of the environment of the aviation enterprise's management

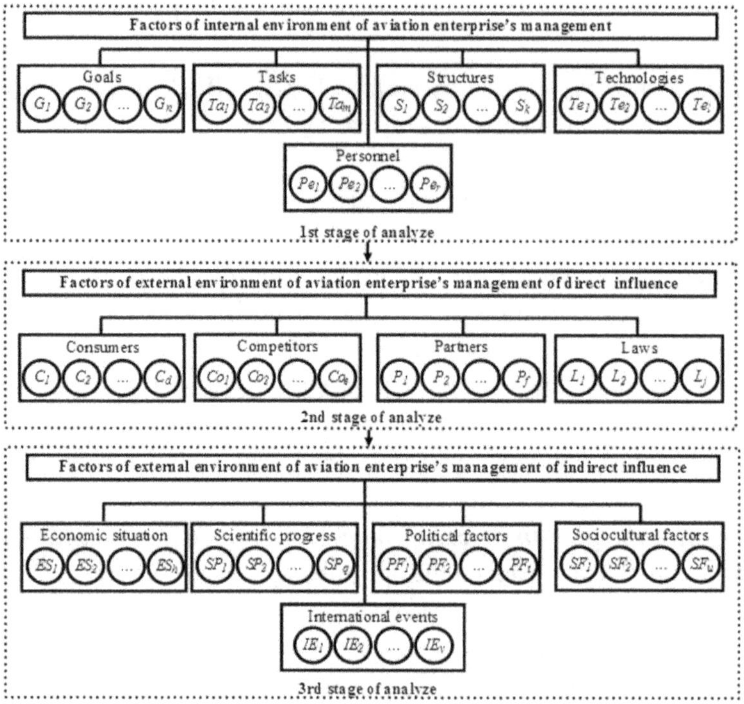

This gave possibilities to consider structural hierarchy, heterogeneity, dynamic instability of factors and to define conditions for their assessment by means of equation (20):

$$\overline{F} = \overline{F}_{ie} \bigcup \overline{F}_{eedi} \bigcup \overline{F}_{eeii}, \tag{20}$$

where

$\overline{F}_{ie} = \left\{\overline{G}, \overline{Ta}, \overline{S}, \overline{Te}, \overline{Pe}\right\}$ is a set of factors of internal management environment of aviation enterprise (goals $\overline{G}$, tasks $\overline{Ta}$, structures $\overline{S}$, technologies $\overline{Te}$, personnel $\overline{Pe}$);

$\overline{F}_{eedi} = \left\{\overline{C}, \overline{Co}, \overline{Pa}, \overline{L}\right\}$ is a set of factors of external management environment of aviation enterprise of direct influence (consumers $\overline{C}$, competitors $\overline{Co}$, partners $\overline{Pa}$, laws, and authorities $\overline{L}$);

$\overline{F}_{eeii} = \left\{\overline{ES}, \overline{SP}, \overline{PF}, \overline{SF}, \overline{IE}\right\}$ is a set of factors of external management environment of aviation enterprise of indirect influence (economic situation $\overline{ES}$, scientific and technical progress $\overline{SP}$, political factors $\overline{PF}$, sociocultural factors $\overline{SF}$, international events $\overline{IE}$).

Figure 12 shows the graphical interpretation of the method of inhomogeneous factors generalization of internal and external management environment using set-theoretical approach.

Table 5. Decomposition of factors of internal management environment of aviation enterprise $\overline{F}_{ie}$

Nº	Factors	Parameters	Coding
1	Goals	Safety of aviation activity	G_1
2	$\overline{G} = \{G_1, G_2, ..., G_n\}$	The regularity of aviation activity	G_2
3		Economic efficiency of aviation activity	G_3
4	Tasks	Conservation of human life and health	Ta_1
5	$\overline{Ta} = \{Ta_1, Ta_2, ..., Ta_m\}$	Protection from acts of unlawful interference	Ta_2
6		Environmental control	Ta_3
7		Effective resource exploitation	Ta_4
8		Protection from informative hazards	Ta_5
9		Prevention of failures in work	Ta_6
10		Profit earning	Ta_7
11		Quality aviation services production	Ta_8
12	Structures	Linear structure	S_1
13	$\overline{S} = \{S_1, S_2, ..., S_k\}$	Functional structure	S_2
14		Divisional structure	S_3
15		Design structure	S_4
16		Matrix structure	S_5
17		Process structure	S_6
18	Technologies	Aircraft performance characteristics	Te_1
19	$\overline{Te} = \{Te_1, Te_2, ..., Te_l\}$	The level of aircraft dilapidation	Te_2
20		The level of special techniques dilapidation	Te_3
21		The level of buildings and constructions dilapidation	Te_4
22		Technological operations on development of aviation services	Te_5
23		Technological operations on the commanders, inspectors, and instructors realization of aviation services	Te_6
24		Technological operations on ground servicing	Te_7
25		Technological operations on maintenance	Te_8
26		Technological operations on airport servicing	Te_9
27		Technological operations on air navigation servicing	Te_{10}
28		Flight crew and passenger cabin crew	Pe_1
29		Commanders, inspectors, and instructors	Pe_2
30		Experts who are carrying out the regulation of airspace usage and air traffic service	Pe_3
31		Experts who are carrying out organization and maintenance of the aircraft and other types of flight servicing	Pe_4
32	Aviation personnel	Experts on air traffic service	Pe_5
33	$\overline{Pe} = \{Pe_1, Pe_2, ..., Pe_r\}$	Experts who carried out organization and performance of design and experimental, experimental, research and development works while flight tests of aviation techniques	Pe_6
34		Experts who are carrying out supervision and control of flights safety and those who perform the investigation of aviation events	Pe_7
35		Experts who are carrying out the analysis and control of the flight validity of the aircraft during development, testing, certification and a mass production	Pe_8
36		Experts of aviation safety	Pe_9
37		Aviation experts	Pe_{10}

Table 6. Decomposition of factors of external management environment of direct influence $\overline{F}_{eedi}$

Nº	Factors	Parameters	Coding
1	Consumers	First-class passengers	C_1
2	$\overline{C} = \{C_1, C_2, ..., C_d\}$	Business-class passengers	C_2
3		Economy-class passengers	C_3
4		Baggage clientage	C_4
5		Corporate clients	C_5
6		VIP-clients	C_6
7		The customer of aviation works	C_7
8	Competitors	Airlines – limited liability company	Co_1
9	$\overline{Co} = \{Co_1, Co_2, ..., Co_e\}$	Airlines – a private joint stock company	Co_2
10		Airlines – a public joint-stock company	Co_3
11		Airlines – government companies	Co_4
12		Airlines – private companies	Co_5
13		Foreign airlines	Co_6
14	Partners	Members of the aviation alliance	Pa_1
15	$\overline{Pa} = \{Pa_1, Pa_2, ..., Pa_f\}$	Flights sales agencies	Pa_2
16		Air navigation service providers	Pa_3
17		Airports	Pa_4
18		Handling companies	Pa_5
19		Cleaning companies	Pa_6
20		Fuel servicing companies	Pa_7
21		Catering companies	Pa_8
22		Organization for aircraft maintenance	Pa_9
23		Auto transportation enterprises	Pa_{10}
24		Hotels	Pa_{11}
25		Constitution of Ukraine	L_1
26		Laws of Ukraine	L_2
27		International agreements of Ukraine	L_3
28		Aviation legislation of Ukraine	L_4
29		International conventions and proceedings	L_5
30		ICAO Standards and Recommended Practices (SARPs)	L_6
31		Joint Aviation Requirements (JAR)	L_7
32		Eurocontrol SAfety Regulatory Requirements (ESARRs)	L_8
33	Laws	Resolutions of the Verkhovna Rada of Ukraine	L_9
34	$\overline{L} = \{L_1, L_2, ..., L_j\}$	Edicts of the President of Ukraine	L_{10}
35		Resolutions, the State of investment activities in transport domain ordinance of the Cabinet of Ministers of Ukraine	L_{11}
36		State standards of Ukraine	L_{12}
37		Orders of State Aviation Administration of Ukraine	L_{13}
38		Aviation rules of Ukraine	L_{14}
39		Sectoral guidance documents	L_{15}
40		Aviation rules of the former USSR	L_{16}
41		State standards of USSR	L_{17}
42		Sectoral guidance documents (sectoral standards, guidance documents)	L_{18}

Table 7. Decomposition of factors of external management environment of indirect influence $\overline{F}_{eeii}$

N°	Factors	Parameters	Coding
1	Economic situation	Cost of living	SE_1
2	$\overline{ES} = \{ES_1, ES_2, ..., ES_h\}$	Minimum salary	SE_2
3		Minimum pension	SE_3
4		Average earnings	SE_4
5		Inflation rate	SE_5
6		Bank interest rate	SE_6
7		Unemployment rate	SE_7
8		State of investment activities in the transport domain	SE_8
9		Development of tourism	SE_9
10	Scientific and technical progress	Level of moral depreciation of aircraft	ST_1
11	$\overline{SP} = \{SP_1, SP_2, ..., SP_q\}$	Level of moral depreciation of special techniques	ST_2
12		Level of moral depreciation of buildings and constructions	ST_3
13		The progressiveness of technological schemes on development and realization of aviation services	ST_4
14		The progressiveness of technological schemes from the ground, technical, airport, air navigation services	ST_5
15	Political factors	Stability of political regime	PF_1
16	$\overline{PF} = \{PF_1, PF_2, ..., PF_t\}$	Level of corruption	PF_2
17		Crime rate	PF_3
18		Level of state regulation in aviation sphere	PF_4
19	Sociocultural factors	The social structure of society	SF_1
20	$\overline{SF} = \{SF_1, SF_2, ..., SF_u\}$	National and household traditions of the population	SF_2
21		The popularity of air traffic	SF_3
22	International events	Global economic situation	IE_1
23	$\overline{IE} = \{IE_1, IE_2, ..., IE_v\}$	Strategic-military situation	IE_2
24		Influence of certain states	IE_3
25		Nature-effect	IE_4
26		State of raw material and natural resources	IE_5

Method of expert estimation was used in order to define the level of influence of factors of internal and external management environment of aviation enterprise on aviation activity (Shmelova, Sikirda, Assaul, & Stasiuk, 2015). Respondents from among pilots and ATCO of different age categories, with the different professional experience, filled in the offered questionnaires and defined individual preferences concerning the influence of factors of internal and external management environment of the aviation enterprise on aviation activity. The pairwise comparison method and ranking were applied; the consensus of expert opinion was confirmed by the coefficient of the variation, Kendal's concordat coefficient, and Spearman's rating correlation coefficient. Priority influence of factors on aviation activity was established by comparison of weight coefficients. As an example, Figure 13 and Figure14 show

Figure 12. Graphical interpretation of the method of inhomogeneous factors generalization of internal and external management environment of aviation enterprise

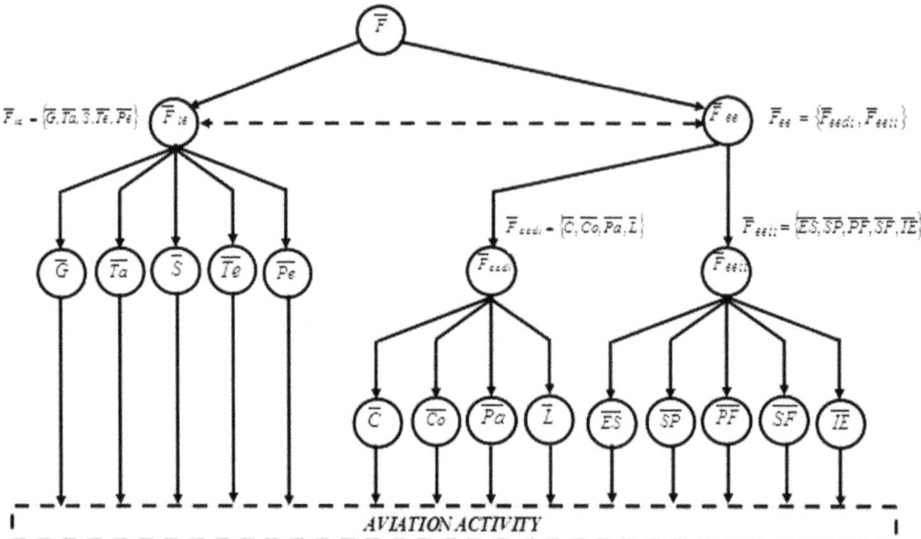

the influence of goals and international events of internal and external management environment correspondingly on aviation activity.

There were defined, that the greatest influence has the level of safety of aviation activity (factor of internal management environment of aviation enterprise); global economic situation (factor of external management environment of aviation enterprise).

FUTURE RESEARCH DIRECTIONS

It is our belief, that the methods of analysis of H-O ANS of DM can be used in fields other than aviation. The algorithms and methodology for analysis of situation development under influence DM by H-O can be useful in any technogenous production. Technogenous production is a complex system that contains interrelated technical, economic and social objects. Technogenous production has a multilevel hierarchical structure and a high level of risk. Recently, there are frequent and common emergencies such as disaster, accident, crash on hydraulic engineering, chemical and military industries, gas and oil pipelines, nuclear power plants and transport. The main strategic problem of mankind on the path to sustainable development is the safety and stability of technogenous production in complex systems.

It is necessary to build models for enterprises of forecasting the development of the situation with using stochastic models: GERT network, decision tree, and Markov's network; reflexive theory of moving to positive or negative choice.

Figure 13. Influence of factors of internal management environment of aviation enterprise

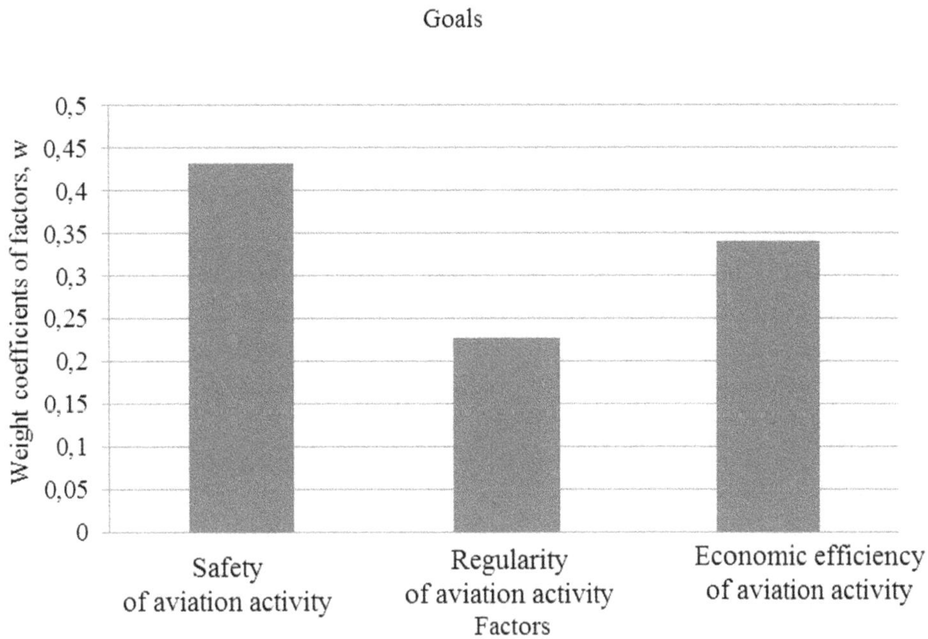

Figure 14. Influence of factors of external management environment of indirect influence of aviation enterprise

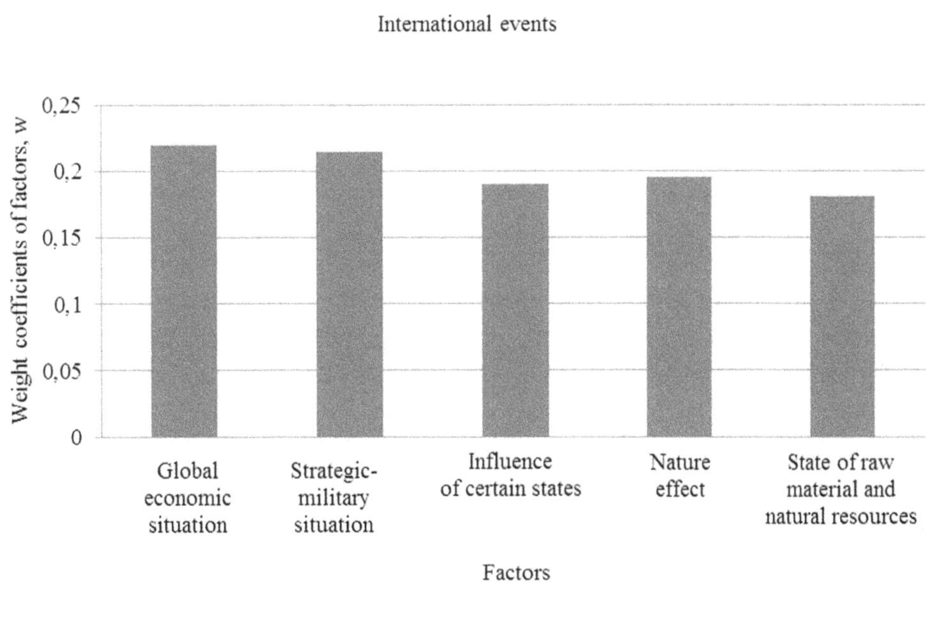

What's Next?

On next steps, we are planning to consider another different complex STS, such as hydraulic engineering, chemical and military industries, gas and oil pipelines, nuclear power plants, transport, etc. It is necessary to analyze all the factors influencing the DM by operators in these systems in order to predict the development of the technogenous catastrophe and prevent it. The enhancement and improvement of new DM models, safety balance models from the modern stage of evolution of human factor's models, for the next stage of development too.

System research of the internal and external environment of the enterprises, quantitative assessment, aggregation, the choice of the principles of the convolution, normalization and priorities of the respective indicators of the enterprises will lead to the solving of the multicriterial task of optimizing the enterprises' management, which enables to reduce the negative impact of risk factors as well as increase opportunities for optimal organizational performance.

CONCLUSION

So, the ANS is presented as a complex STS. The influence on the DM by H-O of the ANS of the professional factors (knowledge, skills, abilities, and experience) as well as the factors of non-professional nature (individual-psychological, psychophysiological and socio-psychological) has been defined. The network analysis of the actions of an aircraft crew and ATCO in the flight emergencies has been made using stochastic models DM by H-O ANS in-flight emergencies. The numerical indices of an expected risk in the stochastic models DM by H-O ANS under risk and uncertainty have been defined. The scenarios of developing a flight situation in case of selecting either the positive or negative pole under the pressure of the external environment, the preceding experience of H-O and the intentional selection (intention) in accordance with the reflexive theory have been obtained. The results of the evaluation of non-professional factors are determining the influence of social factors on the H-O by defining preferences; diagnostics of individual psychological qualities of the H-O ANS in the situation; monitor the emotional state of the H-O ANS for the timely diagnosis of potentially dangerous transition of mental activity and determine the stability of the ANS by professional activities. On the basis of the reflexive theory of bipolar choice, the expected risks of DM of the ANS's operator have been studied and the influence of the external environment, previous experience and intention of the H-O have been identified. The methods for analysis of DM by the H-O in ANS using stochastic networks have been developed.

The authors have obtained the results of the evaluation of professional and non-professional factors: determination the social-psychological impact on DM of H-O by identifying the system of H-O's preferences; diagnosing individual-psychological qualities of H-O ANS in the development of flight situation. In future work, we are planning to research new methods of monitoring of the psychophysiological factors as emotional state H-O ANS for timely diagnosis transition to a potentially dangerous mental activity and determine the stability of STS in the performance of professional actives.

Factors of internal and external management environment of aviation enterprise which influence aviation activity were determined from the point of the systematic approach. Aviation enterprise is a complex of basic elements such as goals, tasks, structures, technologies, personnel, and also environment – political, economic, cultural, market which interacts and to which it has to adapt. The internal environment of aviation enterprise is a source of life force and includes the potential which gives possibilities to exist and survive in a defined time period but also it can be a source of problems. The external environment is a source of resources needed for the maintenance of its internal potential at the necessary level in order to achieve aviation enterprise's goals. Inhomogeneous factors of internal and external management environment of aviation enterprise were generalized using the set-theoretical approach. This gave possibilities to define that the level of safety of aviation activity has the greatest influence among factors of the internal environment and global economic situation – among factors of the external environment.

Proposed methodology of evaluation of the influence of the inhomogeneous factors of management environment of aviation enterprise on flight safety will allow developing the safety passports. They will be applied by the aviation authorities during the certification inspections of subjects of aviation activity for comparison of aviation enterprise's normative and actual indicators and determination of their organizational performance.

REFERENCES

Aviation Accident Statistics. (2017). *National Transportation Safety Board*. Retrieved from www.ntsb.gov/aviation/ aviation.htm

Baxter, G., & Sommerville, I. (2011). Socio-technical systems: From design methods to systems engineering. *Interacting with Computers*, *23*(1), 4–17. doi:10.1016/j.intcom.2010.07.003

Bertsch, V., Treitz, M., Geldermann, J., & Rentz, O. (2007). Sensitivity analyses in multi-attribute decision support for off-site nuclear emergency and recovery management. *International Journal of Energy Sector Management*, *1*(4), 342–365. doi:10.1108/17506220710836075

Carayon, P. (2006). Human factors of complex sociotechnical systems. *Applied Ergonomics*, *37*(4), 525–535. doi:10.1016/j.apergo.2006.04.011 PMID:16756937

Clegg, C. W. (2000). Sociotechnical principles for system design. *Applied Ergonomics*, *31*(5), 463–477. doi:10.1016/S0003-6870(00)00009-0 PMID:11059460

Flueler, T. (2006). *Decision making for complex socio-technical systems: Robustness from Lessons Learned in Long-Term Radioactive Waste Governance. In Environment & Policy* (Vol. 42). Dordrecht, The Netherlands: Springer. doi:10.1007/1-4020-3529-2

International Civil Aviation Organization. (1998). Human Factors Training Manual (1st ed.). Doc. ICAO 9683-AN/950. Montreal, Canada: Author.

International Civil Aviation Organization. (2002). Human Factors Guidelines for Safety Audits Manual (1st ed.). Doc. ICAO 9806-AN/763. Montreal, Canada: Author.

International Civil Aviation Organization. (2004). *Cross-Cultural Factors in Aviation Safety: Human Factors Digest Nº 16. Circ. ICAO 302-AN/175*. Montreal, Canada: Author.

International Civil Aviation Organization. (2005). *Global Air Traffic Management Operational Concept. Doc. ICAO 9854*. Montreal, Canada: Author.

International Civil Aviation Organization. (2009). *Manual on Global Performance of the Air Navigation System. Doc. ICAO 9883*. Montreal, Canada: Author.

International Civil Aviation Organization. (2013a). Safety Management Manual (SMM) (3rd ed.). Doc. ICAO 9859-AN 474. Montreal, Canada: Author.

International Civil Aviation Organization. (2013b). *State of Global Aviation Safety*. Montreal, Canada: Author.

International Civil Aviation Organization. (2014). *Safety Report*. Montreal, Canada: Author.

Keating, C., Fernandez, A. A., Jacobs, D. A., & Kauffmann, P. (2001). A methodology for analysis of complex sociotechnical processes. *Business Process Management Journal*, *7*(1), 33–50. doi:10.1108/14637150110383926

Kharchenko, V., Shmelova, T., & Sikirda, Y. (2011a). Graphanalytical models of decision-making by air navigation system's human-operator. *Proceedings of the National Aviation University*, *1*, 5–17.

Kharchenko, V., Shmelova, T., & Sikirda, Y. (2011b). Methodology for analysis of decision making in air navigation system. *Proceedings of the National Aviation University*, *3*, 85–94.

Kharchenko, V., Shmelova, T., & Sikirda, Y. (2012a). *Decision-making by an operator in the air navigation system: A monograph*. Kirovograd: KFA of NAU.

Kharchenko, V., Shmelova, T., & Sikirda, Y. (2012b). Modeling of behavioral activity of air navigation system's human-operator in flight. *Proceedings of the National Aviation University*, *2*, 5–17.

Kharchenko, V., Shmelova, T., & Sikirda, Y. (2016). *Decision-making in socio-technical systems: A monograph*. Kyiv: NAU.

Kharchenko, V., Shmelova, T., & Sikirda, Y. (2018). The methodology of Research and Training in Air Navigation Socio-technical System. *Proceedings of the National Aviation University*, *1*(74), 8–23.

Kuchar, J. K., & Yang, L. C. (2000, December). A review of conflicts detection and resolution modeling methods. *IEEE Transactions on Intelligent Transportation Systems*, *1*(4), 179–189. doi:10.1109/6979.898217

Lefebvre, B. (2008, January-June). Intentional-reflective model of the agent. *Reflexive Processes and Control, 1,* 69-78.

Lefebvre, B., & Adams-Webber, J. (2002, January-June). Functions of fast reflexion in bipolar choice. *Reflexion Processes and Control, 1,* 29-40.

Lefebvre, V. A. (2001). Algebra of conscience (2nd enlarged ed.). Dordrecht, The Netherlands: Kluwer Publishers.

Leychenko, S., Malishevskiy, A., & Mikhalic, N. (2006). *Human factors in aviation: a monograph in two books. Book 1st*. Kirovograd: YMEKS.

Makarov, R., Nidziy, N., & Shishkin, G. (2000). *Psychological foundations of didactics in-flight education: A monograph*. Moscow: MAPCHAK.

Mescon, M., Albert, M., & Khedouri, F. (2008). *Management: A monograph*. Moscow: Vilyams.

Shmelova, T., & Sikirda, Y. (2012). Analysis of human-operator's decision-making in the air navigation system. *Radioelectronic and Computer Systems*, *7*(59), 319–324.

Shmelova, T., & Sikirda, Y. (2016). Calculation the scenarios of the flight situation development using GERT's and Markov's networks. In *Proceedings of the 5th International Scientific and Practical Conference "Management of High-Speed Moving Objects and Professional Training of Operators of Complex Systems"*. Kirovograd: KFA NAU.

Shmelova, T., & Sikirda, Y. (2017). Model of optimization of transport flows taking into account combinatorics and regression analysis methods. In *Proceedings of the 13th International Scientific and Technical Conference "AVIA-2017"*. Kyiv: NAU.

Shmelova, T., Sikirda, Y., Assaul, A., & Stasiuk, O. (2015). Structural Analysis of Management Environment of Aviation Enterprise from the Point of Systematic Approach. *Proceedings of the National Aviation University*, *2*, 3–12.

Shmelova, T., Sikirda, Y., & Assaul, Y. (2015). Influence of Factors of Management Environment of Aviation Enterprise on the Level of Aviation Safety. *Technological Audit and Production Reserves*, *2/3*(22), 17–24. doi:10.15587/2312-8372.2015.41440

Shmelova, T., Sikirda, Y., & Jafarzade, T. (2013). Models of flight situations development while decision-making by air navigation system's human-operator. *Information Processing Systems*, *8*(115), 136–142.

Shmelova, T., Sikirda, Y., & Sunduchkov, K. (2013). A sociotechnical analysis of air navigation system. *Science and Technics of Air Forces of Armed Forces of Ukraine*, *4*(13), 34–39.

Sikirda, Yu., Shmelova, T., & Tkachenko, D. (2017). Evaluation of the organizational factors influence on flight safety in air traffic control. *Collections of Scientific Works of the Kharkiv National University of Air Force, No.*, *3*(52), 39–44.

KEY TERMS AND DEFINITIONS

Air Navigation Socio-Technical System (ANSTS): A complex large-scale, high-tech man-machine system that requires complex interactions between their human and technological components; the operations in socio-technical systems generally involve high-risk/high-hazard activities; the consequences of safety breakdowns are often catastrophic in terms of loss of life and property.

Air Navigation System (ANS): A complex of organizations, personnel, infrastructure, technical equipment, procedures, rules, and information that is used to provide airspace users with safe, regular, and efficient air navigation service.

ATC: Air traffic control.

ATCO: Air traffic control's operator.

DM: Decision making.

GERT: Graphical evaluation and review technique.

H-O: Human-operator.

ICAO: International Civil Aviation Organization.

SMS: Safety management system.

Socio-Technical System (STS): The large-scale, high-technology systems, because they require complex interactions between their human and technological components; the operations in socio-technical systems generally involve in high-risk/high-hazard activities; the consequences of safety breakdowns are often catastrophic in terms of loss of life and property.

Chapter 4
A Sociotechnical Systems Approach Applying a Novel Taxonomy to a Survey for the Assessment of Safety Performance

Toivo Niskanen
Ministry of Social Affairs and Health, Finland

ABSTRACT

The aim of this chapter was to explore a sociotechnical systems approach applying a novel taxonomy with respect to safety performance. The study applied a combination of qualitative and quantitative methodologies. Workers (n = 120) and managers (n = 85) were asked to complete a questionnaire survey (Appendix). The following hypotheses were supported: "Activities of the management" had positive impacts on five aggregated variables, namely "near-accident investigation and instructions" (H1), "occupational safety and health (OSH) training" (H2), "operations, technical processes, and the safe use of chemicals" (H3), "use of personal protective equipment" (H4), and "measuring, follow-up, and prevention of major accidents" (H5). By undertaking a statistical evaluation and then devising a novel taxonomy, it was possible to gain detailed insights into diverse aspects of a high-risk industry's work with regard to complex sociotechnical systems. When applying the current approach through participatory cooperation, organizations may acquire new perspectives on their safety performance.

INTRODUCTION AND BACKGROUND

Sociotechnical systems consist of the job design, software and/or hardware, an internal environment, an external environment, and/or an organizational design that fits into the work system (Hendrick, 2002a). Sociotechnical systems theory is based on several core constructs, including joint causation, joint optimization, and joint design (Kleiner, 1999). Joint causation refers to the belief that personnel and technological

DOI: 10.4018/978-1-5225-7192-6.ch004

subsystems are jointly affected by, or open to, the environment (Kleiner, 1999), while joint causation results in the related sociotechnical system's theoretical concept of joint optimization (Kleiner, 1999). Such joint optimization is operationalized through joint design, a human-centered approach to function and task allocation/design, including consistently accessing the expertise of workers, and attending to the organization's sociotechnical characteristics (Hendrick, 2002a, 2002b). Checkland (2000) defined the difference between "hard" and "soft" systems thinking as follows: "Hard" systems thinking assumes that systems exist as entities in the real world that can be characterized by their well-defined objectives and also engineered to better meet those objectives, while "soft" systems thinking recognizes that different people may or may not perceive a particular situation to be problematic and, when exploring the situation, the inquiry process can be structured as a system of learning (Checkland, 2000).

Sociotechnical systems theory concerns itself with the design of complete work systems. Hence, models have been proposed to help guide researchers in terms of identifying organizational-level variables that influence both safety and safe behavior (Robertson et al., 2015). The application of the theoretical constructs results in the management of change via organizational development within the sociotechnical system (Appelbaum, 1997). A sociotechnical system cannot be treated as a static entity but should instead be recognized as a dynamic process for decreasing the risk of accidents, which is continually adapting in order to achieve its ends and react to changes in both itself and its environment (Leveson et al., 2006, pp. 96-97). Decisions and strategies intended to manage the external environment within sociotechnical systems must be congruent as well as compatible with the organization's structure (Appelbaum, 1997). Sociotechnical systems can be characterized as follows (Maier, 1998): the elements of the system are themselves sufficiently complex to be considered systems; when operating together, the systems produce functions and fulfill purposes not produced or fulfilled by the elements alone; the elements possess operational independence; and the elements possess managerial independence.

The key issue associated with sociotechnical systems is the need to design work so that the two parts of a given system yield positive outcomes with joint optimization (Appelbaum, 1997). Wilson (2014) concluded that the systems thinking approach must involve an understanding of the interfaces between people and all other elements, and it must use this as the basis for devising a system that takes into account inter-related or coupled activities or entities (e.g., hardware, software, spaces, communities, and people) with a joint purpose. Switching from a traditional work design or organization to one based on the principles of sociotechnical system requires a transitional structure for managing the change process, innovation, and learning (Appelbaum, 1997). The key advantage of considering a situation using the sociotechnical systems framework is that one can consider, in a principled and organized manner, how all the components and subsystems of the system behave and interact (Dul et al., 2012; Vincent, Ward, & Langdon, 2012). The work system is composed of people in the form of a personnel and social subsystem, as well as technology in the form of a technological subsystem. The personnel subsystem defines the ways in which "individuals" perform tasks, while the technological subsystem defines the "tasks" that need to be performed (Hendrick, 2002a; Kleiner, 2008). Robertson et al. (2015) argued that the causality across system levels represents a critical challenge in terms of understanding risks to safety, particularly in highly complex sociotechnical systems. Indeed, optimizing one subsystem and fitting the second to it results in the sub-optimization of the joint work system (Kleiner, 1999).

The growth of different types of frameworks, models, and methods can be seen as evidence of the success of sociotechnical systems theory (Robertson et al., 2015). Although research interest in the analysis of work system models (e.g., Carayon et al., 2014), the systems thinking approach (e.g., Wilson, 2014), and "System of Systems" (e.g., Siemieniuch & Sinclair, 2014) has increased in recent years, no efforts involving statistical analysis have previously been made to determine the proactive dimensions of occupational safety and health (OSH) that underpin the concepts of sociotechnical exchange and relationships. Robertson et al. (2015) argued that "we might indeed expect to find such a rich diversity of sociotechnical systems, given the widely varying properties and characteristics of different work domains and their associated safety challenges." The present study attempts to address this knowledge gap by concentrating on a sociotechnical systems approach involving the fixed analysis of quantitative and qualitative methods. The focus of the present study is on the application of sociotechnical systems in order to explore how the different levels of OSH measures can represent a basis for obtaining a better understanding that would in turn allow for the analysis of both top-down and bottom-up processes within organizations.

Robertson et al. (2015) argued that the research with respect to the invariant characteristics of sociotechnical systems could also be expanded to examine other aspects of the external environment, including safety and industry standards. The present study attempts to address the knowledge gap by concentrating on sociotechnical systems' relationships in terms of promoting the organizational objectives associated with safety performance. Given that workplace safety is one of many goals that organizations need to achieve, future research should explore how a sociotechnical systems approach may help to balance different organizational objectives (Robertson et al., 2015). Little research has previously been conducted with the aim of gathering evidence-based data concerning OSH so as to validate the scientific theories that underpin the sociotechnical systems approach. Research involving a combined quantitative and qualitative analysis is especially scarce. The present taxonomy in the qualitative study will specify the fundamental dimensions and processes that are involved in the sociotechnical systems' relationships, define their integrative dynamics, and finally, clarify the stages at which these dimensions take on greater or lesser significance in quantitative terms.

Carayon et al. (2015) argued that due to the conceptual overlap between sociotechnical systems and the broader class of complex adaptive systems, we should further explore these connections. When we seek to embrace sociotechnical advances in relation to safety improvement, we need to improve, evolve, and arguably generate a revolution in our evaluative and analytic techniques (Carayon et al., 2015). A better understanding of systemic structures could then facilitate the design of more effective safety culture interventions, while the application of systems thinking concepts will improve the overall effectiveness of safety performance and management (Goh, Brown, & Spickett, 2010). The focus of the present study was on the development and application of the OSH taxonomy to data obtained from a survey concerning decreasing the risk of accidents by adopting a systems-thinking approach. The present systematic view utilizes the theories inherent within sociotechnical systems and applies such concepts in the framework of this analysis.

THE AIM OF THE RESEARCH

The aim of the present article was to examine how OSH managers and workers' OSH representatives in the Finnish chemical industry evaluate the impact of "Activities of the management" on different aggregated variables, namely:

- Instructions
- OSH training
- Operations, technical processes, and safe use of chemicals
- Use of personal protective equipment (PPE)
- Measuring, follow-up, and prevention of major accidents (the variables from the online questionnaire are presented in Appendix) in the joint framework of sociotechnical systems

The main objectives of the study focus on the following: introduce a procedure for applying the developed novel qualitative taxonomy based on sociotechnical systems using data from a survey and adopting a sociotechnical approach; demonstrate that a qualitative taxonomy based on sociotechnical systems makes it possible to interpret the OSH survey's data; and show that the developed novel qualitative taxonomy based on technical systems is able to extract OSH domain knowledge from the data collected in an internet survey questionnaire.

RESEARCH MATERIALS

The online questionnaire survey of workers' OSH representatives and OSH managers working in the Finnish chemical industry was conducted from September to October 2011. The data were collected via an online internet survey. Although the data were collected in 2011, the survey was conducted in an environment that still exists today; thus, it is not believed that the responses would differ significantly if they were collected in 2018.

A register of OSH managers is kept by the Finnish Centre for Occupational Safety. A total of 258 usable e-mail addresses were obtained from this register for the chemical industry, and 85 OSH managers voluntarily participated in the survey. Their response rate was 33%. Their average age was 49 years; 28% were female and 72% were male. The enterprises represented in the sample were divided into the following size classes: 1–9 workers (8%), 10–19 (17%), 20–49 (23%), 50–249 (35%), and 250 or more workers (17%). The durations of the OSH managers' experience in the role were as follows: less than 1 year (3%), 1–5 years (53%), and over 5 years (44%).

The workers' elected OSH representatives provide their e-mail information to the register of the Finnish Industrial Union TEAM (blue-collar workers). A total of 120 e-mail addresses were obtained from this register. The clerical workers' OSH representatives list their e-mail addresses in the register of the Finnish Trade Union PRO (clerical workers). A total of 228 e-mail addresses were obtained from this register. Consequently, a total of 348 e-mail addresses were obtained, and 120 workers' OSH representatives voluntarily participated in the survey (i.e., the response rate was 34%). Their average age was 43 years; 29% were female and 71% male. The enterprises represented in this part of the sample were divided into the following size classes: 1–9 workers (2%), 10–19 (8%), 20–49 (23%), 50–249 (44%), and 250 or more workers (23%). The durations of the workers' OSH representatives' experience in the role were as follows: less than 1 year (2%), 1–5 years (46%), and over 5 years (52%).

RESEARCH METHODS AND DATA ANALYSIS

Combined Qualitative and Quantitative Methods and Analysis

Several authors (e.g., Teddlie, 2005) have stated that pragmatism represents the best paradigm for justifying the use of a mixed methods research approach. In terms of a basic definition of mixed methods research, it is a combination of "elements of qualitative and quantitative research approaches" (Johnson, Onwuegbuzie, & Tumer, 2007). Such an approach involves the use of qualitative and quantitative viewpoints, data collection, analysis, and inference techniques for the purposes of breadth and depth of understanding, and corroboration (Johnson et al., 2007). Mixed methods research legitimizes the use of multiple approaches when answering research questions, and it is inclusive, pluralistic, complementary, and eclectic (Johnson et al., 2007). The goal of any analysis of sociotechnical systems is a comprehension accounting of the "joint optimization of the social and technical systems," that is, the different sub-systems or different system components (Carayon et al., 2015). Joint optimization involves interactions among system components, as well as between the system and its external environment (Hendrick, 2002a, 2002b).

Quantitative Methods and Analysis

The utilized questionnaire survey form was developed specifically for this study. The OSH managers and workers' OSH representatives were asked to evaluate how well the different OSH measures had been achieved by marking their responses on a four-point Likert scale as follows: 1=strongly disagree, 2=somewhat disagree, 3=somewhat agree, or 4=strongly agree.

Hypotheses H1–H5 were tested using a regression analysis.

Hypothesis 1: For workers and managers, "Instructions" would have positive impacts on the "Activities of the management" in the joint framework of sociotechnical systems.
Hypothesis 2: For workers and managers, "OSH training" would have positive impacts on the "Activities of the management" in the joint framework of sociotechnical systems.
Hypothesis 3: For workers and managers, "Operations, technical processes, and the safe use of chemicals" would have positive impacts on the "Activities of the management" in the joint framework of sociotechnical systems.
Hypothesis 4: For workers and managers, the "Use of personal protective equipment (PPE)" would have positive impacts on the "Activities of the management" in the joint framework of sociotechnical systems.
Hypothesis 5: For workers and managers, "Measuring, follow-up, and prevention of major accidents" would have positive impacts on the "Activities of the management" in the joint framework of sociotechnical systems.

A regression analysis determines how one dependent variable associates with several independent variables. Once a regression model had been constructed, this study confirmed the goodness of fit of the model as well as the statistical significance of the estimated parameters. The checks of the goodness of fit included the R^2 and hypothesis testing, while the statistical significance was checked using t-tests of the individual parameters as presented with SAS version 9 (SAS software, 2005). The coefficient of determination (R^2) is calculated in a regression analysis in order to indicate the percentage of the dependent variable that can be predicted by the independent variables. This level of accuracy in the prediction of the dependent variable will change based on which independent variables are included in the model. In terms of the statistical consistency of the responses, Cronbach's alpha was estimated as a coefficient of reliability or consistency (SAS software, 2005). It should be noted that the alpha coefficient is included as one of the statistics reported with differential decisions, but not with absolute decisions (Cronbach & Shavelson, 2004). It is commonly accepted that Cronbach's $\alpha>0.6$ = questionable, $\alpha>0.7$ = acceptable, and $\alpha>0.8$ = good (Cronbach & Shavelson, 2004).

Qualitative Methods and Analysis

A Qualitative Taxonomy of the Sociotechnical Systems Approach Developed for This Study

The present study focuses on the characteristics described by Wilson (2014). These sub-categories are based on the systems thinking approach, and they were developed for the present study based on a taxonomy originating from a sociotechnical systems approach. The sub-categories of these taxonomies are as follows: (1) Systems and Organization: work systems, rules, and leadership; (2) People and Performance: communication, processes, and competencies; and (3) Equipment and Interfaces: technology, procedure, and performance. The present taxonomy thus consists of the three different sections. In this qualitative systems analysis, the taxonomy is classified into sub-categories. Further, the author of this article, as the investigator, has selected in the present novel taxonomy of sociotechnical systems (Figures 1, 2, and 3) the two most important individual sub-categories that exert the greatest impacts with respect to safety performance.

The present novel qualitative taxonomy was applied and its constructs of indicators were presented in a resilience engineering (RE) -related approach (Niskanen, 2018). This novel taxonomy can also be applied in the sociotechnical systems. The both theoretical (RE and sociotechnical) approaches seek to shift the focus of safety research from a traditional reductionist approach towards a focus on the broader levels of social, ecological, organizational and technological factors that create and sustain functional (or dysfunctional) work systems and environments. The novel qualitative taxonomy provides information about what is happening in intermediate stages of the processes of the in the sociotechnical systems. Starting from the level of the whole system, the present qualitative approach leads to the development of operational details and specific steps that can be taken on a concrete level. The developed taxonomy techniques place a strong emphasis on the analysis of the relationships between strategy, process, procedures and function within the inherent features of systems.

Each section aims to reveal important aspects and properties of sociotechnical systems. The study began by explaining and exemplifying event-based constraints, which are the contextual factors that sociotechnical systems must cope with. Wilson (2014) showed that the systems thinking approach involves four phases: (1) identifying and assessing the links between the entities, which may be of state, form, function, and causation; (2) conceptualizing any system of interest as existing within a boundary and thus a defined context, having inputs and outputs that may connect in many to many mappings; (3) treating the system as holistic, with the whole usually being greater (more useful, powerful, functional, etc.) than the sum of its parts; and (4) recognizing that the system changes and modifies both its state and the interactions within it in light of circumstances and events, thereby revealing emergent properties.

As the external environment, when operating under joint causation, may be the most influential subsystem in terms of determining whether sociotechnical systems will be successful or not, achieving valid organization and environmental fit as well as joint optimization is essential (Kleiner, 1999). The transformation is accomplished through a joint effort on the part of many different entities of systems (Edwards & Jensen, 2014), with technology, facilities, and formal and informal organizations (structures, procedures, and processes), workers (qualifications, competencies, attitudes, and values), and managers all serving as examples of the entities typically used in the problem-solving processes associated with safety performance.

Systems and Organization

For (A) "Systems and Organizations" the sub-categories of the taxonomy are as presented in Figure 1:

1. **Work Systems:** Sociotechnical systems, Strategy, Interactions, Goal-oriented, Self-regulation, Proactive, Adaptation, Integrity, and Robustness.
2. **Rules:** Legislation, Command-and-control, Guidelines, Rule hierarchy, Collaboration, Mobilization, Controlling, Monitoring, and Anticipation.
3. **Leadership:** Policy, Operations, Change management, Pro-active, Flexibility, Adaptation, Re-balance, Adjustments, and Mobilization.

The decisions concerning the boundaries of "Systems and Organization" with regard to the identification of the entities "work systems," "rules," and "leadership" comprising the system are perceived as outcomes, that is, entities closely related to the system's key performance indicators (e.g., organization), while some are seen as entities to be implemented during the design process (Edwards & Jensen, 2014). Performance management systems also have a direct influence on the first-line management behavior manifested by two mechanisms (Edwards & Jensen, 2014), namely (1) standard operating procedures are expected as a managerial response to an employee, and (2) induced practice. The organizational conditions of "Systems and Organization" can be represented in the process through the transitions between different individuals and their tasks, coordination and communication across the process, and other temporal aspects of the process (Carayon et al., 2014).

A good system identification should be able (Jackson, 2003, p. 96): (1) to identify the purpose(s) to be pursued; (2) determine the relevant system for achieving the purpose(s) (technology and human factors as "system in focus"); (3) specify the entity of which the system in focus is a part (wider systems, environments); and (4) specify the viable parts of the system in focus ("unfolding complexity"). These are the parts that "produce" the system in focus. Any solution within sociotechnical systems involves iterating through the following series of steps: identify priority hazards, isolate the causes, determine the mechanisms, and develop countermeasures to protect workers from these hazards (Carayon et al., 2015).

Figure 1. "Systems and organizations": The sub-categories of the taxonomy of sociotechnical systems

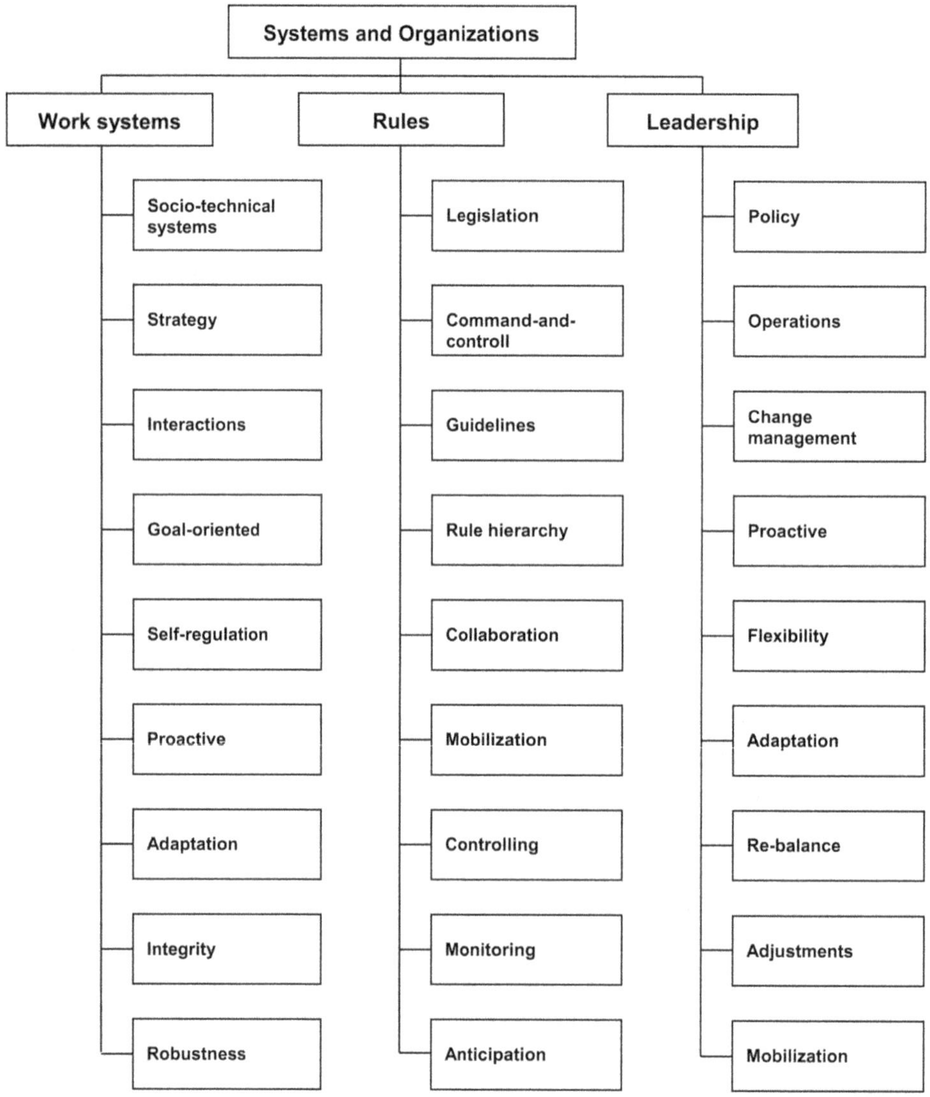

People and Performance

For (B) "People and Performance," the sub-categories of the taxonomy are as follows (Figure 2):

1. **Communication:** Interactions, Information exchange, Management of meaning, Sense-making procedures, Flexibility, Anticipation, and Adaptation.
2. **Processes:** Procedures, Operations, Indicators, Flexibility, Relationships, Micro- and macro-levels, and Control and interactions.
3. **Competencies:** Evaluation, Knowledge, Comprehension, Application, Exchange, Analysis, and Synthesis.

"People and Performance" should be consistent with the principles of sociotechnical systems, including, for example, both technical and social objectives, which will likely include participatory ergonomics and should predict multidimensional performance improvements in "communication," "processes," and "competencies" (applied from Kleiner, 1999). The work system model can be used as a guide when asking questions about an event (Carayon et al., 2014). For instance, (1) who was involved [Person], (2) what were they doing [Tasks], (3) what tools/technologies were they using [Tools/Technologies], (4) where did the event take place [Environment], and (5) what organizational conditions contributed to the event [Organization]. Edwards and Jensen (2014) found that technology, facilities, formal and informal organizations (structures, procedures, and processes), workers (qualifications, competencies, attitudes, and values), and (layers of) managers can all be designated as examples of those entities typically used in this kind of problem-solving process.

Equipment and Interfaces

For (C) "Equipment and Interfaces," the sub-categories of the taxonomy are as follows (Figure 3):

1. **Technology:** Sociotechnical system, Application, Expertise, Flexibility, Implementation, Control, and Self-regulation.

Figure 2. "People and performance": The sub-categories of the taxonomy of sociotechnical systems

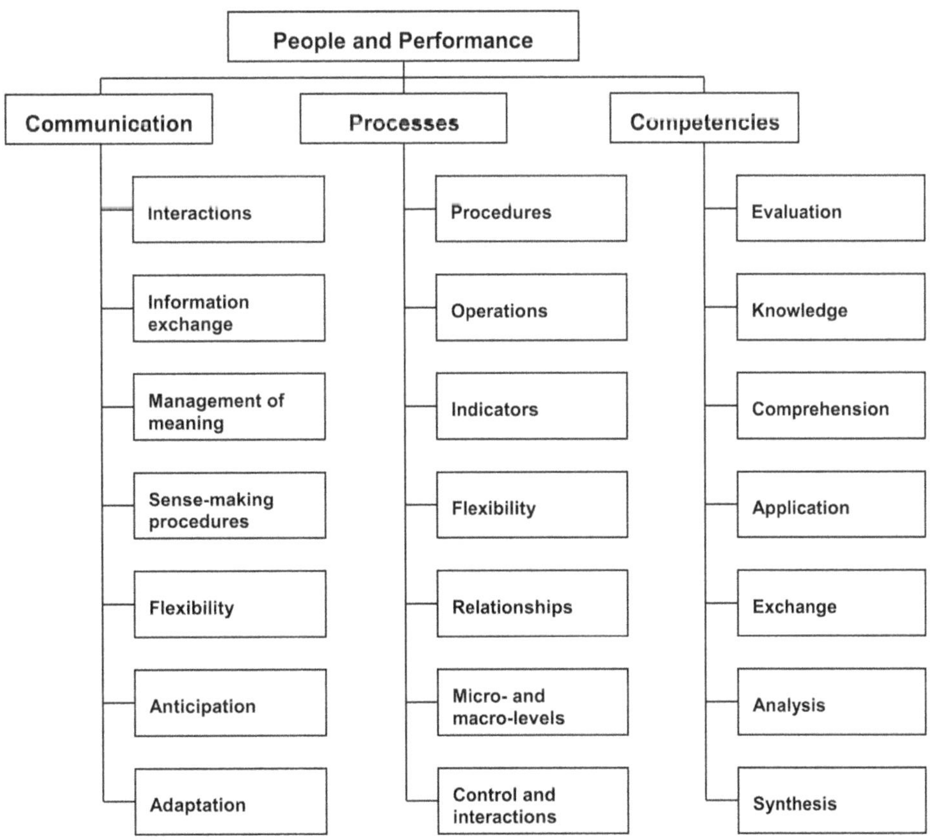

2. **Procedure:** Best practices, Adaptation, Relationships, Flexibility and balance, Calibration, Up-to-dateness, and Preparedness.
3. **Performance:** Hierarchical activities, Control, Flexibility, Cognitive performance, Procedures, Experience, and Anticipation.

The "Equipment and Interfaces"-related systems approach is needed to support operating systems as they attempt to promote excellent safety performance in terms of "technology," "procedure," and "performance". The outputs of this variety of safety performance prevent errors and deviations in the normal workflow, control information systems, provide feedback from systems and processes, implement resilient technology, redesign workstation systems, and integrate systems mechanisms. By understanding the causal forces stemming from the environment, changes to the personnel and technological subsystems could be jointly designed, which should result in improved perceived and objective safety performance (applied from Kleiner, 1999). The inputs of the systems are typically a combination of material and non-material (e.g., interfaces, knowledge) objects (Edwards & Jensen, 2014).

Figure 3. "Equipment and Interfaces": The sub-categories of the taxonomy of sociotechnical systems

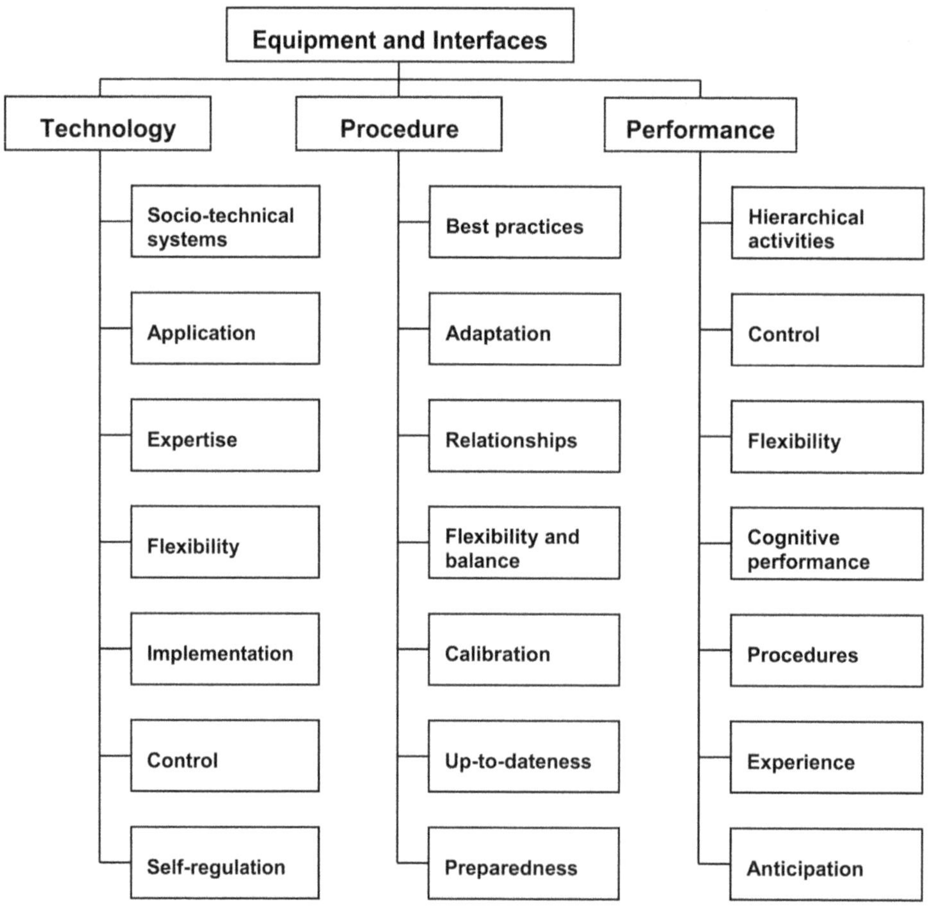

REFERENCE RESULTS AND HOW THEY FIT HYPOTHESES H1–H5

The Independent Variable "Activities of the Management"

Hale and Borys (2014) indicated that one fundamental aspect of any system of management is its need to view rule sets as dynamic in order to (1) cope with diversity and exceptions to practices and rules, and (2) place the management focus on the activities involving monitoring, flexibility, and change. The management take steps to organize activities in such a way that they make better use of resources, information, and equipment; they plan activities, assign tasks, determine resource requirements, and coordinate interrelated activities (Dulac & Leveson, 2004). The activities of the management are likely to involve task behaviors such as setting challenging yet realistic OSH goals and deadlines, developing specific action plans for proactive OSH, determining ways to overcome obstacles, organizing the work efficiently, and emphasizing OSH performance (applied from Yukl, 2002, p. 83).

Grote (2014) showed that the activities associated with the management of uncertainty can serve as a common framework due to the pervasiveness of uncertainty both as a concept in organizational and design research and as a day-to-day challenge in business operations. Management activities tend to be institutionalized through policies, plans, procedures, and processes, and therefore they are not easily and readily adaptable to the natural and inevitable variations that occur in the work being conducted and the hazards being encountered (Wachter & Yorio, 2014). The activities associated with the management of uncertainty have to occur in ways that foster an appropriate balance between stability and flexibility in organizational functioning, while still taking into due consideration the nature of sociotechnical and environmental contingencies (Grote, 2014).

The Dependent Aggregated Variable "Instructions" (Question 1, Appendix, Referring to Hypothesis 1)

Effective communication channels, for example, concerning instructions, are required between the hierarchical levels of each control structure providing the information necessary to impose constraints, and they act as a measuring channel with which to obtain feedback (Leveson, 2005). The safety measures are presented in instructions for accident prevention and the safe performance of operations (Bellamy, Geyer, & Wilkinson, 2008). It can prove useful to define deviation from normal performance when the instructions refer to activities in well-structured technical systems for which a correct action sequence can be obtained from the functional requirements (Rasmussen, Pejtersen, & Goodstein, 1994). Instructions show how an organization executes deviation investigations and safety measures, for example, implemented by sub-contractors in order to lead to outcomes that maintain a stable relationship with the environment (McDonald, 2006, p. 157). The occurrence of a "human error" is a signal that there is a mismatch somewhere in the coupling to the environment (Rasmussen et al., 1994). Sometimes, such an event is observed directly and locally within a closed loop as a deviation from the normal behavior of an individual (Rasmussen, 1997). Therefore, this study generated the following hypothesis:

Hypothesis 1 (H1): For workers and managers, "Instructions" would have positive impacts on the "Activities of the management" in the joint framework of sociotechnical systems.

The Dependent Aggregated Variable "OSH Training" (Question 2, Appendix, Will Be Evaluated in Hypothesis 2)

On-the-job training is mainly aimed at increasing individual knowledge regarding risks and risk protection measures (Grote, 2012). The engineering processes should ensure that the OSH competencies of the staff are optimal, and it is good practice to promote competent staff to OSH roles in the most effective and flexible way (Siemieniuch & Sinclair, 2014). The decision process used within an organization will determine the extent to which any decision incorporates their expertise and knowledge (Yukl, 2002, p. 83). Competence here refers to the criteria and constraints determining the choice between different options (Rasmussen & Svedung, 2000). There should be both a downward reference channel providing the information necessary to impose constraints on the level below and a measuring channel intended to provide feedback about how effectively the constraints have been enforced (Leveson et al., 2006, p. 100). The capacity for flexibility and change allows for learning and adaptation to new and variable demands (Grote, 2014). From a control perspective, two problems must be resolved during training (Rasmussen, 1986). First, the activity must be synchronized with the behavior of the environment in order for it to achieve the desired goal (Rasmussen, 1986). Second, the activity must be optimized so as to create a smooth and efficient pattern within the boundary of the task (Rasmussen, 1986). These data and the arguments discussed above lead to the following hypothesis:

Hypothesis 2 (H2): For workers and managers, "OSH training" would have positive impacts on the "Activities of the management" in the joint framework of sociotechnical systems.

The Dependent Aggregated Variable "Operations, Technical Processes, and the Safe Use of Chemicals" (Question 3, Appendix, Will Be Considered in Hypothesis 3)

The most important aspects of a technical system are the control measures themselves, the equipment, and the process controls, which are the necessary measures of major accident prevention and the safe boundary of operation (Bellamy et al., 2008). In a sociotechnical system, the OSH control system should implement assessments of safe performance, while the boundaries for safe operations should be defined by risk analyses and rules governing system operation (Vicente, 2002). The high levels of routine, standardization, and formalization that create stability generally enhance both predictability and control, and they reduce the need for ad hoc coordination (Grote, 2014). In each control loop, at each level of the sociotechnical control structure, hazardous behavior results from the inadequate enforcement of constraints on the process controlled at the level below (Dulac & Leveson, 2004). Operations "as imagined" are not always as amenable to capture or analysis as the idealized control model. Thus, the starting position would be an understanding of the gap between the system as imagined and the system as actual operation requires (Dekker, 2006, p. 89). In the operation of modern, integrated, large-scale systems, it is becoming increasingly important to control for the possibility of "human error" (Rasmussen et al., 1994). The above discussion leads to the following hypothesis:

Hypothesis 3 (H3): For workers and managers, "Operations, technical processes, and the safe use of chemicals" would have positive impacts on the "Activities of the management" in the joint framework of sociotechnical systems.

The Dependent Aggregated Variable "Use of Personal Protective Equipment (PPE)" (Question 4, Appendix, Will Be Evaluated in Hypothesis 4)

Workers' engagement in safety may act systematically to reduce the probability of human errors from occurring by making workers more involved with, and aware of, their tasks and the use of PPE, as well as any error traps that could be present (Wachter & Yorio, 2014). The flexibility of skilled performance is based on the abilities of human beings to adopt and adjust to the use of PPE so that it will be well suited for specific purposes (applied from Rasmussen et al., 1994). Descriptions of the physical processes and actions involved in the use of PPE are required in order to provide necessary information and evaluate its effectiveness, for example, the safe use of chemicals, exposure limit values for chemicals, ergonomics of equipment use, etc. (applied from Rasmussen et al., 1994). Managers who supervise the use of PPE require extensive knowledge about the best training methods to be used in the performance of the training process. It is important to have technical knowledge regarding both PPE and work processes prior to starting (1) to plan and organize the use of PPE during work operations, (2) to direct and train employees with specialized activities, and (3) to monitor and evaluate their OSH performance. The following hypothesis was hence constructed:

Hypothesis 4 (H4): For workers and managers, the "Use of personal protective equipment (PPE)" would have positive impacts on the "Activities of the management" in the joint framework of sociotechnical systems.

The Dependent Aggregated Variable "Measuring, Follow-Up, and Prevention of Major Accidents" (Question 5, Appendix, Referring to Hypothesis 5)

Safety must be designed into the human factors system, while safety when the system is operating depends partly on the original system design and partly on ensuring that there is effective control over operations (Leveson et al., 2006, p. 99). Dulac and Leveson (2004) showed that as the complexity of the engineered systems has increased, the hazard analysis techniques have continued to lag behind the state-of-the-art engineering practices. When safety incidents (i.e., employee injuries and near misses) do occur, organizations can investigate those accidents with the ultimate goal of reducing the probability of such an event occurring again (Wachter & Yorio, 2014). When organizations need to achieve a more comprehensive overview of accidents, risk management tools and models need to treat the systems as dynamic processes that are continually adapting in order to achieve their ends and react to changes in both themselves and their environment (Leveson, 2002). Risk management that is directly related to the management of uncertainty is suggested as another opportunity for bringing human factors and ergonomics expertise to the fore due to pervasive efforts across all business sectors to better handle risks (Grote, 2014). The following hypothesis was thus devised:

Hypothesis 5 (H5): For workers and managers, "Measuring, follow-up, and prevention of major accidents" would have positive impacts on the "Activities of the management" in the joint framework of sociotechnical systems.

RESEARCH RESULTS

For workers' OSH representatives (n=120) and OSH managers (n=85), the results with respect to the Hypotheses 1-5 (H1-H5) concerning how the "Activities of the management" influenced the different dependent aggregated variables are presented in Table 1.

In this qualitative systems analysis, the taxonomy is classified into sub-categories. In this context, the author of the present article, as the investigator, selected the two most important sub-classes with respect to the impacts of the "Activities of the management" on Questions 1 to 5 in the following:

1. The sub-classes for "Systems and Organizations" (see the sub-classes in Figure 1) within the following sub-categories: (1) Work systems, (2) Rules, and (3) Leadership.
2. The sub-classes for "People and Performance" (figure 2) within the following sub-categories: (1) Communication, (2) Processes, and (3) Competencies.

Table 1. For the workers' OSH representatives (W) and OSH managers (M), the values of R-squared and the parameter estimates in the regression analysis with respect to the impacts of the "Activities of the management" (independent) on the different dependent aggregated variables (Questions 1–5, Appendix)

Independent Variable "Activities of the Management"										
Dependent Aggregated Variables			R^2	Parameter Estimate	Standard Error	t-Value	p-Value	Standard Estimate	Tests of H1-H5
H1	Question (Q) 1: "Instructions" (A) "Systems and Organizations" (Fig, 1) (1) Work systems: Sociotechnical systems; Proactive. (2) Rules: Guidelines; Monitoring. (3) Leadership: Operations; Change management. (B) "People and Performance" (Fig, 2) mm (1) Communication: Interactions: Information exchange (2) Processes: Procedures; Control and interactions. (3) Competencies: Evaluation; Knowledge. (C) "Equipment and Interfaces" (Fig, 3) (1) Technology: Expertise; Implementation. (2) Procedure: Best practices; Preparedness. (3) Performance: Control; Flexibility; Procedures.	W	0.23	1.62	0.28	5.74	<.0001	0.43	***
		M	0.12	1.10	0.33	3.37	0.0011	0.35	**

Independent Variable "Activities of the Management"										
Dependent Aggregated Variables			R^2	Parameter Estimate	Standard Error	t-Value	p-Value	Standard Estimate	Tests of H1-H5
H2	Q2: "OSH training" (A) "Systems and Organizations" (Fig, 1) (1) Work systems: Goal-oriented; Self-regulation. (2) Rules: Guidelines; Rule hierarchy; Collaboration. (3) Leadership: Policy; Operations. (B) "People and Performance" (Fig, 2) (1) Communication: Information exchange; Sense-making procedures. 2) Processes: Operations; Control and interactions. (3) Competencies: Knowledge; Comprehension. (C) "Equipment and Interfaces" (Fig, 3) (1) Technology: Expertise; Implementation. (2) Procedure: Best practices; Preparedness. (3) Performance: Cognitive performance; Experience.	W	0.43	2.40	0.26	9.15	<.0001	0.65	***
		M	0.45	1.99	0.24	8.21	<.0001	0.67	***
H3	Q3: "Operations, Technical processes, Safe use of chemicals" (A) "Systems and Organizations" (Fig, 1) (1) Work systems: Sociotechnical systems; Self-regulation. (2) Rules: Legislation; Guidelines; Rule hierarchy. (3) Leadership: Policy; Operations; Change management. (B) "People and Performance" (Fig, 2) (1) Communication: Management of meaning; Sense-making procedures. (2) Processes: Procedures; Control and interactions. (3) Competencies: Evaluation; Synthesis. (C) "Equipment and Interfaces" (Fig, 3) (1) Technology: Sociotechnical system; Implementation. (2) Procedure: Best practices; Adaptation. (3) Performance: Control; Cognitive performance.	W	0.41	3.02	0.36	8.44	<.0001	0.65	***
		M	0.22	2.09	0.45	4.64	<.0001	0.47	***

Independent Variable "Activities of the Management"									
Dependent Aggregated Variables			R^2	Parameter Estimate	Standard Error	t-Value	p-Value	Standard Estimate	Tests of H1-H5
H4	Q4: "Use on personal protective equipment (PPE)" (A) "Systems and Organizations" (Fig, 1) (1) Work systems: Self-regulation; Proactive. (2) Rules: Legislation; Command-and-control; Guidelines. (3) Leadership: Policy; Operations; Adaptation. (B) "People and Performance" (Fig,2) (1) Communication: Interactions: Management of meaning; Sense-making procedures. (2) Processes: Procedures; Control and interactions. (3) Competencies: Comprehension; Application. (C) "Equipment and Interfaces" (Fig, 3) (1) Technology: Application; Implementation. (2) Procedure: Best practices; Up-to-dateness. (3) Performance: Control; Anticipation.	W	0.39	1.83	0.23	8.00	<.0001	0.62	***
		M	0.25	1.35	0.27	4.99	<.0001	0.50	***
H5	Q5: "Measuring, Follow-up, Prevention of major-accidents" (A)"Systems and Organizations" (Fig, 1) (1) Work systems: Strategy; Goal-oriented. (2) Rules: Collaboration; Monitoring. (3) Leadership: Policy; Mobilization. (B) "People and Performance" (Fig, 2) (1) Communication: Interactions: Information exchange. (2) Processes: Indicators; Control and interactions. (3) Competencies: Evaluation; Analysis. (C) "Equipment and Interfaces" (Fig, 3) (1) Technology: Application; Implementation. (2) Procedure: Adaptation; Preparedness. (3) Performance: Hierarchical activities; Anticipation.	W	0.26	2.01	0.38	5.96	<.0001	0.51	***
		M	0.15	1.46	0.40	3.63	0.0005	0.38	***

** $p < 0.01$.
*** $p < 0.001$.

3. The sub-classes for "Equipment and Interfaces" (figure 3) within the following sub-categories: (1) Technology, (2) Procedure, and (3) Performance.

In an additional analysis concerning the effects on Q1–Q5 (Questions 1–5 in Appendix), the determining independent variables, namely (2) "In my workplace, the requirements of the OSH legislation are promoted in a concrete manner" and (3) "In my workplace, the collaboration between the employer and workers regarding OSH matters promotes workplace safety in a concrete manner," gave the following results:

With respect to the effects of the "OSH legislation" for the workers' OSH representatives, the aggregated variables in Q1–Q5 were statistically highly significant. For the OSH managers, the effects of the "OSH legislation" were as follows: Q1 (0.0184*), Q2 (0.0010**), Q3 (0.2840), Q4 (0.0005***), and Q5 (0.0285*). The most important individual categories, which gave the most important effects in this taxonomy for Q1–Q5, were as follows:

Q1: "Instructions"

(A) "Systems and Organizations": (1) Work systems: Self-regulation, Adaptation; (2) Rules: Legislation, Command-and-control; and (3) Leadership: Operations, Proactive.

(B) "People and Performance": (1) Communication: Information exchange, Sense-making procedures; (2) Processes: Procedures, Control and interactions; and (3) Competencies: Comprehension, Application.

(C) "Equipment and Interfaces": (1) Technology: Implementation, Control; (2) Procedure: Best practices, Flexibility and balance; and (3) Performance: Hierarchical activities, Control.

Q2: "OSH Training"

(A) "Systems and Organizations": (1) Work systems: Goal-oriented, Self-regulation; (2) Rules: Legislation, Command-and-control; and (3) Leadership: Policy, Adaptation.

(B) "People and Performance": (1) Communication: Management of meaning; (2) Processes: Relationships, Control and interactions; and (3) Competencies: Comprehension, Application.

(C) "Equipment and Interfaces": (1) Technology: Application, Implementation; (2) Procedure: Up-to-datedness, Preparedness; and (3) Performance: Hierarchical activities, Control.

Q3: "Operations, Technical Processes, and the Safe Use of Chemicals"

(A) "Systems and Organizations": (1) Work systems: Strategy, Interactions, Self-regulation; (2) Rules: Legislation, Command-and-control; and (3) Leadership: Operations, Adaptation.

(B) "People and Performance": (1) Communication: Sense-making procedures, Adaptation; (2) Processes: Indicators, Control and interactions; and (3) Competencies: Comprehension, Application.

(C) "Equipment and Interfaces": (1) Technology: Implementation, Self-regulation; (2) Procedure: Best practices, Adaptation; and (3) Performance: Control, Procedures, Experience.

Q4: "Use of Personal Protective Equipment (PPE)"

(A) "Systems and Organizations": (1) Work systems: Self-regulation, Proactive; (2) Rules: Guidelines, Rule hierarchy; and (3) Leadership: Policy, Adaptation, Mobilization.
(B) "People and Performance": (1) Communication: Anticipation, Adaptation; (2) Processes: Micro- and macro-levels, Control and interactions; and (3) Competencies: Comprehension, Application.
(C) "Equipment and Interfaces": (1) Technology: Implementation, Control; (2) Procedure: Best practices, Up-to-datedness; and (3) Performance: Control, Procedures.

Q5: "Measuring, Follow-Up, and Prevention of Major Accidents"

(A) "Systems and Organizations": (1) Work systems: Strategy; (2) Rules: Collaboration, Monitoring; and (3) Leadership: Policy, Operations.
(B) "People and Performance": (1) Communication: Anticipation, Adaptation; (2) Processes: Micro- and macro-levels, Control and interactions; and (3) Competencies: Evaluation, Comprehension.
(C) "Equipment and Interfaces": (1) Technology: Implementation, Self-regulation; (2) Procedure: Up-to-datedness, Preparedness; and (3) Performance: Control, Procedures.

With respect to the effects of the "collaboration between the employer and workers" for the workers' OSH representatives, the aggregated variables in Q1–Q5 were statistically highly significant. For the OSH managers, the effects of the "collaboration between the employer and workers" were as follows: Q1 (0.1140), Q2 (0.0011**), Q3 (0.0278*), Q4 (0.0270*), and Q5 (0.0350*). The most important individual categories, which gave the most important effects in this taxonomy for Q1–Q5, were as follows:

Question (Q) 1: "Instructions"

(A) "Systems and Organizations": (1) Work systems: Interactions, Self-regulation; (2) Rules: Guidelines, Collaboration; and (3) Leadership: Policy, Adjustments, Mobilization.
(B) "People and Performance": (1) Communication: Interactions, Information exchange; (2) Processes: Relationships, Control and interactions; and (3) Competencies: Application.
(C) "Equipment and Interfaces": (1) Technology: Implementation, Self-regulation; (2) Procedure: Best practices, Relationships, Flexibility and balance; and (3) Performance: Hierarchical activities, Anticipation.

Q2: "OSH Training"

(A) "Systems and Organizations": (1) Work systems: Self-regulation, Robustness; (2) Rules: Legislation, Collaboration, Monitoring; and (3) Leadership: Operations, Mobilization.
(B) "People and Performance": (1) Communication: Interactions, Information exchange; (2) Processes: Relationships, Control and interactions; and (3) Competencies: Application, Exchange.
(C) "Equipment and Interfaces": (1) Technology: Expertise, Self-regulation; (2) Procedure: Best practices, Relationships; and (3) Performance: Hierarchical activities, Anticipation.

Q3: "Operations, Technical Processes, and the Safe Use of Chemicals"

(A) **"Systems and Organizations":** (1) Work systems: Interactions, Self-regulation; (2) Rules: Legislation, Collaboration, Anticipation; and (3) Leadership: Operations, Proactive.
(B) **"People and Performance":** (1) Communication: Interactions, Sense-making procedures; (2) Processes: Relationships, Control and interactions; and (3) Competencies: Comprehension, Synthesis.
(C) **"Equipment and Interfaces":** (1) Technology: Implementation, Self-regulation; (2) Procedure: Best practices, Relationships; and (3) Performance: Hierarchical activities, Procedures.

Q4: "Use of Personal Protective Equipment (PPE)"

(A) **"Systems and Organizations":** (1) Work systems: Interactions, Proactive; (2) Rules: Rule hierarchy, Collaboration; and (3) Leadership: Policy, Operations.
(B) **"People and Performance":** (1) Communication: Interactions, Adaptation; (2) Processes: Procedures, Relationships; and (3) Competencies: Comprehension, Exchange.
(C) **"Equipment and Interfaces":** (1) Technology: Implementation, Self-regulation; (2) Procedure: Best practices, Preparedness; and (3) Performance: Hierarchical activities, Cognitive performance.

Q5: "Measuring, Follow-Up, and Prevention of Major Accidents"

(A) **"Systems and Organizations":** (1) Work systems: Strategy, Robustness; (2) Rules: Guidelines, Collaboration; and (3) Leadership: Policy, Operations.
(B) **"People and Performance":** (1) Communication: Interactions, Sense-making procedures; (2) Processes: Procedures, Control and interactions; and (3) Competencies: Evaluation, Comprehension.
(C) **"Equipment and Interfaces":** (1) Technology: Control, Self-regulation; (2) Procedure: Best practices, Up-to-datedness; and (3) Performance: Hierarchical activities, Cognitive performance, Procedures.

DISCUSSION, SOLUTIONS, AND RECOMMENDATIONS

Theoretical Implications

In the present study, Hypotheses H1–H5, which were postulated in the framework of sociotechnical systems, were confirmed for both workers (n=120) and OSH managers (n=85) employed in the chemical industry in the following taxonomy: A) "Systems and Organizations" (Figure 1): (1) Work systems, (2) Rules, and (3) Leadership; (B) "People and Performance" (Figure 2): (1) Communication, (2) Processes, and (3) Competencies; and (C) "Equipment and Interfaces" (figure 3): (1) Technology, (2) Procedure, and (3) Performance. Hypotheses H1–H5 were verified when the "Activities of the management" had effects on the following variables (see Table 1): H1: "Instructions" and H2: "OSH training" within the following framework of sociotechnical systems: H3: "Operations, technical processes, and the safe use of chemicals;" H4: "Use of personal protective equipment (PPE);" and H5: "Measuring, follow-up, and prevention of major accidents".

The present results support those of Robertson et al. (2015), who found that a framework of sociotechnical systems for workplace safety could help in terms of bridging the gap between the applied theories and the practice of workplace safety. The sociotechnical approach is acutely sensitive to context (Robertson et al., 2015), that is, it not only seeks to illuminate the immediate organizational and technical factors that impact safe work performance, but also the less immediate and more diffuse influences of organizational policy, communications, and decision making as they "flow down" to influence performance.

With respect to the gathering of data, this study supports the findings of Stringfellow (2010), who concluded that by conducting a taxonomy assessment, one can develop a causal factor taxonomy that can be used as a framework for understanding how best to prevent accidents and anticipating how system behavior may be adequately controlled. This study also supports the results of other investigators (e.g., Murphy, Robertson, & Carayon, 2014), who noted that the purpose of the foundational theoretical framework of sociotechnical systems theory is to establish a "harmonized" work system that improves numerous aspects of organizational performance and effects. In addition, this study supports the results of Murphy et al. (2014), who found that there exists a need to emphasize the importance of a multilevel analysis that simultaneously examines the impact of the different work system design interfaces in order to integrate a sociotechnical system model.

Practical Implications

The present results support those of Robertson et al. (2015), since the taxonomy of sociotechnical systems could be used by safety practitioners in order to understand the role of the broader socio-organizational context in fostering safety, as well as to devise more powerful, sustainable solutions for enhancing safety The results of this study showed that the present novel taxonomy mechanisms for the collection and analysis of data would be beneficial in terms of enabling managers and workers to better understand and monitor whether or not priority measures of best practices are being met. The strength of the taxonomy systems approach lies in its explanatory power when examining performance and behaviors. Another strength concerns its ability to highlight new parts of the system that could be better utilized so as to achieve safer performance in proactive measures providing avenues outside of traditional warning devices.

A sociotechnical systems approach enables the behavior of all managers and workers to be understood, as opposed to providing an understanding of only individual user groups. A comprehensive review of the various approaches of sociotechnical systems serves to generate valuable insights that could lead to innovative approaches for improving safety (Carayon et al., 2015). A systems-thinking approach involves organizing complexity into a coherent story that illuminates both the causes of problems and how they can be remedied in enduring ways (Senge, 1994, p. 128). The feedback perspective suggests that everyone shares responsibility for the problems generated by a system. Yet, this does not necessarily imply that everyone involved can exert equal leverage in terms of changing the system (Senge, 1994, p. 78). Learning always involves new understandings and new behaviors, that is, "thinking" and "doing" (Senge, 1994, p. 374).

This study supports the findings of prior investigations (e.g., Karsh, Waterson, & Holden, 2014) that noted how research concerning the micro-, meso-, and macro-levels is important in terms of helping to determine the extent and design of workplace-related improvements and their justifications. Stringfellow (2010, p. 110) found that a process capable of lessening the gap between the classification of human and organizational errors in order to prevent them from occurring is needed, and even though applying a taxonomy may prove useful in relation to accident analysis and investigation, it is only part of the solution. Furthermore, Stringfellow (2010) concluded that useful safety engineering methods must identify how flaws in what he referred to as the mental models of humans or organizations can occur, as well as what will lead to mismatches between reality and the mental model that adversely affects safety. This study identified a need to emphasize a multilevel analysis that simultaneously examines the impact of the different work system design interfaces in order to integrate a sociotechnical systems model.

Generalization of the Present Results

This quantitative study was based on self-reported data. Therefore, the present results may be subject to potential inaccuracies related to the inability of respondents to accurately recall information. King and Bruner (2000) found that so-called desirable responding is most likely to occur when individuals respond to socially sensitive questions. Chung and Monroe (2003) stated that social desirability concerns "the tendency of individuals to deny socially undesirable actions and behaviors and to admit to socially desirable ones." Further, Arnold and Feldman (1981) highlighted how subjective assessment methodologies are more likely to result in the appearance of a social desirability bias in questions related to job and organizational characteristics.

The present study makes its arguments explicit and allows readers to consider the generalization claim. On the basis of the detailed contextual descriptions of the qualitative results, the reader must judge for him/herself whether the findings may be generalized to novel situations. Kvale and Brinkmann (2009) indicated that validity within a qualitative study is related to whether a method actually examines what it is intended to investigate. The generalizability of the results refers to the question of whether or not the findings of the research are applicable to other situations. Eriksson and Kovalainen (2008) showed that one can utilize four different criteria when evaluating the reliability, validity, and replicability of a qualitative investigation, namely the fruitfulness, quality of interpretation, quality of transcription, and usefulness of the findings.

FUTURE RESEARCH DIRECTIONS

This study identified a number of areas of research worthy of further development. The best way to validate a measure of the design of sociotechnical systems would be to test the scales in order to refine the parameters. Future studies could hence benefit from the adoption of research designs that permit the examination of causality, with longitudinal designs and intervention investigation likely being the most appropriate type of set-up. It would be interesting to explore whether an application of sociotechnical dynamics could promote productivity, reliable and safe task performance, and workers' participation in the highly hazardous systems of the chemical industry. The present results support the conclusion of Carayon et al. (2015) that future research concerning sociotechnical systems for workplace safety should also address the "active" role of workers. This study showed that the effective control of workplace risks would require the systematic assessment of the safety impacts, the consequent identification of areas where risks need to be better controlled, and the joint optimization of technology, organization, and human factors for a learning organization.

CONCLUSION

In a framework of sociotechnical systems, it is important to identify all the potentially significant parts, as well as to understand how each part interacts with the others in mediating the excellent performance of organizations. A combination of both qualitative and quantitative research methodologies established a balanced work system that provides measures intended to improve organizational performance. An assessment of the qualitative taxonomy can be applied as a framework for understanding how best to promote safety performance. The constructionist approach of the present qualitative taxonomy assessment emphasized the joint optimization of technology, organization, and human factors, as well as the shared patterns of meaning making and coordinated interactions within organizations. Top management should be empowered to initiate and manage improvements in sociotechnical systems, as well as to encourage managers to support "bottom-up" practices. The different forms of employees' participation are likely to be effective when their managers have sufficient skills in terms of managing sociotechnical systems, facilitating proactive problem solving with regard to safety measures, and dealing with proactive safety measures in collaboration.

Within sociotechnical systems, the upper technical systems and human factors of organizations exert influences on their sub-systems. In this context, work systems of production methods need to be jointly optimized if they are to attain maximum efficiency in relation to achieving the system's primary safety task. The practices and processes implemented through safety management systems provide the necessary "first steps" in sociotechnical systems in terms of attaining excellent safety performance. The present results showed that many of the safety practices most conducive to fostering an organization's adherence to good safety procedures are embedded into the disciplines necessary for building a learning organization.

ACKNOWLEDGMENT

The author is grateful for the help with the questionnaire received from Kyösti Louhelainen, Pirjo Korhonen, Niina Kallio, Milja Koponen, and Eija-Riitta Hyytinen of the Finnish Institute of Occupational Health (FIOH). The author also wishes to thank Maria L. Hirvonen from the FIOH for her assistance with the data analysis. The author is very grateful to the two anonymous reviewers for their important comments, which improved substantially the quality of the chapter.

REFERENCES

Appelbaum, S. H. (1997). Socio-technical systems theory: An intervention strategy for organizational development. *Management Decision*, *35*(6), 452–463. doi:10.1108/00251749710173823

Arnold, H. J., & Feldman, D. C. (1981). Social desirability response bias in self-report choice situations. *Academy of Management Journal*, *24*(2), 377–385. doi: 10.5465/255848

Bellamy, L. J., Geyer, T. A. W., & Wilkinson, J. (2008). Development of a functional model which integrates human factors, safety management systems and wider organizational issues. *Safety Science*, *46*(3), 461–492. doi:10.1016/j.ssci.2006.08.019

Carayon, P., Hancock, P., Leveson, N., Noy, I., Sznelwar, L., & van Hootegem, G. (2015). Advancing a sociotechnical systems approach to workplace safety - developing the conceptual framework. *Ergonomics*, *58*(4), 548–564. doi:10.1080/00140139.2015.1015623 PMID:25831959

Carayon, P., Wetterneck, T. B., Rivera-Rodriguez, A. J., Shoofs Hundt, A., Hoonakker, P., Holden, R., & Gurses, A. P. (2014). Human factors systems approach to healthcare quality and patient safety. *Applied Ergonomics*, *45*(1), 14–25. doi:10.1016/j.apergo.2013.04.023 PMID:23845724

Checkland, P. B. (2000). Soft Systems Methodology: A Thirty Year Retrospective. *Systems Research and Behavioral Science*, *17*(1), 11–58. doi:10.1002/1099-1743(200011)17:1+<::AID-SRES374>3.0.CO;2-O

Chung, J., & Monroe, G. S. (2003). Exploring social desirability bias. *Journal of Business Ethics*, *44*(4), 291–302. doi:10.1023/A:1023648703356

Cronbach, L. J., & Shavelson, R. J. (2004). My current thoughts on coefficient alpha and successor procedures. *Educational and Psychological Measurement*, *64*(3), 391–418. doi:10.1177/0013164404266386

Dekker, S. W. A. (2006). Resilience engineering: Chronicling the emergence of confused consensus. In E. Hollnagel, D. D. Woods, & N. Leveson (Eds.), *Resilience Engineering: Concepts and Precepts* (pp. 77–92). Aldershot, UK: Ashgate Publishing Limited.

Dul, J., Broder, R., Buckle, P., Carayon, P., Talzon, P., Marras, W., ... Van Der Doelen, B. (2012). A strategy for human factors/ergonomics: Developing the discipline and the profession. *Ergonomics*, *55*(4), 377–395. doi:10.1080/00140139.2012.661087

Dulac, N., & Leveson, N. (2004). An approach to design for safety in complex systems. *Int. Conference on System Engineering (INCOSE '04)*. Retrieved March 22, 2018, from http://sunnyday.mit.edu/papers/incose-04.pdf

Edwards, K., & Jensen, P. L. (2014). Design of systems for productivity and well being. *Applied Ergonomics*, *45*(1), 26–32. doi:10.1016/j.apergo.2013.03.022 PMID:23631941

Eriksson, P., & Kovalainen, A. (2008). *Qualitative Methods in Business Research*. London: Sage Publications. doi:10.4135/9780857028044

Goh, Y. M., Brown, H., & Spickett, J. (2010). Applying systems thinking concepts in the analysis of major incidents and safety culture. *Safety Science*, *45*(3), 302–309. doi:10.1016/j.ssci.2009.11.006

Grote, G. (2014). Adding a strategic edge to human factors/ergonomics: Principles for the management of uncertainty as cornerstones for system design. *Applied Ergonomics*, *45*(1), 33–39. doi:10.1016/j.apergo.2013.03.020 PMID:23622735

Hale, A., & Borys, D. (2014). Working to rule, or working safely? Part 1: A state of the art review. *Safety Science*, *55*(1), 207–221.

Hendrick, H. W. (2002a). An overview of macroergonomics. In H. W. Hendrick & B. M. Kleiner (Eds.), *Macroergonomics: Theory, methods, and applications* (pp. 1–23). Mahwah, NJ: Lawrence Erlbaum Associates. doi:10.1201/b12477-2

Hendrick, H. W. (2002b). Macroergonomics methods: Assessing work system structure. In H. W. Hendrick & B. M. Kleiner (Eds.), *Macroergonomics: Theory, methods, and applications* (pp. 45–66). Mahwah, NJ: Lawrence Erlbaum Associates. doi:10.1201/b12477-4

Hendrick, H. W. (2008). Applying ergonomics to systems: Some documented "lessons learned". *Applied Ergonomics*, *39*(4), 418–426. doi:10.1016/j.apergo.2008.02.006 PMID:18374303

Jackson, M. C. (2003). *Systems Thinking: Creative Holism for Managers*. West Sussex, UK: John Wiley & Sons Ltd. Retrieved March 22, 2018, from http://citeseerx.ist.psu.edu/viewdoc/download?doi=10.1.1.476.2858&rep=rep1&type=pdf

Johnson, R. B., Onwuegbuzie, A. J., & Tumer, L. A. (2007). Toward a definition of mixed-methods research. *Journal of Mixed Methods Research*, *1*(1), 112–133. doi:10.1177/1558689806298224

Karsh, B.-T., Waterson, P., & Holden, R. J. (2014). Crossing levels in systems ergonomics: A framework to support 'mesoergonomic' inquiry. *Applied Ergonomics*, *5*(1), 45–54. doi:10.1016/j.apergo.2013.04.021 PMID:23706573

Kleiner, B. M. (1999). Analysis and Design for Improved Safety and Quality Performance. *Occupational Safety and Ergonomics, 5*(2), 217-245.

Kleiner, B. M. (2008). Macroergonomics: Work system analysis and design. *Human Factors*, *50*(3), 461–467. doi:10.1518/001872008X288501 PMID:18689054

Kvale, S., & Brinkmann, S. (2009). *Interviews: Learning the Craft of Qualitative Research Interviewing* (2nd ed.). London: Sage Publications.

Leveson, N. (2002). *Model-based Analysis of Socio-Technical Risk*. Technical Report. Engineering Systems Division, Massachusetts Institute of Technology. Retrieved March 22, 2018, from http://sunnyday.mit.edu

Leveson, N. G. (2005). *Safety in integrated systems health engineering and management*. NASA Ames Integrated System Health Engineering and Management Forum (ISHEM). Retrieved from http://sunnyday.mit.edu

Maier, M. W. (1998). Architecting principles for systems-of-systems. *Systems Engineering*, *1*(4), 267–284. doi:10.1002/(SICI)1520-6858(1998)1:4<267::AID-SYS3>3.0.CO;2-D

McDonald, N. (2006). Organizational Resilience and Industrial Risk. In E. Hollnagel, D. D. Woods, & N. Leveson (Eds.), *Resilience Engineering: Concepts and Precepts* (pp. 155–180). Aldershot, UK: Ashgate Publishing Limited.

Murphy, L. A., Robertson, M. M., & Carayon, P. (2014). The next generation of macroergonomics: Integrating safety climate. *Accident; Analysis and Prevention*, *68*, 16–24. doi:10.1016/j.aap.2013.11.011 PMID:24368052

Niskanen, T. (2018). A Resilience Engineering -related approach applying a taxonomy analysis to a survey examining the prevention of risks. *Safety Science*, *101*, 108–120. doi:10.1016/j.ssci.2017.08.016

Rasmussen, J. (1986). *Information Processing and Human-Machine Interaction: An Approach to Cognitive Engineering. North Holland series in system science and engineering 12*. New York: Elsevier Science Publishing Co. Inc.

Rasmussen, J. (1997). Risk management in a dynamic society: A modelling problem. *Safety Science*, *27*(2-3), 183–213. doi:10.1016/S0925-7535(97)00052-0

Rasmussen, J., Pejtersen, A. M., & Goodstein, L. P. (1994). *An Introduction to Systems Engineering*. New York: John Wiley & Sons, Inc.

Rasmussen, J., & Svedung, I. (2000). *Proactive Risk Management in a Dynamic Society*. Karlstad: Räddningsverket.

Reynaldo, J., & Santos, A. (1999). Cronbach's alpha: A tool for assessing the reliability of scales. *Journal of Extension*, *37*(2), 11–14.

SAS. (2005). *SAS User's Guide. Version 9*. Cary: SAS Institute Inc.

Senge, P. M. (1994). *The Fifth Discipline: the Art & Practice of the Learning Organization* (2nd ed.). New York: Bantam Doubleday/Currency Publishing Group.

Siemieniuch, C. E., & Sinclair, M. A. (2014). Extending systems ergonomics thinking to accommodate the sociotechnical issues of Systems of Systems. *Applied Ergonomics*, *45*(1), 85–98. doi:10.1016/j.apergo.2013.03.017 PMID:24011651

Stringfellow, M. V. (2010). *Accident analysis and hazard analysis for human and organizational factors* (Doctoral dissertation). Massachusetts Institute of Technology. Retrieved March 22, 2018, from http://sunnyday.mit.edu/safer-world/MaggieStringfellowDissertation.pdf

Teddlie, C. (2005). Methodological issues related to causal studies of leadership: A mixed methods perspective from the USA. *Educational Management Administration & Leadership*, *33*(2), 211–217. doi:10.1177/1741143205051054

Thies, C. G. (2002). A pragmatic guide to qualitative historical analysis in the study of international relations. *International Studies Perspectives*, *3*(4), 351–372. doi:10.1111/1528-3577.t01-1-00099

Vicente, K. J., & Rasmussen, J. (1992). Ecological Interface Design: Theoretical Foundations. *IEEE Transactions on Systems, Man, and Cybernetics*, *22*(4), 589–606. doi:10.1109/21.156574

Vincent, C., Ward, J., & Langdon, P. (2012). Unraveling complex system. *The Ergonomist*, *506*, 12–13.

Wachter, J. K., & Yorio, P. L. (2014). A system of safety management practices and worker engagement for reducing and preventing accidents: An empirical and theoretical investigation. *Accident; Analysis and Prevention*, *68*, 117–130. doi:10.1016/j.aap.2013.07.029 PMID:23993683

Wilson, J. R. (2000). Fundamentals of ergonomics in theory and practice. *Applied Ergonomics*, *31*(6), 557–567. doi:10.1016/S0003-6870(00)00034-X PMID:11132041

Wilson, J. R. (2014). Fundamentals of systems ergonomics/human factors. *Applied Ergonomics*, *45*(1), 5–13. doi:10.1016/j.apergo.2013.03.021 PMID:23684119

Wreathall, J. (2011). Monitoring - A Critical Ability in Resilience Engineering. In E. Hollnagel, J. Paries, D. Woods, & J. Wreathall (Eds.), *Resilience Engineering in Practice: A Guidebook* (pp. 61–68). Surrey, UK: Ashgate Publishing Limited.

Yukl, G. (2002). *The Nature of Leadership* (5th ed.). London: Prentice-Hall International Inc.

APPENDIX

Questionnaire for Workers' OSH Representatives (N=120) and OSH Managers (N=85) Employed in the Chemical Industry

The following statements were provided and the respondents answered on a four-point Likert scale: 1= Strongly disagree, 2= Somewhat disagree, 3= Somewhat agree, or 4= Strongly agree.

In terms of exploring Hypotheses H1–H5, the following statement (independent variable) was tested in relation to questions Q1–Q5:

1. "The 'activities of the management' of the company promote safety in a concrete manner in the workplace."

As additional questions in relation to H1–H5, the following statements (independent variables) were described in the results:

2. "The requirements of the 'OSH legislation' are promoted in a concrete manner in the workplace."
3. "The 'collaboration between the employer and workers' regarding OSH matters is promoted in a concrete manner in the workplace."

Question 1 (Aggregated Variable): "Instructions"

(a) Deviations from safety are assessed (Cronbach's Alpha [α] for W is 0.79 and for M is 0.56); (b) Safety measures are included in instructions (α for W is 0.79 and for M is 0.59); and (c) Safety measures are implemented by sub-contractors (α for W is 0.79 and for M is 0.58).

Question 2 (Aggregated Variable): "OSH Training"

(a) Management is trained in safety (α for W is 0.85 and for M is 0.85); (b) Workers are trained in safety (α for W is 0.85 and for M is 0.85); (c) Safety measures are part of on-the-job training (α for W is 0.85 and for M is 0.85); and (d) Risks related to chemicals and implementing safe working measures have been explained to workers (α for W is 0.87 and for M is 0.87).

Question 3 (Aggregated Variable): "Operations, Technical Processes, and the Safe Use of Chemicals"

(a) Employer uses chemicals that cause the least harm (α for W is 0.86 and for M is 0.86); (b) Employer uses production methods that cause the least harm (α for W is 0.86 and for M is 0.86); (c) Employer actively implements choices to replace dangerous chemicals (α for W is 0.86 and for M is 0.86); (d) General air ventilation system functions well (α for W is 0.86 and for M is 0.86); (e) Local exhaust ventilation is used in workstations (α for W is 0.85 and for M is 0.86); and (f) Efficiency of ventilation is monitored (α for W is 0.85 and for M is 0.86).

Question 4 (Aggregated Variable): "Use of Personal Protective Equipment (PPE)"

(a) PPE has been selected based on risk assessments (α for W is 0.85 and for M is 0.85); (b) Workers participate in the choice of PPE (α for W is 0.86 and for M is 0.85); (c) Use of PPE is mandated (α for W is 0.84 and for M is 0.84); and (d) Storage and maintenance of PPE are implemented properly (α for W is 0.84 and for M is 0.85).

Question 5 (Aggregated Variable): "Measuring, Follow-Up, and Prevention of Major Accidents"

(a) Risk of exposure to chemicals is monitored through regular measurements (α for W is 0.81 and for M is 0.81); (b) Exposure of workers to chemicals is monitored with biological monitoring (α for W is 0.84 and for M is 0.84); (c) Exposure limits are not exceeded (α for W is 0.81 and for M is 0.81); and (d) Plans, training, and guidance all promote safety measures intended to prevent major accidents (α for W is 0.81 and for M is 0.82).

Chapter 5
Considerations of the Mental Workload in Socio-Technical Systems in the Manufacturing Industry: A Literature Review

Manuel Alejandro Barajas Bustillos
Autonomous University of Ciudad Juárez, Mexico

Aide Aracely Maldonado-Macías
Autonomous University of Ciudad Juárez, Mexico

Jorge Luis García-Alcaraz
Autonomous University of Ciudad Juárez, Mexico

Juan Luis Hernández Arellano
Autonomous University of Ciudad Juárez, Mexico

Liliana Avelar Sosa
Autonomous University of Ciudad Juárez, Mexico

ABSTRACT

As cognitive tasks have displaced physical tasks in today's manufacturing industry, this sector can demand high levels of mental workload from workers. In certain situations, there is a high cognitive load, which affects operators reducing their attention to the task and causing them mental fatigue and distractions, resulting in errors that generate economic costs or even injuries to workers. This literature review aims to provide a comprehensive understanding the use of mental workload in the manufacturing sector. The methodology consisted of conducting a search in four databases. In the search, a combination of keywords was used, classifying each journal according to the mental workload evaluation means, the type of evaluation, and the area of application. Articles not focusing on the manufacturing area were discarded. Of the total of 3839 articles found, 12 have been selected. Regarding the methods used for mental load assessment, the analytic techniques were found to be the most frequently used.

DOI: 10.4018/978-1-5225-7192-6.ch005

INTRODUCTION

Currently, due to industries' growing automation, this sector's modern socio-technical systems feature great complexity (González Munõz & Gutiérrez, 2006). Despite such automation, there are still tasks in industry where people are necessary and even indispensable (Autor, 2015). For example, whenever automation has been viable, simple tasks which previously required several workers with a low level of qualification have turned into complex tasks requiring fewer workers with a higher level of knowledge, that is, with complex cognitive skills that involve decision making, improvisation, handling of information, and failure diagnosis, among others (Beaudry, Green, & Sand, 2013), so as to complement the automated systems. On the other hand, there are also tasks that have not been automatized successfully, but where, ironically, the level of complexity has increased (Mital & Pennathur, 2004). This has occurred in tasks that require people's mental abilities, or in cognitive tasks such as those mentioned above, which cause workers' mental load to tend to increase and exceed their capacities and limitations to perform their work efficiently and safely. On the other hand, a low workload level is also undesirable as it can cause workers to fall into complacency (González Munõz & Gutiérrez, 2006) and to generate errors derived from distractions or lack of attention.

Based on the foregoing, it can be said that mental workload evaluation is a key point in research and, in general, in all those activities that seek higher levels of comfort, satisfaction, efficiency and safety in the workplace (Rubio, Díaz, Martín, & Puente, 2004).

Therefore, the main objective of this work is to determine the state of the art of the assessment and consideration of mental workload into modern socio-technical systems of the manufacturing industry by means of a systematic literature review. Additional objectives of this work are to analyze important characteristics of actual available mental workload evaluation and assessment techniques such as training time, administration time, tools needed, advantages and disadvantages, among others. Also, once some selection criteria are established among the articles found, the most frequently used and widely applied techniques must be identified. Finally, to offer recommendations about the pertinence of the use and evaluation of mental workload in the manufacturing environment and operations is convenient to propose future research trends.

BACKGROUND

Whenever any type of work is done, the body undergoes two types of loads: a physical one and a mental one. The physical load is constituted by the biomechanical, physiological and caloric requirements that the worker's body demands, while the mental load refers to the cognitive processing requirements necessary for good performance at work (González Munõz & Gutiérrez, 2006) although currently there is no consensus as to which factors are most closely associated with mental load (Rubio, Díaz, & Martín, 2001).

Mental Workload Evaluation and Assessment

There are different techniques for evaluating mental work load. According to various authors (Dalmau & Ferrer, 2004; González Munõz & Gutiérrez, 2006; Stanton, Hedge, Brookhuis, Salas, & Hendrick, 2004; Stanton, Salmon, & Rafferty, 2013; Young, Brookhuis, Wickens, & Hancock, 2015), these techniques can be grouped into:

- Analytical.
- Primary task performance measures.
- Secondary task performance measures
- Physiological measures.
- Subjective-rating techniques.

Analytical techniques are those that are carried out through mathematical models and computer simulation. Because of modern computerized systems, this type of techniques has seen a boom over the years.

Primary task performance measures involve measuring the worker's ability to perform the primary task being analyzed. The expectation is that a worker's capacity decreases as the mental work load increases. Some of the advantages of this type of measurements are: their sensitivity to variations in the workload and their ease of use. Some of the disadvantages, on the other hand, are that performance can be affected during low workload tasks and that the level of mental workload can be affected by the worker's experience and skill.

On the other hand, Secondary task performance measures focus on analyzing the worker's ability to perform an additional task to the primary. This type of measurements is based on the assumption that as a worker's mental load increases, his/her ability to perform the secondary task will decrease due to the reduction of the remaining mental capacity, so that the performance of the secondary task will be affected. The main advantages of this type of measurements are the following: it features greater sensitivity in evaluating the mental load during short periods, since secondary tasks, unlike primary ones, are executed in short periods of time; also, it allows for the evaluation of a reserved mental capacity. Its main disadvantage lies in the lack of sensitivity to minor variations in the mental workload.

Next, physiological measures are based on the fact that certain cognitive activities have been associated with physiological variations in people, that is where the variation in heart rate, eye movement and brain activity can be affected by an increase or decrease in the mental workload. The main advantage of these measurements is that they do not interfere with the performance of the primary task. The disadvantages are that this type of measurement is affected by the worker's health and that the equipment needed to carry it out is costly.

Finally, subjective-rating techniques analyze the worker's perception of the complexity of the task. The techniques are generally carried out after the task has been executed and can be classified as one-dimensional or multidimensional, depending on the workload dimensions that they evaluate. When one-dimensional techniques are used, the data obtained are easy to analyze, whereas multidimensional techniques entail a higher level of diagnosis. The advantages of this type of techniques are its ease of use, its speed of application and its low cost. One of its main disadvantages is that, often, the mental work load ratings correlate with the worker's performance in the task under analysis. Also, participants are likely to forget certain parts of the task where variations in their workload may have occurred. Lastly, this type of techniques is the most widely used for evaluating mental workload.

To evaluate the mental load perceived by the subject, one-dimensional scales are used, which contemplate a single load factor, or multidimensional scales, which break down the load factor into other factors. Also, the scales can be classified into absolute or relative. The first ones are based solely on the task under study, while the second, on their comparison to another standardized task in which the assessment is relative or multiple. These scales are usually administered immediately after the completion of the task.

Criteria for Mental Workload Assessment

According to Eggemeier, Wilson, Kramer, and Damos, cited in (Rubio et al., 2004), there are several criteria that any mental workload evaluation technique must meet:

- **Sensitivity:** Refers to a technique's ability to detect changes in the levels of difficulty in the task or in its demands.
- **Diagnosticity:** An index should identify not only when a load varies, but also the cause of this variation. The diagnostic power of a technique results from its ability to detect the type of attentional resources that produce the mental load in a specific task. It is this information that makes the implementation of better solutions possible.
- **Selectivity / Validity:** The index should be selectively sensitive only to differences in the capacity demands and should not reflect changes in factors such as the physical load or the emotional stress, which may not be associated with the load.
- **Intrusiveness:** The index should not interfere with the completion of the task whose load is being evaluated.
- **Reliability:** Like any other behavior measure, a load index must be reliable. If the mental load is evaluated at different times (i.e. changes in mental load that occur during the course of a task or operation), it is important for the index to offer a reliable and consistent estimation of the mental load.
- **Implementation Requirements:** Includes aspects such as the time, the implementation and the software necessary for data collection and analysis, as well as the degree of operator training required to obtain valid results.
- **Subject Acceptance:** Refers to the perception that subjects have of the validity and usefulness of the procedure.

Mental Workload Assessment Techniques

Among subjective techniques, those that stand out are:

- Modified Cooper-Harper (MCH) Scale.
- Bedford Scale.
- Subjective Workload Dominance (SWORD) and Pro-SWORD
- Subjective Workload Assessment Technique (SWAT) and Pro-SWAT
- NASA-Task Load index (Nasa-TLX)
- Workload Profile (WP)
- Multiple Resources Questionnaire (MRQ)
- Subjective Mental Workload Scale (SCAM)

Modified Cooper-Harper (MCH) Scale

It was developed by Cooper and Harper (1969) and modified by Wierwille and Casali (1983) and is the oldest one-dimensional, subjective evaluation technique. It was originally developed to evaluate the mental load associated with aircraft piloting tasks and other related activities through a scale scored from

1 to 10, along with a decision tree that would help the operator estimate the mental load. Originally, this scale was used to obtain pilots' subjective ratings on the level of control of an aircraft; therefore, the result of the scale was based on the capacity to control the aircraft and also on the level of entry required by the pilot to maintain adequate control. The Modified Cooper Harper's scale is based on the assumption that there is a direct relationship between the level of difficulty in controlling the aircraft and the pilot's workload (Stanton et al., 2013).

Bedford Scale

Like the MCH, it was developed for piloting tasks and related activities (Ellis & Roscoe, 1982; Roscoe & Ellis, 1990). It is based on the assumption that the mental load of a pilot is determined by the integration of the mental and physical efforts required to meet the demands of specific elements of the piloting task. Essentially, it analyzes the remaining attentional capacity and involves the use of a hierarchical decision tree to evaluate the participants' workload through evaluating the available capacity while performing a task. The scale is usually filled in after the task has been completed, but it can also be administered during its execution (Stanton et al., 2013).

Subjective Workload Dominance (SWORD) and Pro-SWORD

This technique was introduced by Vidulich (1989) and is a derivation of the Analytic Hierarchy Process (AHP), which is a technique used to deal with complex decisions. SWORD uses task comparison by means of pairs, in order to obtain mental workload ratings for individual tasks. The SWORD technique is performed after the execution of the task and calls for participants to rate the predominance of a task over another in terms of their mental workload. Pro-Sword was developed from SWORD by Vidullch, Ward, & Schueren (1991) to be used in predictive environments. When used predictively, tasks are classified according to their predominance before and after the test begins to check the sensitivity of the predictions (Stanton et al., 2013).

Subjective Workload Assessment Technique (SWAT) and Pro-SWAT

Developed by Reid and Nygren (1988), this technique was used originally to evaluate the mental workload in the US Air Force, evaluating mainly pilots in the cabin environment. It is based on the assumption that the mental workload is a multidimensional construct, which is why time load, mental effort load and stress load are all taken into account. It is administered at the time of completing the task. After an initial assessment procedure, participants are asked to rate each dimension using a scale of 1 to 3. Next, a score is calculated for each dimension and a global score for mental workload of between 1 and 100 is obtained. On the other hand, Pro-SWAT is a derivation of SWAT which is used in a predictive way. Pro-SWAT requires evaluators who are knowledgeable about a task to create a mental projection of the operation of the defined system, imagining themselves performing the task, and then reporting on the magnitude of the workload "experienced" at specific times (Eggleston & Quinn, 1984).

NASA-Task Load Index (Nasa-TLX)

NASA TLX is a multidimensional, subjective evaluation technique presented by Hart & Staveland (1988). According to several authors, it is one of the most widely used subjective techniques in evaluating mental workload (Bommer & Fendley, 2018; Rubio et al., 2004; Stanton et al., 2013; Young et al., 2015). It identifies six dimensions of subjective load: mental demand, physical demand, temporary demand, effort, performance, and frustration level. In general, this instrument involves two steps in sequential order. During the first step, a comparison is drawn, by pairs, among the 6 dimensions that make up this technique; then an assessment is made, on a scale of 1 to 100 with intervals of 5 each, of the role that each dimension has played in the evaluation. In general, this technique is performed immediately after the task has been performed.

Workload Profile (WP)

This technique was elaborated by Tsang & Velazquez (1996) based on the Multiple Resources Model (MRM) proposed by Wickens (1984). It is a multidimensional, subjective technique that makes use of the following dimensions: perceptual/central processing, response selection and execution, spatial processing, verbal processing, visual processing, auditory processing, manual output, and speech output. Before its application, participants must be instructed regarding the dimensions that the MRM handles. Afterwards, once the task has been performed, participants evaluate the resources used during the execution of the task, using a range from 0 to 1, where 0 means that there is no demand for resources and 1, that the task uses up all available resources.

Multiple Resources Questionnaire (MRQ)

The MRQ was developed by Boles & Adair (2001). According to Phillips & Boles (2004), it is based on the Expanded Multiple Resources Theory (EMRT), an evolution from the MRM which was proposed by Boles & Law (1998). It is a multidimensional technique whose dimensions, based on the available processes, include: auditory emotional, auditory linguistic, facial figural, facial motivation, manual, short term memory, spatial attentive, spatial categorical, spatial concentrative, spatial emergent, spatial positional, spatial quantitative, tactile figural, visual lexical, visual phonetic, temporal visual and vocal. For its administration, it is required that the participants be instructed on the principles of the MRQ before executing of the task. After task execution, participants are to evaluate each one of the dimensions, based on a scale of the 0 to 4, thus indicating the level of resources they used for each dimension evaluated.

Subjective Mental Workload Scale (SCAM)

The SCAM is a subjective evaluation technique developed by Rolo González, Díaz Cabrera, and Hernández Fernaud (2009) in 2009. The development of the scale was based as well on the review of the theoretical literature as on the analysis of other instruments in use, taking into consideration three major dimensions or sources of mental load: temporal demands of the task, processing resources demanded by the task and aspects of an emotional nature. The scale developed is characterized by being multidimensional and subjective, based on the approach of the high relevance of the worker's perceptions, competence, and training, which may affect his assessment of the difficulty or level of effort of the task, or what is also known as "workload experience".

Each of the abovementioned techniques has a number of advantages and disadvantages, all of which are shown in Table 1 along with several characteristics of the various techniques.

Issues, Controversies, and Problems

Failure to take the mental workload into account for design within the industrial sector can generate some problems which can impact both the worker's health and the company's economy.

Workers exposed to a high mental load tend to lose concentration, which results in lower levels of attention and an increase in their risk levels (Chen, Song, & Lin, 2016) so that accident rates also tend to increase. Having jobs where workers are exposed to high mental workloads, can influence the organization, affecting the relationship between employees and administrators (Bowling, Alarcon, Bragg, & Hartman, 2015).

In light of the above, mental workload has been found to have effects on physical health, causing symptoms or general health complaints that are indicators of coronary heart disease. The reason is that mental workload has effects on the variability of the heart rate. As the load increases, the heart rate decreases while at the same time, the respiratory rate increases. In addition, the effects of mental workload on musculoskeletal disorders have been demonstrated by various studies and manifest mainly in discomfort in the neck and shoulders (González Munõz & Gutiérrez, 2006).

On the other hand, both overload and a sub-load can have effects on workers' mental health. That is because each individual has an optimal workload band. Constant upward or downward deviations from this band will probably cause stress (Dalmau & Ferrer, 2004; González Munõz & Gutiérrez, 2006).

Fluctuations in workload can also result from an irregular flow of work, which is not under the worker's control. This is not only restricted to the pace in the production lines; many times, it depends on the climatic conditions; also, a great variety of jobs are governed by the work station demands or the production needs. Finally, certain workers, such as air traffic controllers, firefighters and pilots, must undergo long periods of inactivity, yet suddenly, they must take action when a crisis occurs. This is potentially harmful if the employee fails to respond appropriately in an emergency (Rubio, et al., 2001).

As was already mentioned, the impact of the mental load on workers can generate health problems or economic losses both for the worker and for the companies. According to Ayaz et al. (2012), the lack of evaluation of the mental workload in the industrial sector impacts mainly on the workers' performance and can result in a series of errors that can, in turn, trigger catastrophic losses.

According to Stanton et al. (2013), there is a great variety of methods to evaluate the load, but their implementation is complicated as there is no defined unit to measure the level of mental load (Rolo González et al., 2009). There is also the fact that most mental load-related jobs were originally developed from models used in the aviation sector (Vidulich & Tsang, 2012); thus, for other sectors, it is necessary to adapt and/or validate mental load evaluation differently.

METHODOLOGY

The methodology followed by this work includes four stages: selection of databases, selection of keywords, depuration of results and final selection.

Table 1. Characteristics of the mental load evaluation techniques

Technique	Training Time	Execution Time	Tools Needed	Advantages	Disadvantages
Primary task performance measures	Low	Low	Simulator, computer	1) They provide a direct performance index. 2) They are particularly effective measuring the workload in long-duration tasks. 3) They can be easily used along with secondary task performance, subjective and physiological measures.	1) They do not always distinguish between workload levels. 2) They are not a reliable measure for isolated measurements.
Secondary task performance measures	Low	Low	Simulator, computer	1) They are sensitive to mental load variations whenever these go undetected by performance measures. 2) Ease of use 3) Indentifying the secondary task is slightly more difficult.	1) They have been found to be sensitive only to big changes in mental load. 2) They are intrusive in measuring the performance of the primary task. 3) Special care must be taken when designing the secondary task in order to maintain the same resources used in the primary task.
Physiological measures	High	Low	Heart rate monitor, follow-up of eye pupil, EEG	1) Several physiological measures have proven to be sensitive to mental load variation. 2) Data are recorded continually during the test. 3) They can be used in real-world applications.	1) Data can be affected by external interference. 2) Equipment is temperature-sensitive and hard to use. 3) The measuring equipment is physically uncomfortable.
MCH Scale	Low	Low	Paper and pencil	1) Quick and easy to use; it requires little training and has a low cost 2) Widely used in various areas 3) The data obtained are easier to analyze than multidimensional instruments.	1) It measures low sophistication loads. 2) Limited to manual tasks 3) Not as sensitive as TLX or SWAT
Bedford Scale	Low	Low	Paper and pencil	1) Quick and easy to use; it requires little training and has a low cost.	1) More complex than one-dimensional techniques 2) The data are collected after the tests. There is already a certain amount of problems due to correlation with performance.
SWORD	Low	Low	Paper and pencil	1) Quick and easy to use; it requires little training and has a low cost. 2) Very effective when comparing the mental workload caused by two or more interfaces.	1) More complex to analyze tan one-dimensional techniques 2) The data are recorded after the tests. There is a certain number of problems due to correlation with performance.
Pro-SWORD	Low	Low	Paper and pencil	1) Quick and easy to train for and administer 2) Wide validation levels	1) Based on data which are approximations of mental load 2) Not widely used
SWAT	Low	Low	Paper and pencil	1) Quick and easy to use; it requires little training and has a low cost 2) Multidimensional 3) Subscales are generic, allowing their use in various areas	1) More complex to analyze tan one-dimensional techniques 2) Several studies suggest that TLX has greater sensitivity to mental load variations 3) MWL classifications may correlate with task performance
Pro-SWAT	High	High	Paper and pencil, software	1) Simple and efficient method to predict mental load 2) May be applied to any task.	1) It is in its first stages of development and has yet to be validated. 2) Little sensitivity to mental load
NASA-TLX	Low	Low	Paper and pencil	1) Quick and easy to use; it requires little training and has a low cost 2) Better performance than SWAT 3) Features a generic scale, which allows it to be administered in any environment.	1) More complex to analyze tan one-dimensional techniques 2) Weighing procedure is painstaking 3) Suitable for individual mental load only.
Workload Profile Technique	Medium	Low	Paper and pencil	1) Quick and easy to use; it requires little training and has a low cost 2) Based on a solid theoretical foundation	1) More complex to analyze tan one-dimensional techniques 2) Data ae collected after the tests. 3) More complex than other mental load techniques
MRQ	Medium	Low	Paper and pencil	1) Quick and easy to use; it requires little training and has a low cost 2) Based on a solid theoretical foundation	1) More complex to analyze tan one-dimensional techniques 2) Data ae collected after the tests. 3) More complex than other mental load techniques
SCAM	Low	Low	Paper and pencil	1) Quick and easy to use; it requires little training and has a low cost 2) Features a generic scale, which allows it to be administered in any environment.	1) It is in its first stages of development. 2) More complex to analyze tan one-dimensional techniques. 3) Not widely used

Source: (Boles & Adair, 2001; Rolo González et al., 2009; Stanton et al., 2013)

Selection of Databases

To select the databases where the search would be conducted, those that showed a greater number of indexed journals were searched, focusing mainly on the areas of engineering and manufacturing.

Keyword Selection

Once the databases were identified and based on the advanced search characteristics of each of them as well as on the number of quality results, a series of words were selected which would be used to perform the searches.

Depuration of Results

After the results of the advanced searches in the selected databases were obtained, they were exported to various files for depuration. Once these results were obtained, they were classified according to the following criteria: the mental workload evaluation used, the type of evaluation (physiological, subjective, etc.), and the sector of application (aviation, manufacturing, health centers, education, others), and the number of participants. Items that did not meet at least three of these criteria were eliminated as were those that were duplicated.

Final Selection

After this first depuration, those articles that focused on the manufacturing area were set aside. Once the articles were pre-selected, their entire content was analyzed. The following final criteria for selection was used: each article had to be a peer-reviewed, research article, and its application and/or development had to target manufacturing systems.

RESULTS

Below are the results obtained during the four stages described in the methodology.

Selection of Databases

The databases selected for this study were: ACM DL, IEEE Xplore, ScienceDirect and Web of Science. They were chosen because they met the criteria of number of indexed journals and focused primarily on the engineering and manufacturing sectors.

Selection of Keywords

Because the advanced searches for articles are different in each database, in order to obtain quality results, searches were conducted using different keywords. For the ACM DL and IEEE Xplore databases, the selected keywords were: mental and workload. For ScienceDirect, and in order to reduce the number of articles obtained in the search, the logical operators AND, OR and NOT were used; thus, the keywords were: mental, workload, evaluation, design, manufacturing, industry, nurse, medical and driver. The relationship in which these words and the logical operators were used is shown in Figure 1. Finally, for the Web of Science database, the logical operators AND, OR and NOT were also used. However, due to the databases' characteristics, the keyword "TS =" was added, by means of which it was specified that the search would be conducted within the fields of Abstract, Title or Keywords of a registry. Additionally, it was specified in the search options that the search ought to include scientific articles only.

Depuration of Results

A total of 3839 articles were found in the first searches. In the column "First Search" in Table 2, are shown the number obtained from each database. All search results were saved in spreadsheets for analysis.

Once the articles were classified based on the mental workload evaluation used, the type of evaluation of this (physiological, subjective, etc.), the area of application (aviation, manufacturing, health centers, education, others), and the number of participants, and based on the aforementioned inclusion criteria, the number of articles was reduced to 891, this is shown in the column "First Selection" in Table 2. The Figure 2 shown the frequency, by year, of the articles selection. Figure 2 shows the frequency of the first selection of elements, in which you can observe a trend towards the increasing publication of articles of mental workload.

Figure 1. Relationship between keywords and logical operators used for the ScienceDirect database

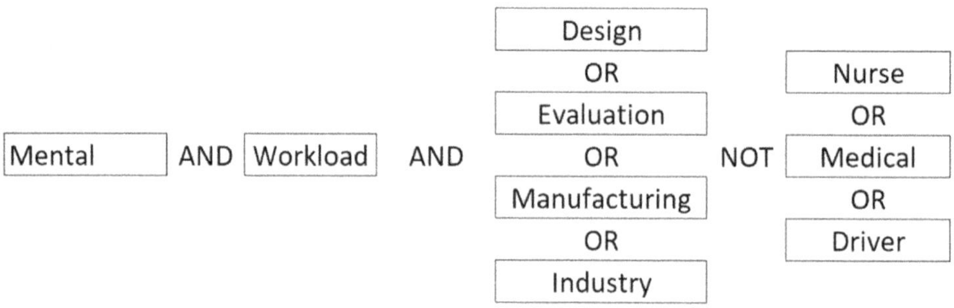

Table 2. Number of articles found by database

Database	Number of Articles			
	First Search	First Selection	Second Selection	Final Selection
ACM DL	56	5	0	0
IEEE Xplore	362	86	4	0
ScienceDirect	3065	658	85	9
Web of Science	356	142	16	3
Total:	3839	891	105	12

Figure 2. Year of publication of the articles of the first selection

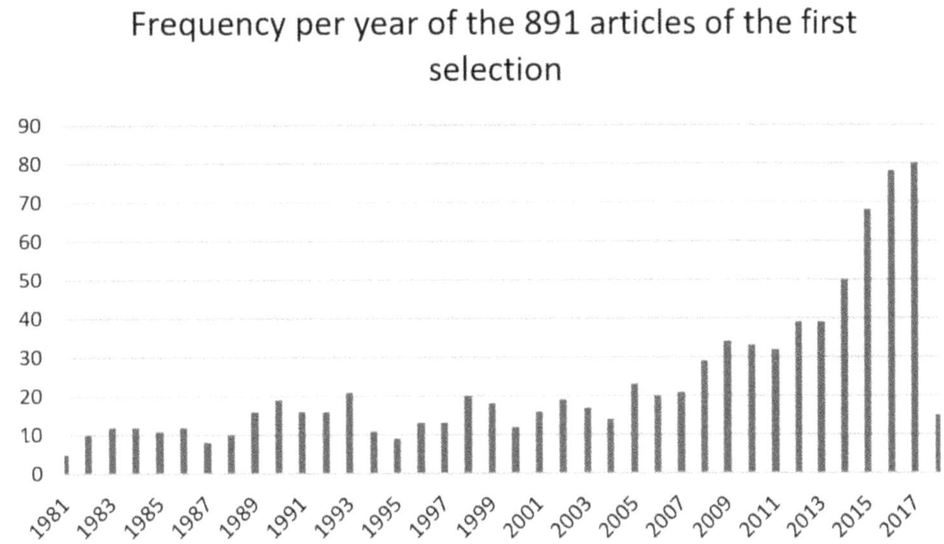

For the next stage, out of the 891 articles obtained from the previous stage, a pre-selection was made of those articles that met the criteria of being focused on the manufacturing sector; thus, the number of articles was reduced to 105. The column "Second Selection" in Table 2 shows the number of preselected articles per database.

Final Selection

For the final selection, in the articles where it was available, the total content of the article was analyzed; this included the Abstract, the introduction, the literature review, the methods, the discussion and the conclusion. At this last stage, 12 articles were selected. The number of articles selected per database is shown in the column "Final Selection" in Table 2, while Table 3 shows the general characteristics of the selected articles.

In five of the articles, various analytical techniques were used; in three of them, NASA-TLX was used; in one, the MCH was used; in two, they did not specify which technique was used; and in another, two techniques were used, one analytical and one subjective (NASA-TLX), this is showed in Table 4. In terms of seniority the oldest article was from the year 1999, while two of them were published this year (2018). Figure 3 shows a graph that contains the year of publication of the articles and where you can see that it is only in the last few years that there has been the greatest interest.

Regarding the area in which the 12 selected articles are focused, it was found that in seven of them the operator's performance is analyzed, in 3 of them is considered the mental workload in the design of the manufacturing systems, 1 is focused on the training area and 1 is focused on planning.

Table 3. Characteristics of the selected articles

Author	Title	Journal	Date	Type of Evaluation	Reference
Shuxin Bi and Gavriel Salvendy	Analytical modeling and experimental study of human workload in scheduling of advanced manufacturing systems	International Journal of Human Factors in Manufacturing	January 1, 1994	Analytical	(Bi & Salvendy, 1994)
Sharon Claxton Bommer and Mary Fendley	A theoretical framework for evaluating mental workload resources in human systems design for manufacturing operations	International Journal of Industrial Ergonomics	January 1, 2018	Subjective (NASA-TLX)	(Bommer & Fendley, 2018)
Anna Brolin, Peter Thorvald and Keith Case	Experimental study of cognitive aspects affecting human performance in manual assembly	Production & Manufacturing Research	January 1, 2017	Subjective (NASA-TLX)	(Brolin, Thorvald, & Case, 2017)
Lin Ding-Yu and Hwang Sheue-Ling	The development of mental workload measurement in flexible manufacturing systems	Human Factors and Ergonomics in Manufacturing & Service Industries	December 7, 1998	Analytical	(Ding-Yu & Sheue-Ling, 1998)
Friedhelm Nachreiner	Standards for ergonomics principles relating to the design of work systems and to mental workload	Applied Ergonomics	August 1, 1995	N/A	(Nachreiner, 1995)
M. Bevilacqua, F. E. Ciarapica, G. Mazzuto and C. Paciarotti	Visual Management implementation and evaluation through mental workload analysis.	IFAC Proceedings Volumes	May 1, 2013	Subjective (MCH)	(Bevilacqua, Ciarapica, Mazzuto, & Paciarotti, 2013)
Celestine A. Ntuen	The application of fuzzy set theory to cognitive workload evaluation of electronic circuit board inspectors	Human Factors and Ergonomics in Manufacturing & Service Industries	June 1, 1999	Analytical	(Ntuen, 1999)
O. J. Sealetsa and A. Thatcher	Ergonomics issues among sewing machine operators in the textile manufacturing industry in Botswana	Work	January 1, 2011	Subjective (NASA-TLX)	(Sealetsa & Thatcher, 2011)
Kuo-Hao Tang, Richard J. Koubek, Nancy J. Lightner and Gavriel Salvendy	Development and validation of a theoretical model for cognitive skills acquisition	International Journal of Industrial Ergonomics	April 1, 1999	Analytical and Subjective (NASA-TLX)	(Tang, Koubek, Lightner, & Salvendy, 1999)
Shezeen Oah, Rudia Na and Kwangsu Moon	The Influence of Safety Climate, Safety Leadership, Workload, and Accident Experiences on Risk Perception: A Study of Korean Manufacturing Workers	Safety and Health at Work	February 9, 2018	N/A	(Oah, Na, & Moon, 2018)
Lin Ding-Yu and Hwang Sheue-Ling	Use of neural networks to achieve dynamic task allocation: a flexible manufacturing system example	International Journal of Industrial Ergonomics	June 27, 1998	Analytical	(Ding-Yu & Sheue-Ling, 1998)
Xiaosong Zhao, Chia-Yu Hsu, Pei-Chann Chang and Li Li	A genetic algorithm for the multi-objective optimization of mixed-model assembly line based on the mental workload	Engineering Applications of Artificial Intelligence	January 2016	Analytical	(Zhao, Hsu, Chang, & Li, 2016)

Figure 3. Year of publication of the final articles selection

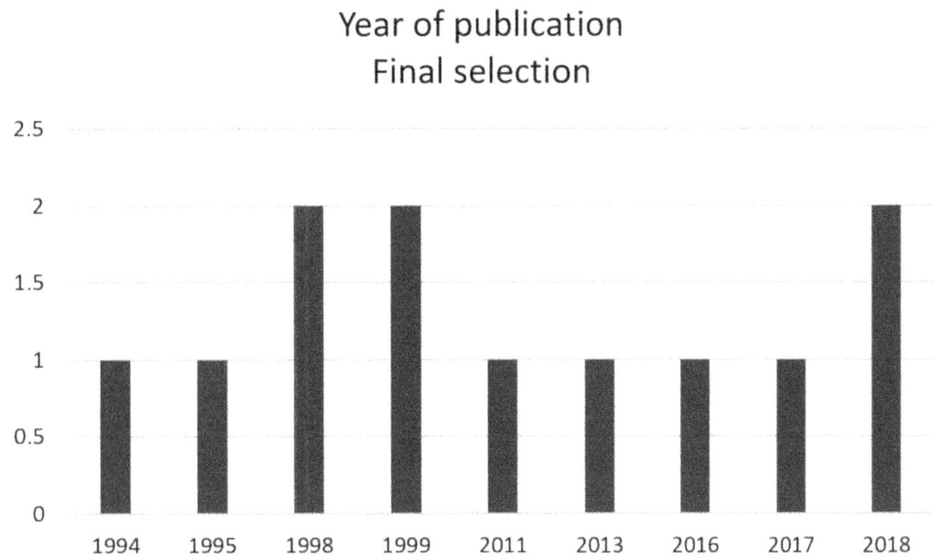

Table 4. Type of evaluation used in the selected articles

Type of Evaluation	Qty	Reference
Analytical	6	(Bi & Salvendy, 1994; Ding-Yu & Sheue-Ling, 1998, 1998; Ntuen, 1999; Tang et al., 1999; Zhao et al., 2016)
Subjective (NASA-TLX)	4	(Bommer & Fendley, 2018; Brolin et al., 2017; Sealetsa & Thatcher, 2011; Tang et al., 1999)
Subjective (MCH)	1	(Bevilacqua et al., 2013)
N/A	2	(Nachreiner, 1995; Oah et al., 2018)

DISCUSSION

Mental work load evaluation has undergone a great change in recent years. It has been gradually incorporated into various sectors, without leaving those where it originally began (aviation sector); thus, its full implementation in the industrial sector's socio-technical systems is now a reality.

During the conduction of this work, only 12 of 3839, which is (approx.) 0.31% percent of the articles analyzed, focused on and/or had been implemented in the manufacturing industry. Returning to Figures 2 and 3, it can be observed that although there has been an increase in the number of articles published per year relating to the mental workload, the application in the manufacturing industry has been marginal. This leaves a research opportunity where assessment and evaluation of the mental workload in the socio-technical systems of the industrial sector can be implemented in a practical and reliable way.

As for the mental workload assessment instrument, and regarding the subjective techniques, these were used in 5 of the 12 selected articles, the most widely used is NASA-TLX, which was used in 4 on these articles. According to the authors of these articles, NASA-TLX was selected because it is the most widely known and used subjective assessment technique (Bommer & Fendley, 2018; Brolin et al., 2017;

Sealetsa & Thatcher, 2011; Tang et al., 1999) which is in line with the opinions of recognized authors (Rubio et al., 2004; Stanton et al., 2004, 2013).

It can also be observed that the use of subjective techniques in the assessment of the mental workload is one of the most commonly used, since, as can be appreciated in Table 1, they present a low level of complexity in their implementation, because it is practically only necessary to use paper and pencil.

In the case of analytical techniques, these are the most commonly used and are used separately for each case and are presented to support a particular model. Also, they are focused mainly on predicting the mental workload, and only one case of study (Ntuen, 1999) use a subjective technique to validate their model.

FUTURE RESEARCH DIRECTIONS

As mentioned in the results section, half of the articles selected in this review were published within the last five years, and two of them were even published as this work was being prepared. Thus, it is necessary to note that mental load evaluation focusing on the manufacturing industry is gradually incorporating the mental workload and that due to the time constraints on the article's search, some articles may have been overlooked.

It should also be kept in mind that, although many of the items discarded in this work were not directly related to manufacturing, in many of them the techniques and methods can be adapted to the socio-technical systems of the manufacturing industry. For example, many of the case studies carried out in the fields of transportation or medicine can be carried out within the industrial sector as well.

CONCLUSION

Due to the various problems derived from the lack of consideration of mental workload in several fields, the relevance of its incorporation into the design and evaluation of current manufacturing systems and processes is highlighted in this chapter. Moreover, the risks of having mental overload or sub load during tasks execution have an impact on workers' performance that tends diminish incurring in work stress, human error, poor quality of life at work and reduced productivity.

As a result of the comprehensive literature review, the number of research articles found in 4 of the main databases is insufficient. In this manner, the field of mental workload research in socio-technical systems is a challenge and represents an opportunity for researchers working in the area of cognitive ergonomics, industrial engineering, industrial psychology among other emerging fields. At the moment a concrete approach for its measurement and evaluation is still pending and represent the most important obstacle for including mental workload assessment and consideration in work and process design in manufacturing and sociotechnical environments.

Additionally, among the articles found for the subjective techniques, the NASA-TLX is distinguished as the most widely used for mental workload and present advantages for its easiness of use, its sensitivity, non-intrusiveness, its diagnosticity, its validity, its subject acceptance and relatively low cost of their implementation.

ACKNOWLEDGMENT

We thank the Autonomous University of Ciudad Juárez (UACJ) and the National Institute of Science and Technology (CONACYT) for the financial support granted through project CONACYT - INS (FRONTERAS CIENCIA) 2016 – 01 - 2433. We also thank Leslie Yaratzed Cedeño Castillo for her outstanding support in the translation and improvement of this paper.

REFERENCES

Autor, D. H. (2015). Why Are There Still So Many Jobs? The History and Future of Workplace Automation. *The Journal of Economic Perspectives*, *29*(3), 3–30. doi:10.1257/jep.29.3.3

Ayaz, H., Shewokis, P. A., Bunce, S., Izzetoglu, K., Willems, B., & Onaral, B. (2012). Optical brain monitoring for operator training and mental workload assessment. *NeuroImage*, *59*(1), 36–47. doi:10.1016/j.neuroimage.2011.06.023 PMID:21722738

Beaudry, P., Green, D. A., & Sand, B. M. (2013). *The Great Reversal in the Demand for Skill and Cognitive Tasks* (Working Paper No. 18901). National Bureau of Economic Research. doi:10.3386/w18901

Bevilacqua, M., Ciarapica, F. E., Mazzuto, G., & Paciarotti, C. (2013). Visual Management implementation and evaluation through mental workload analysis. *IFAC Proceedings Volumes, 46*(7), 294–299. 10.3182/20130522-3-BR-4036.00065

Bi, S., & Salvendy, G. (1994). Analytical modeling and experimental study of human workload in scheduling of advanced manufacturing systems. *The International Journal of Human Factors in Manufacturing*, *4*(2), 205–234. doi:10.1002/hfm.4530040207

Boles, D. B., & Adair, L. P. (2001). The Multiple Resources Questionnaire (MRQ). *Proceedings of the Human Factors and Ergonomics Society Annual Meeting*, *45*(25), 1790–1794. doi:10.1177/154193120104502507

Boles, D. B., & Law, M. B. (1998). A simultaneous task comparison of differentiated and undifferentiated hemispheric resource theories. *Journal of Experimental Psychology. Human Perception and Performance*, *24*(1), 204–215. doi:10.1037/0096-1523.24.1.204 PMID:9483826

Bommer, S. C., & Fendley, M. (2018). A theoretical framework for evaluating mental workload resources in human systems design for manufacturing operations. *International Journal of Industrial Ergonomics*, *63*, 7–17. doi:10.1016/j.ergon.2016.10.007

Bowling, N. A., Alarcon, G. M., Bragg, C. B., & Hartman, M. J. (2015). A meta-analytic examination of the potential correlates and consequences of workload. *Work and Stress*, *29*(2), 95–113. doi:10.1080/02678373.2015.1033037

Brolin, A., Thorvald, P., & Case, K. (2017). Experimental study of cognitive aspects affecting human performance in manual assembly. *Production & Manufacturing Research*, *5*(1), 141–163. doi:10.1080/21693277.2017.1374893

Chen, J., Song, X., & Lin, Z. (2016). Revealing the "Invisible Gorilla" in construction: Estimating construction safety through mental workload assessment. *Automation in Construction*, *63*, 173–183. doi:10.1016/j.autcon.2015.12.018

Cooper, G. E., & Harper, R. P. Jr. (1969). *The use of pilot rating in the evaluation of aircraft handling qualities*. Neuilly-Sur-Seine, France: Advisory Group for Aerospace Research and Development.

Dalmau, I., & Ferrer, R. (2004). Revisión del concepto de carga mental: Evaluación, consecuencias y proceso de normalización. *Anuario de Psicología*, *35*(4), 521–546.

Ding-Yu, L., & Sheue-Ling, H. (1998). The development of mental workload measurement in flexible manufacturing systems. *Human Factors and Ergonomics in Manufacturing & Service Industries*, *8*(1), 41–62. doi:10.1002/(SICI)1520-6564(199824)8:1<41::AID-HFM3>3.0.CO;2-E

Eggleston, R. G., & Quinn, T. J. (1984). A Preliminary Evaluation of a Projective Workload Assessment Procedure. *Proceedings of the Human Factors Society Annual Meeting*, *28*(8), 695–699. 10.1177/154193128402800811

Ellis, G., & Roscoe, A. (1982). *The Airline Pilot's View of Flight Deck Workload: A Preliminary Study Using a Questionnaire*. Rarnborough, UK: Royal Aircraft Establishment.

González Munõz, E. L., & Gutiérrez, R. (2006). La carga de trabajo mental como factor de riesgo de estrés en trabajadores de la industria electrónica. *Revista Latinoamericana de Psicología*, *38*(2), 259–270.

Hart, S. G., & Staveland, L. E. (1988). Development of NASA-TLX (Task Load Index): Results of Empirical and Theoretical Research. In P. A. Hancock & N. Meshkati (Eds.), *Advances in Psychology* (Vol. 52, pp. 139–183). North-Holland. doi:10.1016/S0166-4115(08)62386-9

Mital, A., & Pennathur, A. (2004). Advanced technologies and humans in manufacturing workplaces: An interdependent relationship. *International Journal of Industrial Ergonomics*, *33*(4), 295–313. doi:10.1016/j.ergon.2003.10.002

Nachreiner, F. (1995). Standards for ergonomics principles relating to the design of work systems and to mental workload. *Applied Ergonomics*, *26*(4), 259–263. doi:10.1016/0003-6870(95)00029-C PMID:15677027

Ntuen, C. A. (1999). The application of fuzzy set theory to cognitive workload evaluation of electronic circuit board inspectors. *Human Factors and Ergonomics in Manufacturing & Service Industries*, *9*(3), 291–301. doi:10.1002/(SICI)1520-6564(199922)9:3<291::AID-HFM6>3.0.CO;2-S

Oah, S., Na, R., & Moon, K. (2018). The Influence of Safety Climate, Safety Leadership, Workload, and Accident Experiences on Risk Perception: A Study of Korean Manufacturing Workers. *Safety and Health at Work*. doi:10.1016/j.shaw.2018.01.008

Phillips, J. B., & Boles, D. B. (2004). Multiple Resources Questionnaire and Workload Profile: Application of competing models to subjective workload measurement. *Proceedings of the Human Factors and Ergonomics Society Annual Meeting*, *48*(16), 1963–1967. doi:10.1177/154193120404801636

Reid, G. B., & Nygren, T. E. (1988). The Subjective Workload Assessment Technique: A Scaling Procedure for Measuring Mental Workload. In P. A. Hancock & N. Meshkati (Eds.), *Advances in Psychology* (Vol. 52, pp. 185–218). North-Holland; doi:10.1016/S0166-4115(08)62387-0

Rolo González, G., Díaz Cabrera, D., & Hernández Fernaud, E. (2009). Desarrollo de una Escala Subjetiva de Carga Mental de Trabajo (ESCAM). *Revista de Psicología del Trabajo y de las Organizaciones*, *25*(1), 29–37. doi:10.4321/S1576-59622009000100004

Roscoe, A. H., & Ellis, G. A. (1990). *A subjective rating scale for assessing pilot workload in flight: A decade of practical use*. Rarnborough, UK: Royal Aircraft Establishment.

Rubio, S., Díaz, E., Martín, J., & Puente, J. M. (2004). Evaluation of Subjective Mental Workload: A Comparison of SWAT, NASA-TLX, and Workload Profile Methods. *Applied Psychology*, *53*(1), 61–86. doi:10.1111/j.1464-0597.2004.00161.x

Rubio, S., Díaz, E. M., & Martín, J. (2001). Aspectos metodológicos de la evaluación subjetiva de la carga mental de trabajo. *Archivos de Prevencion de Riesgos Laborales*, *4*(4), 160–168.

Sealetsa, O. J., & Thatcher, A. (2011). Ergonomics issues among sewing machine operators in the textile manufacturing industry in Botswana. *Work (Reading, Mass.)*, *38*(3), 279–289. doi:10.3233/WOR-2011-1131 PMID:21447888

Stanton, N., Hedge, A., Brookhuis, K., Salas, E., & Hendrick, H. W. (Eds.). (2004). *Handbook of Human Factors and Ergonomics Methods*. CRC Press. doi:10.1201/9780203489925

Stanton, N., Salmon, P. M., & Rafferty, L. A. (2013). *Human Factors Methods: A Practical Guide for Engineering and Design*. Ashgate Publishing, Ltd. doi:10.1201/9781315587394

Tang, K.-H., Koubek, R. J., Lightner, N. J., & Salvendy, G. (1999). Development and validation of a theoretical model for cognitive skills acquisition. *International Journal of Industrial Ergonomics*, *24*(1), 25–38. doi:10.1016/S0169-8141(98)00085-7

Tsang, P. S., & Velazquez, V. L. (1996). Diagnosticity and multidimensional subjective workload ratings. *Ergonomics*, *39*(3), 358–381. doi:10.1080/00140139608964470 PMID:8849491

Vidulich, M. A. (1989). The Use of Judgment Matrices in Subjective Workload Assessment: The Subjective Workload Dominance (SWORD) Technique. *Proceedings of the Human Factors Society Annual Meeting*, *33*(20), 1406–1410. 10.1177/154193128903302009

Vidulich, M. A., & Tsang, P. S. (2012). Mental Workload and Situation Awareness. In G. Salvendy (Ed.), *Handbook of Human Factors and Ergonomics* (pp. 243–273). John Wiley & Sons, Inc. doi:10.1002/9781118131350.ch8

Vidullch, M. A., Ward, G. F., & Schueren, J. (1991). Using the Subjective Workload Dominance (SWORD) Technique for Projective Workload Assessment. *Human Factors*, *33*(6), 677–691. doi:10.1177/001872089103300605

Wickens, C. D. (1984). *The Multiple Resources Model of Human Performance: Implications for Display Design*. Urbana, IL: Illinois Univ. at Urbana. Retrieved from http://www.dtic.mil/docs/citations/ADP004516

Wierwille, W. W., & Casali, J. G. (1983). A Validated Rating Scale for Global Mental Workload Measurement Applications. *Proceedings of the Human Factors Society Annual Meeting*, *27*(2), 129–133. 10.1177/154193128302700203

Young, M. S., Brookhuis, K. A., Wickens, C. D., & Hancock, P. A. (2015). State of science: Mental workload in ergonomics. *Ergonomics*, *58*(1), 1–17. doi:10.1080/00140139.2014.956151 PMID:25442818

Zhao, X., Hsu, C.-Y., Chang, P.-C., & Li, L. (2016). A genetic algorithm for the multi-objective optimization of mixed-model assembly line based on the mental workload. *Engineering Applications of Artificial Intelligence*, *47*, 140–146. doi:10.1016/j.engappai.2015.03.005

KEY TERMS AND DEFINITIONS

Cognition: The activity related to the acquisition of knowledge.

Cognitive Ergonomics: The branch of ergonomics that is interested in mental processes such as perception, memory, reasoning, and motor response, to the extent that they affect the interactions between human beings and the other components of a system.

Database: In the academic field, they are repositories of information that may contain references to books, journals, conference proceedings, and theses, among others.

Mental Workload: Refers to the set of strains inflicted on a person by the demands of the mental work that he/she performs.

Socio-Technical Systems: Refers to the integration in the company as it consists of a technical system (equipment and production methods) and a social system (a group of men) that interact with and complement each other.

Stress: Associated with the state that results from events that generate anxiety or overwhelmingness. When a person receives demands that are excessive, he/she will begin to feel anxiety and to experience various symptoms.

Subjective: In the field of mental load assessment, subjective instruments are those in which the mental workload is not measured directly, but indirectly.

Section 2
Macroergonomic Assessments

Chapter 6
Lean Production and Its Impact on Worker Health:
Force and Fatigue-Based Evaluation Approaches

Murray Gibson
Auburn University, USA

Beata Mrugalska
Poznan University of Technology, Poland

ABSTRACT

Lean is currently applied successfully in many industrial sectors. However, its value and impact on human health is not fully understood. To gain a better understanding, this chapter explores how ergonomics force and fatigue evaluation methods can be applied in a manner to enhance lean initiatives. These methods incorporate ergonomic-related variables of force type, force duration, force frequency, and degree of awkward posture, and incorporate the recommended cumulative rest allowance (RCRA) model as a practical fatigue-based metric. These methods and their application are discussed.

INTRODUCTION

In today's global competition, successful companies focus on diversity and inclusion as a source of competitive advantage. The organizations use diverse methods and techniques to improve their productivity, quality, and profitability. In order to achieve this, the work organization system of lean production has been widely adopted throughout the industrial world. Lean originates from the Toyota Production System and has been recognized as doing more with less. Thus, it is oriented toward reduction of unnecessary variations and steps in the work process by the elimination of waste, which can be thought of as a non-value-added aspect of a product or service. Originally, the focus was paid to the elimination of wastes such as defects requiring rework, unnecessary processing steps, movement of materials or people, waiting time, excess inventory, and overproduction. But today Lean systems impact a wide-range of

DOI: 10.4018/978-1-5225-7192-6.ch006

manufacturing related systems including the initial stages of product life cycle such as product development, procurement and manufacturing over to distribution (Womack & Jones, 1996). The most common practices associated with lean production are the following: bottleneck removal (production smoothing), cellular manufacturing, competitive benchmarking, continuous improvement programs, cross-functional work force, cycle time reductions, focused factory production, just-in-time/continuous flow production, lot size reductions, maintenance optimization, new process equipment/technologies, planning and scheduling strategies, preventive maintenance, process capability measurements, pull system/Kanban, quality management programs, quick changeover techniques, reengineered production process, safety improvement programs, self-directed work teams, and total quality management (Shah & Ward, 2003).

It should be clearly pointed out that the misusage of lean techniques can lead to overlooking issues that lead to health problems, injuries, or even accidents. All these consequences lead to the need to spend money on compensation claims, which may be the best example of waste (Pai et. 2009). The awareness of the relationship between worker health and lean among lean practitioners is not always visible. There are some studies which show such an impact in the areas of occupational injuries and illnesses, or job characteristics related to job strain and linked to hypertension and cardiovascular disease (Landsbergis et al., 1999). Furthermore, it has been investigated that lean production can empower workers (Shah & Ward, 2003; Taj & Morosan, 2011) as it requires intensified work pace and increased job demands. Decision authority and skill levels can increase modestly or temporarily, while decision latitude is kept at low level which can lead to job strain and finally, musculoskeletal disorders. In jobs with ergonomic stressors, intensification of labor appears to lead to increases in musculoskeletal disorders (Landsbergis et al., 1999). Lean may result in raised stress levels and therefore increase worker turnover and absenteeism with a negative impact on manufacturing performance (Gill, 2003). However, further studies revealed that this approach in fact increases the stress levels. But not on the shop floor as it was previously suggested, but rather on the managerial level (Sangwan, 2013). Furthermore, when lean was tested against such work organization models as: Taylorism model, human relations model and socio-technology models, on the two dimensions central or decentralized orientation and human factor orientation; it turned out that it outperforms both the Taylorism model and the human relations model (European Foundation Report, 2001).

Ergonomics is defined as the application of theory, principles, data and methods to design a system which accommodates both the human operator's capabilities and limitations in the work process (Wickens, 1992). Ergonomics enables us to better understand the interactions among humans and other elements of a system, allowing the optimization of overall system performance (Naranjo-Flores et. al, 2014). Ergonomics continues to become more and more important in industry due to the positive influence on productivity and efficiency as well as its modern-day focus to reduce or prevent work-related injuries. It should be appreciated that ergonomics and lean are highly inter-related. Ergonomic risks are strongly associated with lean-related wastes; and lean transformation should naturally lead to ergonomic risk reduction (Aqlan et. al, 2013). Despite these facts, many companies have implemented lean-related systems to control productivity and quality, without a clear understanding of the potential of ergonomics to contribute to the objectives of lean.

The purpose of this research is to better understand and communicate the application of modern ergonomics force and fatigue evaluation methods in lean initiatives. To achieve this objective, a systematic search of the literature was be carried out using the following electronic databases: MEDLINE, PubMed, Scopus, Clarivate Analytics (former Web of Science) and Science Direct. All databases were searched from their inception to April 2018. Keywords used in the search strategy were: "lean" AND "force" AND "fatigue" OR "fatigue" AND "calculator" OR "fatigue" AND "calculation" OR "muscle" AND "fatigue" OR "grip" AND "force" AND "fatigue" OR "pinch" AND "force" AND "fatigue" OR "press" AND "force" AND "fatigue". In order to find the supposed papers while reading the titles and abstracts the reference was made to human. This search was supposed to reveal better insight into approaches for the quantification of ergonomic related muscle fatigue, particularly those utilizing calculations-based models.

BACKGROUND

Fatigue is commonly perceived as a non-specific symptom that refers to health conditions of human beings. It is defined as an overwhelming sense of psychological and psychical tiredness, lack of energy and feeling of exhaustion which results in the perception of difficulty performing tasks (Gruet et al., 2013; Tran et al. 2010). If it is accumulated and not resolved, it can cause overwork, chronic fatigue syndrome, overtraining syndrome, and even endocrine disorders, immunity dysfunction, organic diseases and a threat to human health (Wan et al., 2017; Larun et al., 2017). It can be perceived as a multidimensional concept which can be integrated within the World Health Organization's Classification of Functioning, Disability, and Health (WHO-ICF). Following it, the effect of disease on body function and structure, activity and participation of the patient can be shown (Figure 1) (World Health Organization 2001, Vries et al. 2010).

Diverse fatigue classification can be found in the literature; however, the most general approach is to differentiate between objective and subjective fatigue. Objective fatigue is the observable and measurable loss in performance due to the repetition of a physical or mental task, while subjective fatigue is a feeling of early exhaustion, weariness and unwillingness to exert effort (Staub & Bogousslavsky, 2001). Moreover, when we refer to its duration, we can differentiate between acute fatigue and chronic fatigue. Acute fatigue can be quickly relieved by rest or life-style changes, whereas chronic fatigue is a persistent tiredness lasting months, and not ameliorated by rest (Norheim et al., 2011; Silverman et al., 2010; Kroenke et al., 1988). On the other hand, it can also be classified as mental fatigue, which refers to the cognitive or perceptual aspects of fatigue; and physical fatigue, which is related to the performance of the motor system. It is also possible to refer to different levels of the motor pathway and differentiate

Figure 1. Multidisciplinary concept of fatigue
Source: Adopted from (World Health Organization 2001, Vries et al. 2010)

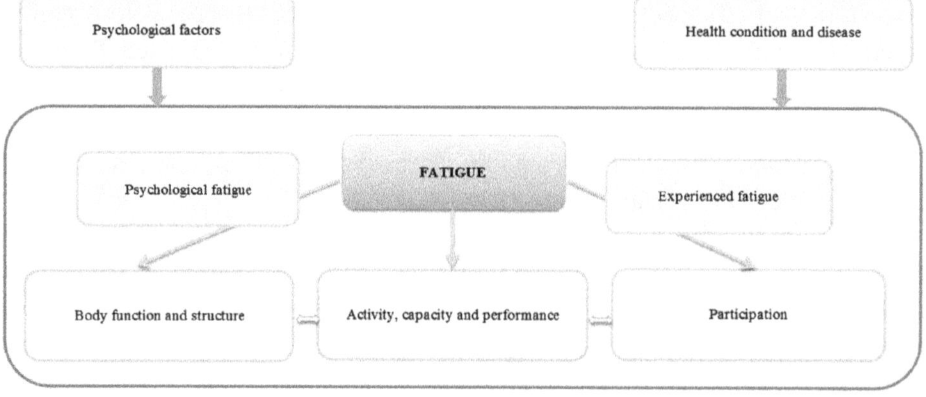

central and peripheral fatigue. Central fatigue originates at the central nervous system (CNS), which decreases the neural drive to the muscle. Peripheral fatigue is produced by changes at or distal to the neuromuscular junction (Gandevia, 2001; Bigland-Ritchie et al., 1978).

Two types of fatigue commonly associated with the application of industrial ergonomics are cognitive (mental) fatigue and whole-body fatigue. Methods of quantifying cognitive fatigue are not widely adopted by ergonomics practitioners in the field. Cognitive fatigue is often "measured" by administering subjective rating scales, resulting in metrics/outputs that are apt to be perceived as "highly subjective" by management. Due to cognitive fatigue being highly dependent upon individual perception, it is difficult to establish reliable (and credible) metrics for cognitive fatigue across jobs and tasks (Beurskens et al., 2016; Montalvo et al., 2016).

Whole body fatigue on the other hand is associated with energy consumed by the body during performance of a physical activity (i.e., materials handling, walking, running, etc.) (Kim et al., 2016). Whole body fatigue can be measured indirectly by measuring oxygen consumption by the body (i.e., ml O_2 per kg of bodyweight per min.). Then oxygen consumption is converted to energy consumption of the body (i.e., 1 liter of O_2 = 5 kcal). Direct measurement of oxygen consumption can be cumbersome, requiring workers to wear equipment that is somewhat invasive. This method is not commonly relied upon in the application of industrial ergonomics (Thomas et al. 2016).

Another method of quantifying energy consumption in the evaluation of whole body fatigue is by table lookup. Using this method, work is classified by distinct type/category of effort such as Light Work (consuming 1-2.5 kcal/min), Moderate Work (consuming 2.5-3.8 kcal/min), etc. Then by measuring or estimating the time spent performing each type/category of activity, a cumulative value of energy consumption is established.

Measuring whole body fatigue is not utilized to a significant extent in the mainstream application of industrial ergonomics. This may be due to most modern jobs not requiring high levels of whole body physical exertion. For example, strenuous materials handling tasks today are likely performed with the assistance of mechanized equipment rather than performed manually. For whatever reasons, reduction and prevention of musculoskeletal disorders are the primary focus of today's ergonomics initiatives rather than whole body fatigue.

A form of fatigue that lends itself to a higher specificity of measurement and practical application is localized muscle fatigue. Localized muscle fatigue concerns the fatiguing of active tissue associated with force applications by the worker. When muscles and other soft tissues become fatigued, this can be detrimental to the performance of physical work. This is depicted in Figure 2.

The authors prefer a practical definition of localized muscle fatigue:

A reduction in human current ability or capacity to apply force due to previous force exertions.

This reduction in ability or capacity to apply force can result in the worker using a higher percentage of his/her maximum strength toward the end of the shift (fatigued state) than was necessary at the beginning of the shift (non-fatigued state). This can also be a reduction in speed (velocity or time duration) at which a force is applied, or similarly a reduction in the speed of bodily movement (i.e., requiring more time to reach, bend, perform a movement sequence, etc.). This is particularly detrimental at a bottleneck station, potentially negatively impacting the productivity rate of the entire line or process when the employee in a bottleneck station becomes fatigued. Or the reduction in ability or capacity to apply force could be

Figure 2. Localized muscle fatigue's impact on human performance

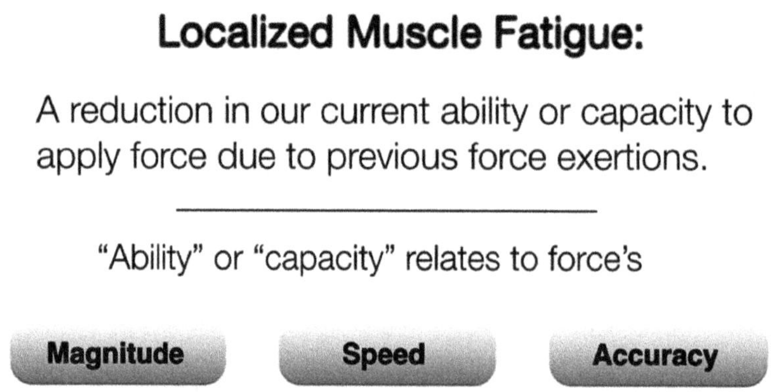

a reduction in movement accuracy, resulting in increased quality defects for tasks requiring high levels of precision and accuracy. All these things can negatively impact work from a performance perspective.

FORCE AND FATIGUE EVALUATION IN LEAN APPLICATIONS: INDUSTRIAL INSIGHTS

Based on the authors' observations in industry, it is common for organizations to implement lean, yet experience considerable difficulty achieving the performance levels predicted by the lean tools of standard work, line balancing, etc. For work with a high manual work content component, it is a common occurrence for visual factory performance boards to display satisfactory productivity levels for the first 2-4 hours of the work shift, but then mysteriously record drops in productivity over the remainder of the work shift. In a similar fashion, it is common to observe satisfactory productivity levels during the first part of the work week, but then observe productivity decline (unexplainably!) toward the end of the work week. The authors believe this breakdown between "model productivity" and actual productivity may be due to lean approaches failing to adequately account for worker fatigue.

Worker fatigue can manifest itself in multiple forms. As previously outlined in this paper, commonly recognized categories of fatigue are cognitive fatigue, whole body fatigue, and localized muscle fatigue. Of these 3 categories of fatigue, the authors believe that the category most readily applied with significant benefits in a lean environment is localized muscle fatigue.

Localized muscle fatigue can be readily modeled within the lean initiative. Lean's standard work breaks jobs down into a sequence of defined tasks and subtasks. These task and subtask times are then used to calculate the cycle time of the job (projected time to complete one work cycle). Standard work can utilize predetermined time systems (such as MOST, MTM, etc.), but it is commonly accomplished through simple stopwatch time study. In the simple stopwatch method, a "representative worker" is selected for observation. Then a number of observations are recorded, and averaged to estimate the job cycle time. Most of these time study methods include fatigue allowances (allowances developed decades ago), but these standard work fatigue allowances do not incorporate modern methods involving the quantification of localized muscle fatigue.

After standard work's breakdown of jobs into tasks and subtasks, all that remains to quantify localized muscle fatigue is to identify a the additional ergonomic-related variables of

- **Force Type:** Pinch grip, hand grip, press with thumb, etc.
- **Force Duration:** How many seconds the force is applied – 1.2 s, 3.4 s, etc.
- **Force Frequency or Force "Count":** # times the force is applied per job cycle
- **Documentation of Degree of Awkward Posture:** Wrist flexion, ulnar deviation of wrist, etc.

This detailed ergonomic evaluation detail can be readily collected during the development of lean standard work. The next step is to process this data using an ergonomics evaluation model. An emerging model for evaluating localized muscle fatigue is the Recommended Cumulative Recovery Allowance, RCRA (Gibson & Potvin, 2016).

For a given segment of the worker population (often 25th percentile female strength), the RCRA can be used to calculate the recovery time necessary for muscle groups to recover from individual force exertions. For similar force types (hand grip, pinch, one-arm push, etc.), the recovery time for the task as a whole can be calculated by summing the individual task and subtask level recovery times. If the calculated total model recovery time is greater than the actual in-task recovery time, then in-task recovery is insufficient … and muscle fatigue should be anticipated. On the other hand, if the total model recovery time is less than the actual in-task recovery time, then the worker's muscles have (based on model calculations) sufficient recovery, and the worker should be able to perform the job without experiencing undue muscle fatigue over the course of the work shift.

SOLUTIONS AND RECOMMENDATIONS

RCRA based offshoot models and ensuing macro-ergonomics solutions have tremendous potential to benefit both lean and ergonomics. These lean-benefiting solutions include:

- **Line Balancing:** Utilizing RCRA calculations enable us to better assess localized muscle fatigue in traditional line balancing. Line balancing is a methodology for distributing tasks amongst stations to balance work content. Line balancing utilizes time-based metrics (values developed through standard work); but using the RCRA, it is possible to assign/distribute tasks among workstations in a manner to also prevent (or at least minimize) localized muscle fatigue. This approach can be thought of as "fatigue balancing", where tasks are assigned to stations in the optimal manner to prevent localized muscle fatigue.
- **Ergonomics Risk Assessment:** RCRA metrics complement traditional ergonomics risk assessment methods, helping identify jobs posing significant musculoskeletal injury risks. In additional to injury prevention, keeping job demands within RCRA-based limits also prevents localized muscle fatigue, positively impacting productivity and quality metrics.
- **Individualized Ergonomics Evaluation:** Incorporating individual strength levels (measured directly or estimated with predictive equations) into RCRA-based evaluation models enables ergonomics evaluation with respect to the individual. Instead of a one-size-fits-all approach to ergonomics evaluation, individualized evaluation offers greater "fidelity" in our efforts to identify jobs and job groups where physical job demands are misaligned with individual worker capabilities.

These individualized evaluation approaches have the potential to promote increased diversity and inclusion through more effective work assignment, accommodation, and thoughtful job rotation considering persons with strength limitations (i.e., elderly employees, employees with strength reduction resulting from a prior injury, some female workers with limited upper body strength, etc.).

- **Job Rotation:** Aid in design of more effective job rotation plans, either designing for a specific strength percentile (i.e., 10th percentile female strength, 25th percentile female strength, etc.), or rotation plans utilizing individual strength inputs in the job rotation model logic.
- **Simulated Ergonomics Evaluation:** Once job demands are measured/documented, simulation approaches can be used to predict the impact to RCRA metrics of equipment or process changes (i.e., reducing cycle time, changing to a part design requiring higher assembly force, changing the mix of models run on an assembly line, etc.).

The above force and fatigue based macro-ergonomics solutions, incorporating RCRA-based fatigue evaluation models, have tremendous potential to enable lean practitioners to make better decisions with respect to design of physical work, employee work assignments, work scheduling, etc.; thus enabling lean practitioners to both optimize traditional lean objectives and positively impact ergonomics and the physical wellbeing of workers.

FUTURE RESEARCH DIRECTIONS

The authors believe that future research should explore the application of modern force and fatigue-based macro-ergonomics solutions for the purpose of positively impacting productivity and quality-related metrics, in addition to the long-standing role of ergonomics as a tool to prevent or reduce work-related injuries and associated costs. Macro-ergonomics solutions having such potential are line/"fatigue" balancing, individualized ergonomics evaluation, and simulated ergonomics evaluation. Advancements in this area require both empirical and theoretical research.

CONCLUSION

Lean and ergonomics go hand in hand. Successful integration of ergonomics into lean initiatives results in competitive advantages for companies … predictable productivity levels, prevention of workplace injuries, etc. Besides these obvious benefits, emerging force and fatigue evaluation approaches have tremendous potential to positively impact workforce diversity and inclusion, through better measurement and management of physical work requirements.

ACKNOWLEDGMENT

This research was supported by Saturn Ergonomics Consulting (USA) and Poznan University of Technology [grant number 503217/11/141/DSPB/0575].

REFERENCES

Aqlan, F., Lam, S. S., Testani, M., & Ramakrishnan, S. (2013). Ergonomic Risk Reduction to Enhance Lean Transformation. *Industrial and Systems Engineering Research Conference (ISERC).* Retrieved from https://www.highbeam.com/doc/1P3-3169576661.html

Beurskens, R., Haeger, M., Kliegl, R., Roecker, K., & Granacher, U. (2016). Postural control in dual-task situations: Does whole-body fatigue matter? *PLoS One*, *11*(1), e0147392. doi:10.1371/journal.pone.0147392 PMID:26796320

Bigland-Ritchie, B., Jones, D. A., Hosking, G. P., & Edwards, R. H. (1978). Central and peripheral fatigue in sustained maximum voluntary contractions of human quadriceps muscle. *Clinical Science and Molecular Medicine*, *54*, 609–614. PMID:657729

de Vries, J. M., Hagemans, M. L. C., Bussmann, J. B. J., van der Ploeg, A. T., & van Doorn, P. A. (2010). Fatigue in neuromuscular disorders: Focus on Guillain-Barr syndrome and Pompe disease. *Cellular and Molecular Life Sciences*, *67*(5), 701–713. doi:10.100700018-009-0184-2 PMID:20196238

Gandevia, S. C. (2001). Spinal and supraspinal factors in human muscle fatigue. *Physiological Reviews*, *81*(4), 1725–1789. doi:10.1152/physrev.2001.81.4.1725 PMID:11581501

Gibson, M., & Potvin, J. R. (2016). *An equation to calculate the recommended cumulative rest allowance across multiple subtasks*. Association of Canadian Ergonomists Conference, Niagara Falls, NY.

Gruet, M., Temesi, J., Rupp, T., Levy, P., Millet, G. Y., & Verges, S. (2013). Stimulation of the motor cortex and corticospinal tract to assess human muscle fatigue. *Neuroscience*, *231*, 384–399. doi:10.1016/j.neuroscience.2012.10.058 PMID:23131709

Kim, J. S., Gračanin, D., Yang, T., & Quek, F. (2015). Action-transferred navigation technique design approach supporting human spatial learning. *ACM Transactions on Computer-Human Interaction*, *22*(6), 30. doi:10.1145/2811258

Kroenke, K., Wood, D. R., Mangelsdorff, A. D., Meier, N. J., & Powell, J. B. (1988). Chronic fatigue in primary care. Prevalence, patient characteristics, and outcome. *Journal of the American Medical Association*, *260*(7), 929–934. doi:10.1001/jama.1988.03410070057028 PMID:3398197

Landsbergis, P. A., Cahill, J., & Schnall, P. (1999). The impact of lean production and related new systems of work organization on worker health. *Journal of Occupational Health Psychology*, *4*(2), 108–130. doi:10.1037/1076-8998.4.2.108 PMID:10212864

Larun, L., Brurberg, K. G., Odgaard-Jensen, J., & Price, J. R. (2017). Exercise therapy for chronic fatigue syndrome. *BJPsych Advances*, *23*(3), 144–144. doi:10.1192/apt.23.3.144 PMID:28444695

Montalvo, F., Kozachuk, J., Rupp, M. A., Michaelis, J. R., McConnell, D. S., & Smither, J. A. (2016). *Examining methods to induce cognitive fatigue*. Academic Press.

Naranjo-Flores, A. A., & Ramírez-Cárdenas, E. (2014). Human factors and ergonomics for lean manufacturing applications. In J. L. García-Alcaraz, A. A. Maldonado-Macías, & G. Cortes-Robles (Eds.), *Lean Manufacturing in the Developing World: Methodology, Case Studies and Trends from Latin America* (pp. 281–299). Springer International Publishing Switzerland. doi:10.1007/978-3-319-04951-9_13

Norheim, K. B., Jonsson, G., & Omdal, R. (2011). Biological mechanisms of chronic fatigue. *Rheumatology*, *50*(6), 1009–1018. doi:10.1093/rheumatology/keq454 PMID:21285230

Pai, P., Cudney, E. A., & Murray, S. L. (2009). An Analysis of Integrating of Lean and Safety. *Proceedings of the 30th Annual National Conference of the American Society for Engineering Management: ASEM 2009.*

Shah, R., & Ward, P. (2003). Lean manufacturing: Context, practice bundles, and performance. *Journal of Operations Management*, *21*(2), 129–149. doi:10.1016/S0272-6963(02)00108-0

Silverman, M. N., Heim, C. M., Nater, U. M., Marques, A. H., & Sternberg, E. M. (2010). Neuroendocrine and immune contributors to fatigue. *PM & R*, *2*(5), 338–346. doi:10.1016/j.pmrj.2010.04.008 PMID:20656615

Staub, F., & Bogousslavsky, J. (2001). Post-stroke depression or fatigue. *European Neurology*, *45*(1), 3–5. doi:10.1159/000052081 PMID:11205620

Thomas, K., Goodall, S., Stone, M., Howatson, G., Gibson, A. S. C., & Ansley, L. (2015). Central and peripheral fatigue in male cyclists after 4-, 20-, and 40-km time trials. *Medicine and Science in Sports and Exercise*, *47*(3), 537–546. doi:10.1249/MSS.0000000000000448 PMID:25051388

Tran, Y., Craig, A., Wijesuriya, N., & Nguyen, H. (2010). Improving Classification Rates for Use in Fatigue Countermeasure Devices using Brain Activity. In *Proceedings of 32nd Annual International Conference of the IEEE Engineering in Medicine and Biology Society* (pp. 4460-4463). Buenos Aires, Argentina: IEEE.

Wan, J. J., Qin, Z., Wang, P.-Y., Sun, Y., & Liu, X. (2017). Muscle fatigue: General understanding and treatment. *Experimental & Molecular Medicine*, *49*(10), e384. doi:10.1038/emm.2017.194 PMID:28983090

Wickens, C. D. (1992). *Engineering Psychology and Human Performance* (2nd ed.). New York: Harper Collins.

Womack, J. P., & Jones, D. T. (1996). *Lean thinking: banish waste and create wealth in your corporation.* New York: Simon and Schuster.

World Health Organization. (2001). *International Classification of Functioning, Disability and Health: ICF*. Geneva: WHO.

ADDITIONAL READING

Arezes, P., Carvalho, D., & Alves, A. C. (2010). *Threats and opportunities for workplace ergonomics in lean environments. In 17th International Annual EurOMA Conference-Managing Operations in Service Economics. EurOMA.*

Hamja, A., Hossain, A., Maalouf, M. M., & Hasle, P. (2017). A review paper on Lean and Occupational Health and Safety (OHS) in RMG industry. In *4th International Conference on Mechanical Engineering and Renewable Energy Icmere 2017, Chittagong, Bangladesh 18-20 December 2017.*

Maia, L. C., Alves, A. C., & Leão, C. P. (2012, November). Design of a Lean Methodology for an ergonomic and sustainable work environment in Textile and Garment Industry. In *ASME 2012 International Mechanical Engineering Congress and Exposition* (pp. 1843-1852). American Society of Mechanical Engineers.

Nunes, I. L. (2015). Integration of ergonomics and lean six sigma. A model proposal. *Procedia Manufacturing*, *3*, 890–897. doi:10.1016/j.promfg.2015.07.124

Starheim, L., & Hasle, P. (2017). Lean as a Tool for Local Workplace Innovation in Hospitals. In *Workplace Innovation* (pp. 209–225). Cham: Springer. doi:10.1007/978-3-319-56333-6_13

Walder, J., Karlin, J., & Kerk, C. (2007). Integrated lean thinking & ergonomics: utilizing material handling assist device, solutions for a productive workspace. *MHIA White Paper, USA.*

Yusoff, S. M., Arezes, P., & Costa, N. (2013). The integration of lean manufacturing and ergonomics approach in workplace design. In *International Symposium on Occupational Safety and Hygiene–SHO'*13 (pp. 447-448). Sociedade Portuguesa de Segurança e Higiene Ocupacionais (SPOSHO).

KEY TERMS AND DEFINITIONS

Ergonomics: Process of adjusting surrounding to human.
Force: Strength generated by muscle.
Health: Condition of body, mind, and spirit to perform tasks effectively.
Lean: Create more with less.
Lean Thinking: Management approach how to deliver more while eliminating waste.
Muscle: Fibrous tissue which lets to contract, moves, or keeps the position of parts of the body.
Physical Fatigue: Temporary impossibility of muscle to perform tasks optimally.

Chapter 7
Exceeding the Recommended Energy Limits Due to Age and Gender in Occupational Aerobic Workloads

Cesar Omar Balderrama Armendariz
Universidad Autonoma de Ciudad Juarez, Mexico

Jose de Jesus Flores Figueroa
Universidad Autonoma de Ciudad Juarez, Mexico

Judith Lara Reyes
University of Texas at El Paso, USA

Ludovico Soto Nogueira
Universidad Autonoma de Ciudad Juarez, Mexico

ABSTRACT

The purpose of this chapter is to analyze the physical aerobic work in terms of the metabolic expenditure and compare it with the recommended boundaries of energy found in literature, proposing an alternative to the potential work overload through a compensatory equation introduced in the standard time of the workstation. To support the study, information considering the estimated metabolic expenditure in workers was applied to a novel procedure to reduce the metabolic demand of the task according to age and gender. Results of the study indicated that women older than 30 years exceeded the energy limits from moderate to very heavy load activities, and men older than 40 years exceeded the energy limits in heavy and very heavy workloads. The proposal of compensatory equation statistically reduced the energy loads below the recommended limits of energy. The aerobic workload is a sensitive factor for age and gender groups and can be potential risks for developing cardiovascular diseases as well as some musculoskeletal disorders.

DOI: 10.4018/978-1-5225-7192-6.ch007

INTRODUCTION

Accurate methods are required to measure and control the efficiency of the industrial processes. Time and motion studies (TMS) provide the necessary information to establish proper time standards and obtain balanced workloads that allow workers to reach production rates. Because of the importance of the manufacturing companies to maintain appropriate productivity levels, standard time (ST) is considered as a key activity to control most of the production processes through the calculation of the work rate. In this respect, an increased work rate can expose operators to an elevated risk of musculoskeletal injury (Gooyers & Stevenson, 2012), and considering the aerobic physical capacity, ST might be inappropriate if it is performed by individuals who do not reach this capacity just because the age, gender or combination of both (Balderrama, Flores, & Maldonado, 2015).

Even though there is a fact of a significant difference to produce energy on individuals, just few evidence to consider age and gender was found in the calculation of predetermined times or some other systems used to determine the ST; Murrel (1965) proposed a fixed relationship to compute rest time considering an energy expenditure of 5 kcal/min for males and a 4.2 kcal/min for females no matter the age of the workers, and Mital and Shell (1984) proposed the use of an energy model to predict rest period as a percentage of working duration to compensate physiological fatigue, but without stablish the application of the rest. Regarding rest pauses and line balancing, some intents to consider individual energy production have been experimented; Ayabar, De la Riva, Sanchez & Balderrama (2015), developed a model to estimate de energy consumption trough heart rate using linear regression and determine ST in moderate workload stations. Unfortunately, the study did not consider age and gender in the calculations.

It is important to remark that gender differences are significant in physiological work; in a longitudinal study during 5 years taking indices of work content, health, work ability, functional capacity, and symptoms of stress in 129 employees, women showed less physically able as men for physical work (due for reasons as the musculoskeletal capacity), and the critical age for women in prolonged physical work resulted in less than 50 years (Ilmarinen, 1988). Job demands could be significate in relation with older employees; lower autonomy and higher job demands increased the association of an array of common chronic health problems with sickness absence. These results were obtained taking in consideration work factors, health and sickness absence, and relative excess risk on 8,984 employees (Leijten, Van den Heuvel, Ybema, Robroek, & Burdorf, 2013). Likewise, a study performed in 2007 with 4 year, and 11 year examinations of 612 men using ultrasonography and the association between five measures of energy expenditure, concluded that high energy expenditures at work are associated with an accelerated progression of atherosclerosis even after controlling virtually all known cardiovascular risk factors, especially among older workers and workers with preexisting ischemic heart disease or carotid artery stenosis (Krause, Brand, Kaplan, Kauhanen, Malla, & Tuomainen, 2007). Workloads are then, an important part in the physical development of the worker; a 16-year follow-up study based on assessments of musculoskeletal and cardiovascular load, resulted in the following conclusión: In general, and contrary to what one might think, aging workers with low workload had better physical capacity than the subjects with high workload (Savinainen, Nygård & Ilmarinen, 2004).

With age, some physiological factors tend to change, for example, a research considering 120 men aged from 23 to 60 years old observed in six different type of work, found that the variation in the relative aerobic strain (RAS) is shown to increase in work tasks and is demonstrated that this increment is mainly related to physical work and the exposure of workers to peak loads (Ilmarinen & Rutenfranz, 1980). These variations can include the production of muscle forces and the coordination of motor functions (Ilmarinen, 1984). The emerge of fatigue is one important reason for decreasing performance (Öztürkoğlu & Bulfin, 2012). In our experience, consequences of physical fatigue caused by high workloads can be associated with execution mistakes that will end in generation of scrap, repairs, idle time, process delays, and other manufacturing costs related to the personal performance. Also, a decrease in the efficiency and the product quality may appear at a short or long term.

Taking into consideration the metabolic rate needed to perform a job, an experiment is presented to investigate the industrial workloads performed by people of age and gender and compare them with the recommended energy limits at work extracted from literature. In addition, a time allowance equation is proposed in order to reduce the energy consumption in the participants; the equation (explained in section of: Time Allowance Equation) includes nine physiological factors and is introduced directly into the calculation of the ST.

Aging and Work

It is a noteworthy finding in the literature review that there is a decrease over the years to produce energy, as well as there is a difference between men and women in the capacity to consume (O2). For example, on an average, the physical working capacity of women is about 30% less than men of the same age (Astrand & Astrand, 1978). The decline in physical working capacity from the age of 20 to 65 years is about 30%, calculated in terms of mean value. This mean value for a 65-year-old man is of the same magnitude as the value for a 20-year-old woman (Astrand, 1988). Considered to compute physical shape and an accurate tool according to ISO 8996 (±5%), the VO2max (Maximal Oxygen Consumption) decline begins after reaching the maturity age, at least after 30 years of age (Shvartz & Reibold, 1990). Nevertheless, females often dominate jobs where major physical risk factors such as repetitive work, forceful exertions, static work, and awkward postures prevail (Collins & O'Sullivan, 2010).

Shephard (2000) findings, are specific in terms that "aging is associated with progressive decreases in aerobic power, thermoregulation, reaction speed, and acuity of the special senses. These changes can reduce productivity, particularly in self-paced activities where the physical or mental input of the individual worker is the rate-limiting step in production" (Shephard, 2000).

Many years ago, an energy consumption of 5.33 kcal/min for men and 4 kcal/min for women was suggested as a limit for 8 h shift. This represents one third of average consumption in the United States (Bink, 1962) and is still in use to determine energy loads and rest pauses (Bridger, 2003; Groover, 2007; Niebel & Freivalds, 2003). A workload of 5.33 kcal/min generates an increase in the heart rate (HR) of 40 beats/min (Grandjean, 1988). Other studies have recommended that the requirement for a physical activity should be between 30 (no rest available) and 50% (rest available) of the maximum capacity of the worker (Ilmarinen, 1984; Bridger, 2003; Astrand, 1956; Jorgensen, 1985; Rutenfranz, Ilmarinen, Klimmer, & Kylian, 1990).

Half of the aerobic capacity cannot be surpassed because when a demand of energy is involved, it is necessary to design tasks below its consumption level (Farrer, Minaya, Niño & Ruiz, 1997). At intensities exceeding 50%, carbohydrate metabolism increases and therefore, the degree of exertion gradually increases, HR and pulmonary ventilation per liter of oxygen uptake increases, and body temperature rises above 38 °C (Astrand, 1988).

Taking in consideration the information reviewed, is important to define that if there is a significant difference in energy production between women and men as well as a difference in workers of different ages, then an adult woman should not perform a task with aerobic load for the same amount of time than a young man.

Determine the Metabolic Rate in Tasks

There are different methods used to estimate the energy consumption at work. Direct calorimetry is the most accurate method with minimal error but is often expensive and difficult to be used in the field. Similarly, doubly labeled water is accurate and expensive and is applicable for measuring energy consumption over long periods of time. The most recognized and accurate indirect measurement method is through reading VO2. For the ISO 8996 (2004), VO2 measurement is accurate to ±5%, while the methods based on tables, HR, and classifications are accurate to ±15%. The Firstbeat's beat-by-beat method uses the HR also and reports better accuracy against other methods (Firstbeat Technologies, 2017).

By monitoring the HR, we can approximate the metabolic level of a person when executing a task. Taking into consideration the HR in basal state, it can be subtracted from the HR when an aerobic task is performed. VO2 has a lineal relation with HR when an aerobic task is carried out (Groover, 2007; Astrand, 1960; Ilmarinen, 1992; Manero & Manero, 1991; Smolander, Juuti, Kinnunen, Laine, Louhevaara & Männikkö, 2008). HR is acceptable as an estimator of VO2 only when the test work closely resembles the muscle work in the job (Oja, Ilmarinen, & Louhevaara, 1982), and is a good predictor when heat stress is present (Imbeau, Desjardins, Dessureault, Riel & Fraser, 1995). Bouchard and Trudeau suggest good reliability of the VO2/HR relationship at HR usually found in workplaces, regardless of the time of day, is not affected by a day at work in both genders (Bouchard & Trudeau, 2007).

Balderrama et al. (2010) evaluated an electronic HR monitor equipment to estimate oxygen consumption in aerobic activities. The HR monitor records every second and computes the VO2 through neural nets. The software uses a model developed by the company called Firstbeat Technologies (2017). A validation was performed by means of a comparison between a direct measurement equipment of VO2 and monitoring equipment of brand Sunnto resulting in no significant differences using a 95% of confidence level.

In order to determine how demanding is a work task in terms of VO2, there is a general classification presented by Astrand and Rodah (1977), where a workload is considered as a light load when VO2 is between 0 and 0.5 l/min (0–2.5 kcal/min), a moderate workload is considered from 0.5 to 1.0 l/min (2.5–5.0 kcal/min), a heavy work is considered from 1.0 to 1.5 l/min (5.0–7.5 kcal/min), a very heavy work is considered from 1.5 to 2.0 l/min (7.5–10.0 kcal/min), and an extremely heavy work is considered from more than 2.0 l/min (>10.0 kcal/min).

Work Rate

In industry, it is common knowledge that work rate is an important measurement of labor because the number of parts produced per hour or per day is one way to control the intensity of work. Companies use this rate to program their production volumes and determine its maximum capacity in relation to the number of workers and equipment available. ST is used as the basis for estimating work rate and is composed by the normal time and a group of allowances applied according to the work conditions. The normal time is the average time to perform a work task that can be determined by stopwatch, statistical data, video analysis, or predetermined motion time systems. Normally, it is determined from the measure of average skilled operators. In any case, the obtained time represents a mean time of the cycle time without taking worker's age and gender into consideration.

Allowances adding to the normal time are for personal needs (such as washroom and coffee breaks), unavoidable work delays (such as equipment breakdown or lack of materials), and worker fatigue (physical or mental) (Niebel & Freivals, 1998). In some cases, different compensations are added to the normal time of the job that help to reach the estimated time when factors do not allow to carry out the task in the calculated time (Meyer, 1991). Predetermined motion time systems are not suitable for aging workers and middle age and old woman can have problems to reach the ST in terms of energy consumption (Balderrama, Flores, & Maldonado, 2015).

Several factors can be utilized to establish allowances due to fatigue; according to the International Labor Organization (Kanawaty, 1992), the physiological factors are weight, force or pressure, posture, and restrictive clothing. The psychological factors such as discipline, concentration, mental and visual demands, and monotony can also be taken into consideration. Also, thermal and atmospheric factors of the environment such as temperature, humidity, wind conditions, noise, vibration, light, and visual factors are aspects that affect the personal performance when they are out of the comfort limits, which are considered as significant, in order to add extra concession to the normal time of the job. Unfortunately, there are no records or investigations that consider the age and gender differences as a required element within the ST.

METHODS

Subjects

Five companies dedicated to manufacture auto parts were visited, where a total of 149 workers gave authorization to collect data during 8 months of work in accordance with the institutional occupational standards. Their ages fluctuated between 20 and 68 years and were categorized in five different age ranges. Workloads as moderate, heavy, and very heavy tasks were considered according to Astrand and Rodah (section of: Determine Metabolic rate in Tasks). Considering a previous laboratory study (Balderrama, Flores, & Maldonado, 2015) about energy consumption, light tasks were not considered because they didn't affect any age group. A summary of the participants' data is presented in Table 1. Mean and standard deviation is presented considering five age groups to include 149 participants. 59 men and 90 women with a mean of 53.42 and 43.91 years old respectively. VO2 on rest as well VO2 MAX looked very similar comparing average measures for both genders. Due to some activities are time consuming and interfere with production purposes, rest VO2 was computed by equations according to age and gender (Kanawaty, 1992) and VO2max was taken from local tables (Sandoval & Ramos, 1995).

Instruments

Considering that VO2 is one of the most accurate methods to measure the capacity to generate body energy and the need to operate noninvasive equipment to workers, this study use an estimation of VO2 by means of HR using the Firstbeat technology (2017) as a new HR variability-based method for the estimation of oxygen consumption without individual laboratory calibration; it was examined as a good estimator in Smolander et al. (2008), Balderrama et al. (2010), and Uusitalo (2011). This equipment

Table 1. Participants' data

Age Range	Gender	Number of Participants	Age (Avg)		Weight (Avg)	Height (Avg)	Rest VO2 ml/kg/min	VO2 Max ml/kg/min
20-29	Men	12		26.40	73.00	170.92	3.68	38.90
	Woman	31		25.31	66.74	157.14	3.89	35.76
	Σ	43	Mean	25.60	68.34	160.66	3.84	37.30
			S.D.	2.68	12.75	8.94	0.34	6.60
30-39	Men	14		35.42	76.79	173.64	3.43	31.47
	Woman	23		34.42	74.08	157.38	0.40	5.53
	Σ	37	Mean	34.78	75.03	163.08	3.83	34.44
			S.D.	2.78	13.51	9.11	3.67	32.36
40-49	Men	14		43.24	78.77	172.06	3.83	31.22
	Woman	19		42.47	67.00	158.47	3.49	28.70
	Σ	33	Mean	42.88	73.22	165.69	3.69	29.96
			S.D.	2.80	13.33	8.31	0.44	5.25
50-59	Men	11		53.20	77.50	169.80	3.89	31.22
	Woman	12		54.50	72.50	160.38	3.49	28.70
	Σ	23	Mean	53.85	75.00	165.09	3.69	29.96
			S.D.	2.46	14.89	7.78	0.44	5.25
60-69	Men	8		63.83	69.83	168.33	3.59	28.26
	Woman	5		62.83	69.50	154.33	3.36	24.50
	Σ	13	Mean	63.33	69.66	161.33	3.47	26.38
			S.D.	1.78	14.37	8.65	0.38	3.66
Total	Men	59		53.42	75.35	170.06	3.77	31.31
	Woman	90		43.91	69.96	157.54	3.58	30.96
	Σ	149	Mean	47.71	72.12	162.55	3.65	31.10
			S.D.	2.50	13.77	8.56	0.38	5.23

was utilized to obtain VO2 on field and compute time allowances by means equation 1 described in the next section. Also, a Sunnto Team Pod© with chest straps and Sunnto Monitor© software were used to monitor HR and calculate VO2.

Time Allowance Equation

An equation that considers individual energy production was formulated. The model considers the energy requirement for the task and compare it with the energy available of the worker (according to age and gender). The objective was to reduce the energy consumption below the recommended limits at work adding more time to the execution of the work task. The equation proposed uses the current energy expenditure in work task (GE) minus the resting metabolism rate of the worker (Eir), divided by half the maximum capacity (expenditure limit) of the person according to age and gender (Eid), minus, once again, the resting metabolism rate (Equation. 1).

$$TOL = (GE - Eir)/(Eid - Eir) \quad (1)$$

The units of the model are relative to body weight in ml/kg/min. The necessary factors for calculating allowance factor (TOL) are as follows:

1. Age
2. Gender
3. Body weight
4. Smoking
5. Physical shape of the individual
6. Metabolic rate (Eir)
7. Capacity in terms of: Maximum oxygen consumption of the individual (Eid x 2)
8. Expenditure in work task (GE)
9. Workload level

The age (A), gender (B), body weight (C), smoking (D), and physical shape (E) were obtained directly from the participant and was fed in the Sunnto software that computes the energy consumption. Because it was not possible for most companies to allow direct measurement of the resting metabolic rate and VO2max of the worker directly, we made use of the equations for obtaining basal metabolism (F) published by the International Labor Office (Kanawaty, 1992); these are based on the age ranges and gender of the people. For VO2max (G) in people of different ages, tables of local statistics were used (Sandoval & Ramos, 1995). The tables mark a difference in VO2max according to age range and gender. In the application of equation 1, it is required that conversion units of basal metabolic rate and the maximum capacity for producing energy be the same (e.g. kcal, VO2, Watts, etc.). The 50% of the VO2max is only considered when rests in the occupational activities are available; otherwise, 33% of the VO2max must be used. The energy expenditure in work task (H) was obtained by direct measure of participants using the Sunnto Team Pod. The workload level (I) was obtained after GE (H) calculation according with Astrand and Rodah classification presented the first section of the document. At least, 20 cycles were taken on each activity.

The factor obtained from equation 1 must be multiplied by the normal time (cycle time) of the workstation in order to obtain the ST of the operation. That is, if the normal time is 80 s and the calculated allowance computed by personal data in Equation 1 is 1.033 then the new ST would be: $80 \times 1.033 = 82.64$ s. It is important to remark that equation variables (Eir, Eid, and GE) can be calculated by different methods according to the data and the availability of the resources. ISO 8996 provides different estimation methods regarding metabolic rate.

Procedures

Measurements were taken on repetitive tasks that possessed an aerobic load from a moderate to very heavy level. Physiological factors such as age, gender, weight, smoker, and physical condition level were collected at the beginning of a shift, as well as an explanation on how to wear the chest strap. In the first part of the experiment, production lines were monitored considering the actual ST with no modification in the process, and information that contained unexpected interruptions to the process was discarded. The first collected data were used to classify the workload level.

Data obtained from chest strap was stored in a laptop, while the VO2 was calculated using the software for the estimation of metabolism, which was validated previously (Balderrama et. al, 2010). Thereafter, the average consumption of the oxygen was calculated for each worker (GE) and then subtracted from the allowed oxygen consumption according to his or her age and gender (Eir). To determine the allowed oxygen consumption, international recommendations were taken into consideration to compare whether work tasks exceeded the 50% of VO2max (33% if rest are unavailable). Thus, VO2max was calculated using the local tables that were published by the Institute of Social Security (Sandoval & Ramos, 1995).

After detection of the workstations that exceeded the recommended limit (exceeded the 50% of VO2max), ST was modified by adding the time that includes the calculated allowances. This procedure was developed to evaluate the proposed equation. To implement the new ST increased in time, some strategies have to be taken to avoid affecting the work rates of the shift; more workers to the production line or more stock were introduced to the workstations. All the activities were performed to meet the requirements presented in Time Allowance Equation section. It should be noted that this study decided to disregard light workload activities due to a previous study (Balderrama, Flores, & Maldonado, 2015) that did not give significance of this level of activity.

Statistical Analysis

A total of five analyses were considered in the experiment: (a) a general linear model of analysis of variance (ANOVA) was carried out using the data of the VO2 in the task; the purpose was to determine whether age, gender, and other factors were significant factors that affect the energy expenditure of the task; (b) a paired t-test was conducted in order to detect whether the workers surpassed the recommended limits of the energy consumption in the workstation according with age and gender. Age groups split in male and females were compared with the average energy consumption on each work station; (c) a similar paired t-test test was developed, but this time, the data of VO2max of men from 20 to 30 years was introduced in the analysis; the reason, to simulate the work performed only by young male workers and figure it out whether they surpass the recommended limits of work; allowances were collected to run (d) a normality test to the results; (e) lastly, a paired t-test to analyze whether the new standard time with time allowances introduced, could reduce the energy consumption below the recommended limits. Minitab Software© was utilized to analyze statistical data including all the figures presented.

RESULTS

Results are presented according to the statistical analysis proposed in chapter 2.5 taking the same letters sub-division:

1. The ANOVA in the first part of the experiment reflects that only the factor introduced as workload level was a significant factor with a p value of 0.017. In other words, the variables introduced as age ($p = 0.067$), gender ($p = 0.567$), physical shape ($p = 0.740$), and smoking ($p = 0.100$) do not modify the response variable when a confidence level of 95% is used. Hence, the worker energy consumption (VO2) in workstation is not significantly different from one person to another and is only affected by the intensity of work.

2. The paired t-test conducted to detect whether the VO2 in the task exceeds the allowed VO2, showed information to deduce that there is enough evidence to say that the VO2 required by the task is higher than the allowed limit (exceeded 50% of the person of maximum energy production) of the VO2 with a p = 0.0. The individual paired tests that considered age and gender in the three different workloads reflect that women older than 30 years surpass the recommended energy limits when they perform moderate, heavy, and very heavy tasks (according with Astrand and Rodah, 1977), while men older than 40 years surpass the recommended limits with heavy and very heavy tasks. In addition, women younger than 30 years surpass the limits in heavy and very heavy task, and men older than 40 years surpass the limits only in very heavy tasks. Figure 1 displays the difference of gender, age, and workload level when a limit of energy is considered. Values close to zero represent better energy to perform a task.
3. Results of the conducted simulating the workers having the VO2max of men in age ranging between 20 and 29 years, reflect that statistical evidence exists to deduce that young workers did not exceed the recommended limits of the energetic expenditure in all the considered workloads levels (p = 0.0). Figure 2 presents how the energy consumption can be below the limits if all the participants were younger men.
4. A residual normality test was conducted in order to validate the experiment obtaining a p value of 0.242 indicating normality in data.
5. Paired t-test performed after the application of the allowances indicates that the energy consumption of the workers did not surpass the energy recommended for 50% of their maximum VO2 (p = 0.00). Figure 3 shows a lower tendency of the VO2 after application of time allowances against results at the beginning of the experiment.

Figure 1. VO2 workload and recommended energy comparison

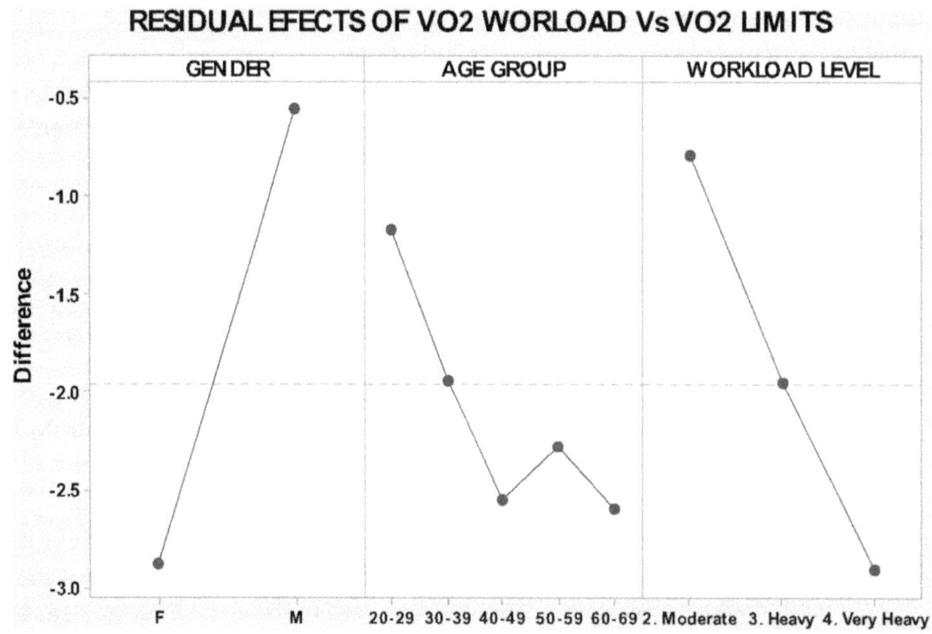

Figure 2. Energy comparison considering 20–29 years energy level

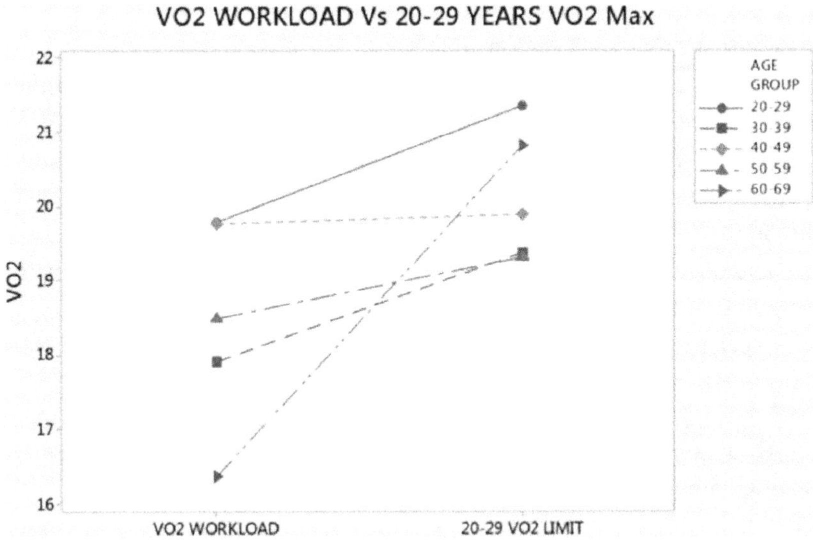

Figure 3. VO2 workload versus VO2 with time allowance

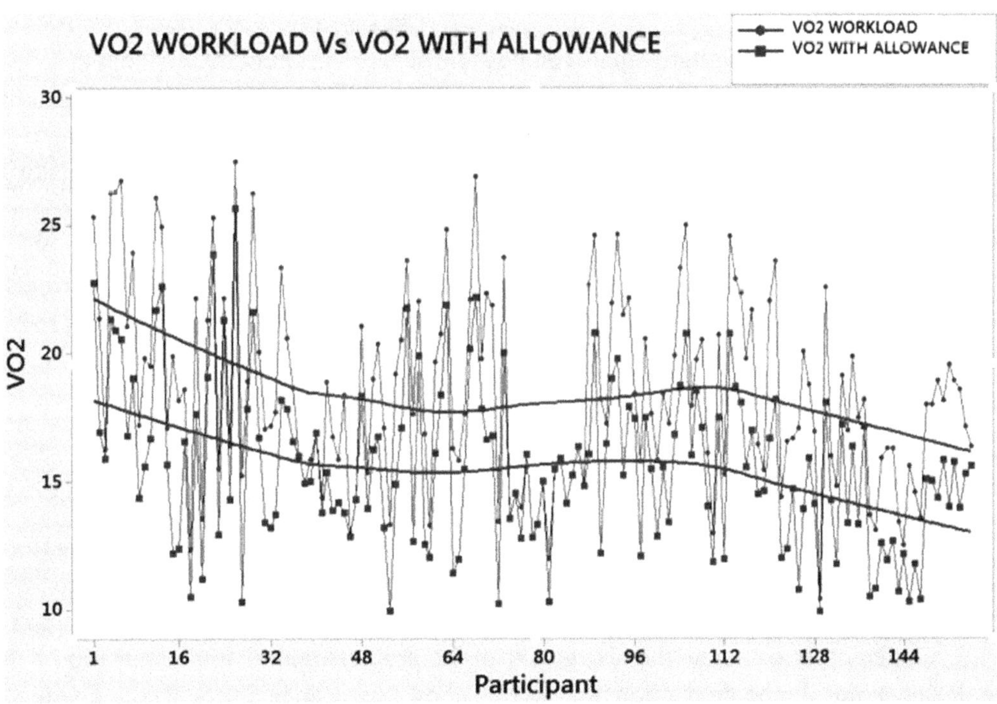

DISCUSSION

This investigation confirms that workstations that possess similar workload, have no significant difference in the worker energy consumption in spite of the physiological conditions of the person. In other words, all persons have similar energy consumption in similar tasks, but the main difference is based on the fact that we have different capacity to generate energy during the day. This result matches up with Louhevaara (1999), Ilmarinen (2001), and the definition of ISO 8996, which states that the metabolic rate can vary from person-to-person, about ±5%, for the same work and under the same working conditions. Also, we confirm, and according to the bibliography presented, that men and women have significant differences in energy production as well as the reduction in energy production after the age of 30 years.

Results of this experiment regarding aerobic workloads that are controlled by ST let us to assume that moderate workloads (2.5–5.0 kcal/min) might be unsuitable for women who are older than 30 years. Heavy (5.0–7.5 kcal/min) and very heavy workloads (7.5–10.0 kcal/min) could be unsuitable for women who are older than 20 years. Similarly, men older than 40 years surpass the recommended limits in workstation over the 7.5 kcal/min.

In accordance with the aforementioned information, workloads determined by the ST are not suitable for all ages and gender, proving that they are only be accepted for specified age ranges. The proposed equation applied to ST reduced the worker energy consumption below the recommended limits suggested in bibliography that mention that workloads should be below of the 50% of the worker capacity to produce energy in one day.

Metabolic capacity of young individuals can be significantly different to carry out tasks with high metabolic requirements compared with adult workers. According to our experience, some companies take strategies to not assign high workloads to adults and women or assign breaks according to manager experience to achieve the recovery of worker, where there is nothing normative or mandatory on this respect. Systems in industry designed to introduce brakes or rest pauses do not consider age and some companies assign pauses according to administrative interests and not when the operator needs it. An allowance of time within the standard time could be a good response to high working demands due the manipulation of the rest by worker each cycle time, achieving a recovery rate throughout the workday in better manner than rest pauses.

CONCLUSION

ST is a valuable planning tool for several important production factors; however, it could affect people depending on their age and gender. The addition of time through a ST allowance helps to accomplish manual aerobic work more equilibrated according to the amount of energy a person can produce. To implement compensatory equation, the acceptance of the problem and its implementation willingness is required. Depending on the production process, the allowances can be introduced individually or according to the characteristics of small working groups. The proposed equation differs from other models because it considers the age and gender of the employee and the feasibility of applying it systematically through ST.

The recommendation of this study is to provide opportunities to the workers to keep energy reserves at the end of the shift in order to succeed on their day-to-day life, to enhance the quality of their lives, and to promote an occupational health program to improve the productivity of the manufacturing companies. Inclusion of the aging people must be one of the commitments of this era due the worldwide increment of old people.

ACKNOWLEDGMENT

We would like to thank The Autonomous University of Juarez City for support the project and give permission in the use of the facilities and equipment during this research.

REFERENCES

Astrand, I. (1960). Aerobic work capacity in men and women with special reference to age. *Acta Physiologica Scandinavica*, *49*(suppl 169), 45–60. PMID:13794892

Astrand, I. (1988). Physical demands in work life. *Scandinavian Journal of Work, Environment & Health*, *14*(1), 10–13. PMID:3393858

Astrand, I., & Astrand, P.-O. (1978). *Aerobic work performance, a review*. New York: Academic Press Inc. doi:10.1016/B978-0-12-261350-0.50015-6

Astrand, P.-O. (1956). Human physical fitness with special reference to sex and age. *Physiological Reviews*, *36*(3), 307–335. doi:10.1152/physrev.1956.36.3.307 PMID:13359126

Astrand, P. O., Rodahl, K., Dahl, H., & Strømme, S. B. (1977). *Textbook of work physiology: physiological bases of exercise*. New York: McGraw Hill.

Ayabar, A., De la Riva, J., Sanchez, J., & Balderrama, C. (2015). Regression model to estimate standard time through energy consumption of workers in manual assembly lines under moderate workload. *Journal of Industrial Engineering*, 1-5.

Balderrama, C., Flores, J., & Maldonado, A. (2015). The effects of age to meet the standard production rate. *Work (Reading, Mass.)*, *51*(4), 827–837. doi:10.3233/WOR-141972 PMID:25425589

Balderrama, C., Ibarra, G., De La Riva, J., & López, L. (2010). Evaluation of three methodologies to estimate the VO2max in people of different ages. *Applied Ergonomics*, *42*(1), 162–168. doi:10.1016/j.apergo.2010.06.017 PMID:20650446

Bink, B. (1962). The physical working capacity in relation to working time and age. *Ergonomics*, *5*(1-4), 25–28. doi:10.1080/00140136208930548

Bouchard, D. R., & Trudeau, F. (2007). Reliability of the assessment of the oxygen/heart rate relationship during a workday. *Applied Ergonomics*, *38*(5), 491–497. doi:10.1016/j.apergo.2006.10.002 PMID:17368556

Bridger, R. (2003). *Introduction to Ergonomics*. London: Taylor and Francis.

Collins, J., & O'Sullivan, L. (2010). Psychosocial risk exposures and musculoskeletal disorders across working age males and females. *Ergonomics in Manufacturing & Service Industries*, *20*(4), 272–286. doi:10.1002/hfm.20220

Farrer, F., Minaya, G., Niño, J., & Ruiz, M. (1997). *Manual de ergonomía*. Madrid: Mapfre.

Firstbeat Technologies. (2017). *VO2 estimation method based on heart rate measurement*. Retrieved from: https://assets.firstbeat.com/firstbeat/uploads/2015/10/white_paper_vo2_estimation.pdf

Gooyers, C. E., & Stevenson, J. M. (2012). The impact of an increase in work rate on task demands for a simulated industrial hand tool assembly task. *International Journal of Industrial Ergonomics*, *42*(1), 80–89. doi:10.1016/j.ergon.2011.11.005

Grandjean, E. (1988). *Fitting the task to the man*. New York: Taylor and Francis.

Groover, M. P. (2007). *Work systems and the methods, measurement, and management of work*. Upper Saddle River, NJ: Pearson Prentice Hall.

Ilmarinen, J. (1984). Physical load on the cardiovascular system in different work tasks. *Scandinavian Journal of Work, Environment & Health*, *10*(6), 403–408. doi:10.5271jweh.2303 PMID:6535243

Ilmarinen, J. (1988). Physiological criteria for retirement age. *Scandinavian Journal of Work, Environment & Health*, *14*(1), 88–89. PMID:3393893

Ilmarinen, J. (1992). Job design for the aged with regard to decline in their maximal aerobic capacity: Part I—guidelines for the practitioner. *International Journal of Industrial Ergonomics*, *10*(1), 53–63. doi:10.1016/0169-8141(92)90048-5

Ilmarinen, J. (2001). Aging Workers. *Occupational and Environmental Medicine*, *58*(8), 546–548. doi:10.1136/oem.58.8.546 PMID:11452053

Ilmarinen, J., & Rutenfranz, J. (1980). Occupationally induced stress, strain and peak loads as related to age. *Scandinavian Journal of Work, Environment & Health*, *6*(4), 274–282. doi:10.5271jweh.2607 PMID:7233116

Imbeau, D., Desjardins, L., Dessureault, P. C., Riel, P., & Fraser, R. (1995). Oxygen consumption during scaffold assembling and disassembling work: Comparison between field measurements and estimation from heart rate. *International Journal of Industrial Ergonomics*, *15*(4), 247–259. doi:10.1016/0169-8141(94)00040-A

ISO-8996. (2004). *Ergonomics of Thermal Environment. Determination of Metabolic Rate*. International Organization of Standardization.

Jorgensen, K. (1985). Permissible Loads Based on Energy Expenditure Measurements. *Ergonomics*, *28*(1), 365–369. doi:10.1080/00140138508963145 PMID:3996374

Kanawaty, G. (1992). *Introduction to Work Study. Ginebra*. International Labor Organization.

Krause, N., Brand, R. J., Kaplan, G. A., Kauhanen, J., Malla, S., Tuomainen, T., & Salonen, J. T. (2007). Occupational physical activity, energy expenditure and 11-year progression of carotid atherosclerosis. *Scandinavian Journal of Work, Environment & Health*, *33*(6), 405–424. doi:10.5271jweh.1171 PMID:18327509

Leijten, F. R. M., Van den Heuvel, S. G., Ybema, J. F., Robroek, S. J. W., & Burdorf, A. (2013). Do work factors modify the association between chronic health problems and sickness absence among older employees? *Scandinavian Journal of Work, Environment & Health*, *39*(5), 477–485. doi:10.5271jweh.3353 PMID:23440271

Louhevaara, V. (1999). Is the physical workload equal for ageing and young blue-collar workers? *International Journal of Industrial Ergonomics*, *24*(5), 559–564. doi:10.1016/S0169-8141(98)00061-4

Manero, R., & Manero, J. (1991). *Dos alternativas para el estudio y promoción de la capacidad física de los trabajadores*. MAPFRE Seguridad.

Meyer, F. (1991). *Motion and Time Study: For Lean Manufacturing*. New York: Prentice Hall.

Mital, A., & Shell, R. L. (1984). A compressive metabolic energy model for determining rest allowances for physical tasks. *Journal of Methods-Time Measurement*, *11*, 2–8.

Murrel, K. F. (1965). *Human performance in industry*. New York: Reinhold.

Niebel, B. W., & Freivalds, A. (1998). *Methods, standards, and work design*. New York: WCB/McGraw-Hill.

Niebel, B. W., & Freivalds, A. (2003). *Methods, standards, and work design*. New York: McGraw-Hill.

Oja, P., Ilmarinen, J., & Louhevaara, V. (1982). Heart rate as an estimator of oxygen consumption during manual postal delivery. *Scandinavian Journal of Work, Environment & Health*, *8*(1), 29–36. doi:10.5271jweh.2499 PMID:7134920

Öztürkoğlu, Y. Y., & Bulfin, R. L. (2012). Scheduling jobs to consider physiological factors. *Human Factors and Ergonomics in Manufacturing & Service Industries*, *22*(2), 113–120. doi:10.1002/hfm.20257

Rutenfranz, J., Ilmarinen, J., Klimmer, F., & Kylian, H. (1990). Work load and demanded physical performance capacity under different industrial working conditions. Fitness for the Aged, Disabled and Industrial Worker. *International Series on Sport Sciences*, *20*, 217–238.

Sandoval, J., & Ramos, J. (1995). *Tablas de Indicadores Antropométricos y Fisiológicos en Trabajadores Mexicanos: Instituto Mexicano del Seguro Social*. IMSS.

Savinainen, M., Nygård, C. H., & Ilmarinen, J. (2004). Workload and physical capacity among ageing municipal employees—a 16-year follow-up study. *International Journal of Industrial Ergonomics*, *34*(6), 519–533. doi:10.1016/j.ergon.2004.06.006

Shephard, R. J. (2000). Aging and productivity: Some physiological issues. *International Journal of Industrial Ergonomics*, *25*(5), 535–545. doi:10.1016/S0169-8141(99)00036-0

Shvartz, E., & Reibold, R. (1990). Aerobic fitness norms for males and females aged 6 to 75 years: A review. *Aviation, Space, and Environmental Medicine*, *6*, 3–11. PMID:2405832

Smolander, J., Juuti, T., Kinnunen, M. L., Laine, K., Louhevaara, V., Männikkö, K., & Rusko, H. (2008). A new heart rate variability-based method for the estimation of oxygen consumption without individual laboratory calibration: Application example on postal workers. *Applied Ergonomics*, *39*(3), 325–331. doi:10.1016/j.apergo.2007.09.001 PMID:17950689

Uusitalo, A., Mets, T., Martinmäki, K., Mauno, S., Kinnunen, U., & Rusko, H. (2011). Heart rate variability related to effort at work. *Applied Ergonomics*, *42*(6), 830–838. doi:10.1016/j.apergo.2011.01.005 PMID:21356531

Chapter 8
Burnout and Obesity in Middle and Upper Management in the Manufacturing Industry of Baja California

Sharon Idali Macias Velasquez
Universidad Autónoma de Baja California, Mexico

Yolanda Angelica Baez-Lopez
Universidad Autónoma De Baja California, Mexico

Aidé Aracely Maldonado-Macías
Universidad Autónoma de Ciudad Juárez, Mexico

Jorge Limon-Romero
Universidad Autónoma de Baja California, Mexico

Diego Tlapa
Universidad Autónoma de Baja California, Mexico

ABSTRACT

Globally, companies are increasingly considering the importance of mental health in workers and their relationship with productivity, which has led to increased research on work stress, which showed that there is a relationship between stress related to work and health disorders, both physical and mental. This chapter addresses the analysis of two of the main consequences that a worker can develop when having work stress: burnout syndrome, measured by the Maslach burnout inventory general survey (MBI-GS) and obesity, through the body mass index (BMI). The study was carried out in 118 people who occupy middle and upper management of the manufacturing industry of Baja California, having as objective to find the relationship that exists between the two variables, using ordinal logistic regression, as well as to characterize the sample using mean difference and hypothesis testing. From this perspective, this chapter can serve as a guide to study the behavior of variables and propose organizational development strategies aimed at reducing and preventing these problems.

DOI: 10.4018/978-1-5225-7192-6.ch008

INTRODUCTION

On a global scale, companies are increasingly considering the importance of their workers' mental health and the relationship it bears with productivity. In general, job strain ranks among the ten main work-related health problems, often resulting in serious mental health disorders (World Health Organization, 2016). However, although it starts by having psychological effects, it is occasionally associated with diseases that affect a person's physical health as well.

Job strain occurs when work demands are in disproportion or exceed workers' capacities and resources. On the other hand, it can also occur when a worker's knowledge and skills to meet such demands do not match the expectations of the company's organizational culture. Thus, the severity of job strain will depend on the magnitude of the demands to be met within a certain period of time, the individual's feeling of control over them, and the decision latitude that he/she has in regards to them (American Institute of Stress, 2016).

Worldwide, Mexico ranks first in job strain, with 75% of Mexican workers suffering from stress. It is followed by China, with 73%, and the United States, with 59% (World Health Organization, 2016). Financially speaking, employers and governments are increasingly becoming aware of the high toll stress takes both on individual companies and on the economy in general (Gamero, 2010), namely poor productivity, the increase in the number of lost work days due to absenteeism, constant job switching, an increase in the rate of work accidents, increasing customer complaints, replacement of absent workers, and training of substitute workers, among others (World Health Organization, 2008). This situation can lead to losses of between 0.5 and 3.5% of countries' Gross Domestic Product (GDP). Particularly in Mexico, stress among workers results in losses of between 5,000 and 40,000 million dollars a year (Noticias Tijuana, 2015).

Therefore, research on the subject of job strain has been increasing, the results of which show a clear relationship between work-related stress and both physical and mental health disorders (Kang, Koh & Cha, 2004; Gonzalez & Landero, 2008; Lazarus & Folkman, 1984). The consequences of suffering from this condition include the appearance of exhaustion, burnout (Aranda, Pando & Salazar, 2015), anxiety and depression (Organization Internacional del Trabajo, 2016), alcohol abuse, poor sleep quality, smoking habits and an unhealthy diet that can lead to weight problems (Norma Oficial Mexicana NOM-043-SSA2-2012) to name a few.

This study analyzes two of the main consequences a worker can suffer when undergoing job strain. The first one is the burnout syndrome. This term refers to a prolonged response to job strain, generally understood as a state of extreme tiredness (Ksiazek, Stefaniak, Stadnyk & Ksiazek J., 2011), which develops from prolonged exposure to high levels of psychosocial risk factors (Aranda et al., 2015). The first professions associated with this syndrome shared the characteristic of being more frequent in the service sector (Ksiazek et al., 2011). This was due to the supplier-client interaction, where the individual offering a service had to deal directly with the one receiving it. However, the scope of such professions has widened to include jobs that fit the characteristics of "high demand". An example of this phenomenon is the presence of the burnout syndrome for long periods of time among company managers, who soon began to have health problems such as cardiorespiratory conditions, obesity, gastritis, ulcers, and sleep difficulties (Gil-Monte, 2008).

The burnout syndrome is characterized by having three dimensions (Maslach, 2003): the exhaustion dimension refers to the feeling of lacking enough mental and physical strength to meet the level of demand that the assigned activity poses. In this dimension, employees believe that they have no source of energy replenishment, which in turn triggers a greater sensation of weakness. Next is the dimension of cynicism, which surfaces in negative and apathetic attitudes towards both people and work activities. At this level, individuals reduce their performance to a minimum, considering practices such as spending less time in the workplace or lowering the amount of energy used in it. The final dimension is that of inefficiency, which refers to the feeling of lack of productivity at work. As a result of all of the above, individuals develop thoughts of professional failure, negative attitudes towards co-workers, and feelings of emotional exhaustion (Gil-Monte & Peiró, 1999).

Secondly, this research work analyzes obesity, which is defined as the abnormal or excessive accumulation of fat that can be harmful to health. In adults, it is present when there is a body mass index (BMI) equal to or greater than 30 kg / m^2, or equal to or greater than 25 kg/m^2 in adults of short height (NOM-043-SSA2-2012). The World Health Organization (WHO) uses the BMI as an indicator to measure obesity; this indicator is obtained by dividing the weight in kilograms divided by the square of the height measured in meters. Obesity has become a health problem due to the fact that since 1980, it has more than doubled around the world. By 2014, more than 1,900 million adults 18 years of age and older were overweight, the equivalent to 39%, and more than 600 million of them were obese (WHO, 2016). The highest prevalence of overweight and obesity occurred in the Region of the Americas (overweight: 62% in both sexes, and obesity: 26%) and the lowest in the Region of Southeast Asia (overweight: 14% in both sexes, and obesity: 3%) (WHO, 2012).

In Mexico, non-contagious diseases such as overweight and obesity have become one of the main problems affecting not only the health system but also the quality of life of Mexicans of all ages. Results of the National Health and Nutrition Survey (2012) show that the prevalence of overweight and obesity in adults in Mexico was of 71.3% (amounting to 48.6 million people). Specifically, the prevalence of obesity in this group was of 32.4% and that of overweight was of 38.8%. Obesity was higher in females (37.5%) than in males (26.8%), unlike overweight, where there was a prevalence of 42.5% in males, while in females there was a prevalence of 35.9%. In the case of obesity, the highest prevalence occurred in the age group between 40 to 49 years in men and between 50 to 59 years in women. Moreover, the problem was also 28.5% higher in urban locations. Finally, the prevalence of overweight, obesity, and morbid obesity was higher in females, and its cost was approximately 67,000 million pesos in Mexico (Encuesta Nacional de Salud y Nutrición de Medio Camino, 2016).

This study focuses on the manufacturing and export service industry (IMMEX, for its acronym in Spanish) in Mexico as this sector contributes importantly to the strengthening of the Mexican economy by creating benefits such as a decrease in unemployment rates, an increase in technological development, increasing participation in international markets and constant training for workers employed in this type of industry (Consejo Nacional de la Industria Maquiladora y Manufacturera de Exportación, 2016). In the period between January 2015 and May 2016, the IMMEX sector showed a growing trend, with an increase in business of 0.47%. The manufacturing companies registered in the program amounted to a total of 6,134 in the year 2016 (March), 5,024 of which belonged to the manufacturing sector and 1,110 of which, to non-manufacturing companies. Of the 32 states in the country, 11 represent 81.8% of IMMEX businesses, namely Baja California, Nuevo Leon, Chihuahua, Coahuila, Tamaulipas, Jalisco, Sonora, Guanajuato, State of Mexico, Querétaro and Puebla. Regarding the employed personnel, 82.6% is represented by the same previous 11 states, although holding different positions. In the indicator of hours worked in the companies, 7 states represent 85.5%: Durango, Michoacán, Tamaulipas, Chihuahua, Baja California, Guanajuato and Sonora (Instituto Nacional de Estadística y Geografía, 2016).

The largest number of IMMEX establishments is located in the northern border region of the country, as a result of this, there is a greater number of hours worked and personnel employed, than in the rest of the states. The above, together with the prevalence of obesity in adults 20 years of age or older in this area, (Encuesta Nacional de Salud y Nutrición de Medio Camino, 2016). It makes evident the need to initiate research in the State of Baja California, mainly in the cities of Tijuana and Mexicali.

The aim of the study is to find out the impact of obesity and the burnout syndrome on the IMMEX of Baja California, as well as the existence of positive or negative relationships between them. The staff under study are middle and upper level managers as research results show that the most stress-generating type of work is that where demands and pressures exceed the worker's knowledge and abilities, opportunities to make decisions or exercise control are scarce, and there is little support from others (Leka, Griffiths & Cox, 2004). The profile of middle managers in industries meets the above characteristics. Holding a college degree gives them the opportunity to start at a medium level and not as operators; thus, the lack of operational experience, the pressures of senior managers and the position's high demands can become a source of stress. On the other hand, they are limited and sometimes hindered in their ability to make decisions; a different scenario would help increase their self-esteem and self-value. Furthermore, the role of senior managers entails constantly facing activities aimed at achieving the company's objectives and having to make decisions that can affect the survival or the disappearance of the company. Thus, the responsibility of the upper management is to be effective and to stay that way (Salgado & Mejía, 2008). Yet having to maintain that effectiveness is precisely what triggers the stress process and turns it into an ever increasing threat (Bittel & Ramsey, 1988) both to the individual and the company as the success of companies is based, largely, on its executives' capacity, ability and well-being.

METHODOLOGY

This section outlines the methodology used in this project, which essentially consisted of five stages: description of the sample, validation of the instruments of measurement, statistics of the sector under study, analysis of difference of means and analysis of association of variables. Each stage contains a detailed description of the methods of analysis used and the partial results obtained.

Phase One: Description of the Sample

A cross-sectional, observational study was conducted along the Mexico-United State border area in the state of Baja California, particularly in the cities of Tijuana and Mexicali. Through the use of a non-probabilistic sampling for convenience (Alaminos & Castejón, 2006), 121 surveys were administered to employees in middle and upper management in the manufacturing industry and export services in seven different companies. Out of the total sample, 42 surveys belonged to the city of Tijuana, with 33 men and 9 women; on the other hand, 79 belonged to the city of Mexicali, with 42 men and 37 women. The ages of the participants ranged from 20 to 60 years with an average value of 35.11 years and a standard deviation of 10.2 years. The respondents mentioned a time of seniority in the company ranging from 1 month to 35 years of service. Figures 1 to 4 show the descriptive statistics of the collected data classified into the respective categories.

Figure 1. Percentages by type of position (Elaborated by the authors)

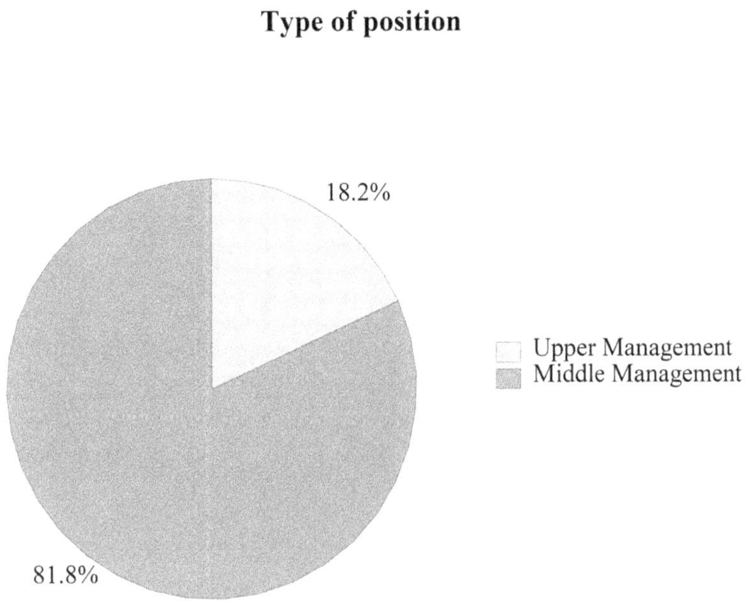

Figure 2. Percentages by type of department (Elaborated by the authors)

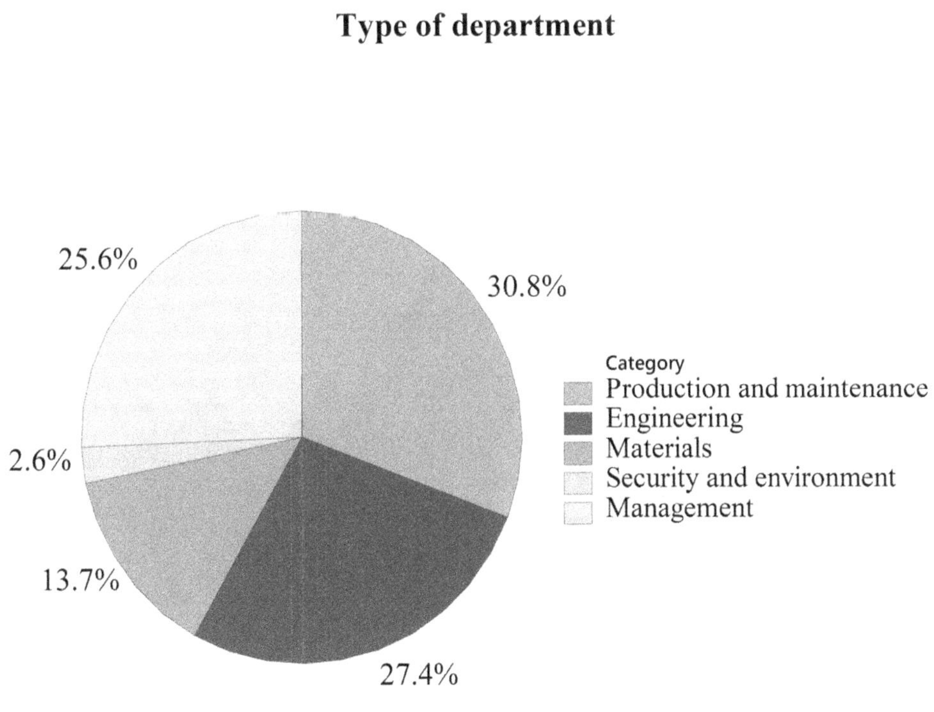

Figure 3. Percentages by working hours (Elaborated by the authors)

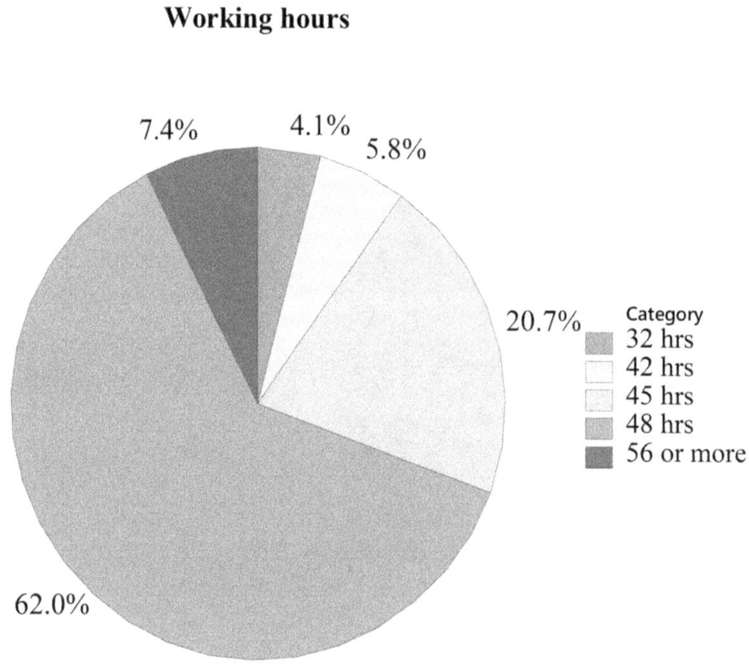

Figure 4. Percentages by type of contract (Elaborated by the authors)

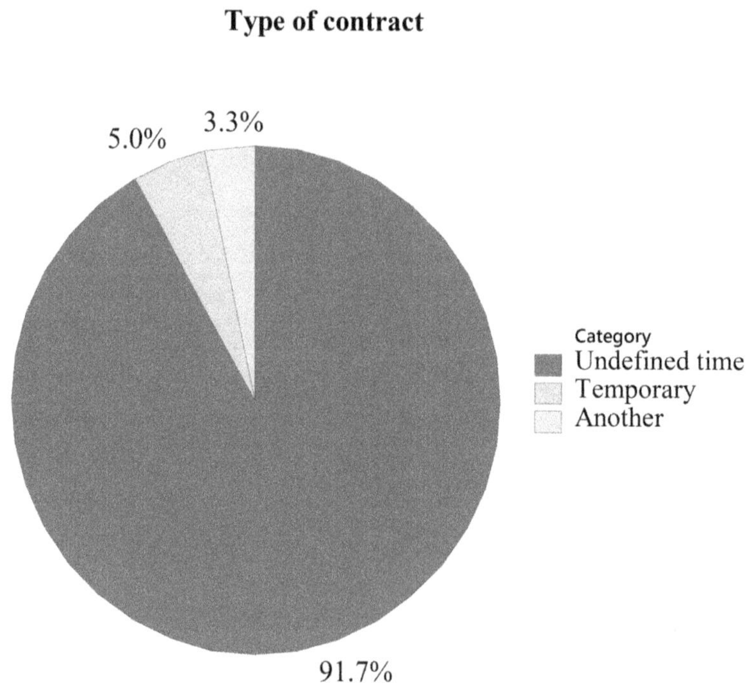

Phase Two: Validation of Instruments of Measurement

To analyze the two thematic variables, occupational stress and obesity, a self-report questionnaire was provided in physical form to each worker, which included a brief introduction explaining the objectives pursued and emphasizing the information's anonymous character as well as the consent required to collect it. The questionnaire was divided into three sections: the first included an introduction and the employee's general data, accompanied by the signature of consent. The second section consisted of the self-report questionnaire called Maslach Burnout Inventory General Survey (MBI-GS), prepared by Maslach, Jackson and Leiter in 1996. This study used the version translated by Moreno, Rodríguez & Escobar (2001), with highly satisfactory reliability indices, which oscillated between 0.84 to 0.89 for all the three dimensions. This instrument is generic as it was addressed to all personnel regardless of their profession. It consisted of a total of 16 items assessing job strain through three dimensions (emotional wear, with 5 items; cynicism, with 5 items; and professional efficiency, with 6 items). The response scale was Likert type of 7 points. The output variable was a worker characterized in one of the five available levels of work stress: nothing, low, medium, high and extreme. The last section included weight and height measurements for the analysis of the obesity variable. To classify each individual into a type obesity, the ratio of Body Mass Index (BMI) established by the World Health Organization (2006) was used, which estimates the degree of obesity by means of the quotient of its weight in kilograms divided by its height squared and expressed in meters (Welborn, Dhaliwal & Bennett, 2003). The convenience of using this method lay in that it was proposed by a highly recognized world organization and is currently the most accepted criterion used in America and Europe (Guzmán, Del Castillo, & García, 2010) to classify individuals according to their weight. The categories are shown in Table 1 for people over 20 years of age.

In order to guarantee the formal validity of the Maslach Burnout Inventory General Survey (MBI-GS), which all questionnaires must show before being used (Cook & Beckman, 2006), an internal consistency analysis was conducted using the Cronbach's α method, with the purpose of finding out the variables of interest in the specific sample and geographical area under study. The results show that the instrument has the necessary validity (Huh, Delorme & Reid, 2006) for data collection in samples of personnel in middle and upper management in the industrial sector of the Baja California Norte border area, featuring an α value of 0.892 for the dimension of emotional exhaustion, 0.816 for professional efficiency, 0.847 for cynicism, and 0.814 for the instrument as a whole since alpha values greater than 0.80 are considered acceptable.

Table 1. Classification of BMI (Secretaría de Salud, 2013)

Category	BMI (Kg/m^2)
Underweight	<18.5
Normal weight	18.5 – 24.99
Overweight	>25
Class I Obesity	30 – 34.99
Class II Obesity	35 – 39.99
Class III Obesity	> 40

Each worker's weight was measured with a TAYLOR brand glass digital scale. To ensure the reliability of the scale, a cross-method reproducibility and replicability study was conducted which focused mainly on determining what part of the variation observed in the process was due to the measurement system used (Portuondo & Portuondo, 2010). The data collection procedure took place as follows:

- Three operators were responsible for taking the measurements.
- A sample of ten people was selected, representing the range of expected value with respect to its weight.
- The three operators measured the ten people the same number of times.
- All measurements were taken with the same instrument, in this case the scale.
- To prevent the order of measurement from altering the results, each operator measured each of the ten people in random order within a replica.
- The procedure was repeated until three replicas are obtained.

The results of the study described above, indicate that the variation of the measurements of the sample was 99.9%, so it is concluded that the difference is due to the weight characteristics of each person and not to the error of measurement by the operator. The variation provided by the device, in this case the scale, is 0.1% repeatability, in addition the system is able to distinguish between 44 different categories, so it is possible to affirm that the instrument is valid (Automotive Industry Action Group, 2010) to perform taking measurements in the population studied.

The depuration and analysis of the weight variable was performed. The database, consisting of 121 weight measurements collected, showed 1 atypical piece of data and 3 missing pieces of data, a reason why it was necessary to replace the missing values and treat the atypical data according to statistical guidelines in order to perform future analyses. Considering that the variable was quantitative, the substitution of missing data was done by means of the average method of two adjacent numbers (Nakai, Chen, Nishimura & Miyamoto, 2014), which was considered the most convenient for this case since the data generated gave continuity to the series of values without altering it. Likewise, the missing height data were replaced by the method of the mean of two adjacent data.

Phase Three: Analysis of the Sector Under Study

Based on the statistics obtained regarding the variables of interest and the participating staff, the following graphs were prepared, where information was organized by gender and city.

Workers were classified into 6 levels using the BMI to determine their degree of nutrition. Thirty-two point eight percent (32.8%) of the total sample was found to have a normal weight, with a higher percentage, 44.3%, suffering from overweight, 19% showing class I obesity, 2.5% featuring class II obesity and 0.8% under te category of class III obesity.

Reagrding job strain, 23% of the total sample of workers fell under the category of low level, 23.8% to the medium level, 19.7% to the quite high level, 13.1% to the extreme level and 19.7 did not show signs of suffering from any kind of job strain.

Figure 5 shows nutrition levels in relation to gender. It can be observed obesity class I occured in 20.83% of men and in 15.21% of women; class II obesity, in 1.38% of men and 2.17% of women; and class III obesity, was not present in both men and women. Overweight had a higher percentage in men,

Figure 5. Overweight category by gender (Elaborated by the authors)

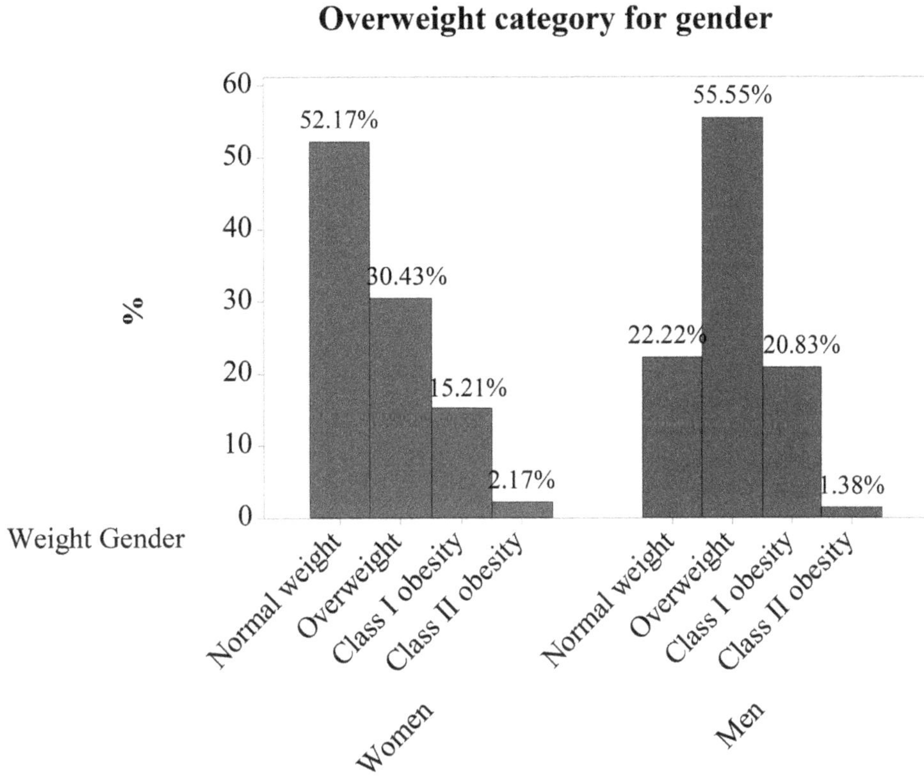

with 55.55%, than in women, with 30.43%. Finally, 22.22% of men are within their normal weight, while 52.17% of women maintain their weight.

The classification of the level of nutrition by cities is featured in Figure 6. For Mexicali, the results were as follows: Featuring the highest percentage, overweight was present in 44.15% of the sample, class I obesity in 15.58%, and class II obesity in 2.59%, while 37.66% featured a normal weight. For the city of Tijuana, the following results were obtained: 48.78% were overweight, 24.39% had class I obesity and 26.82% remained at their normal weight.

Regarding the job strain variable, the percentages that feature this condition are shown by gender in Figure 7. In males, 25% of individuals have a little level of job strain, 26.38% maintain a medium level, 16.66% falls under the category of enough, 11.11% have an extreme level and 20.83% do not have this condition. In females, 21.73% have a little level, 19.56% have a medium level, 26.08% have the condition of enough, 15.21% have this condition at an extreme level and 17.39% of women do not suffer from job strain.

Regarding cities and their level of job strain, the following results are shown in Figure 8: for the city of Tijuana 24.39% of workers have a little level, 31.70% maintain a medium level, 17.07% have the condition of enough, only 7.31% has an extreme level and 19.51% of workers do not have this condition. As for Mexicali, 23.37% presents a little level, 19.48% an medium level, 22.07% a enough level, 15.58% an extreme level, and 19.48% of workers do not have this condition.

Figure 6. Overweight category by city (Elaborated by the authors)

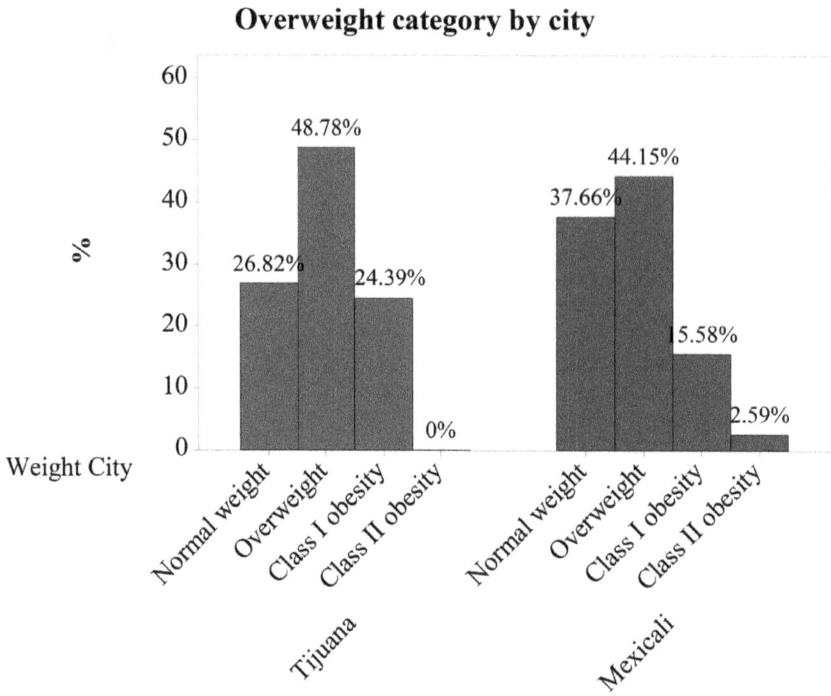

Figure 7. Burnout category by gender (Elaborated by the authors)

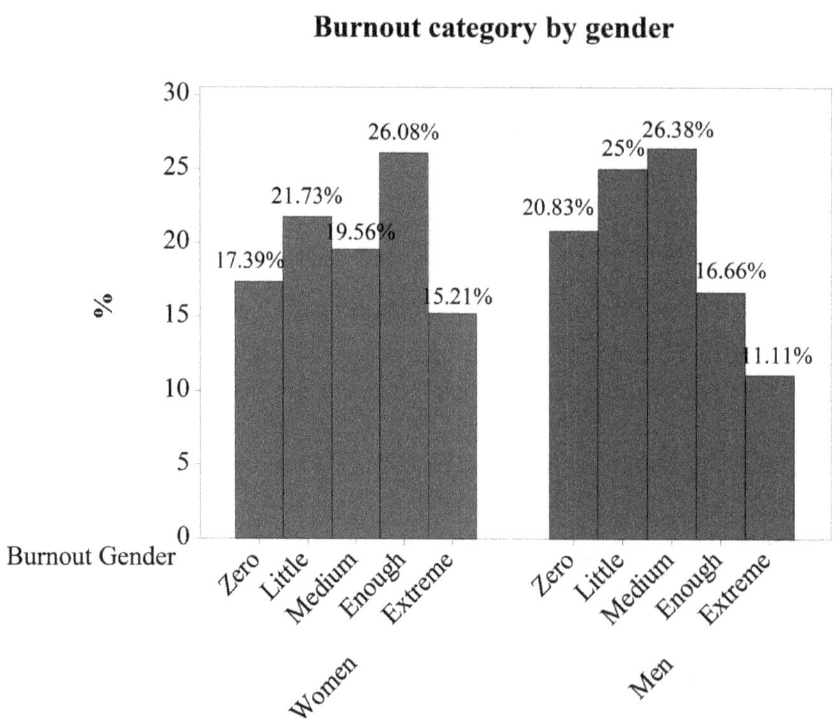

Figure 8. Burnout category by city (Elaborated by the authors)

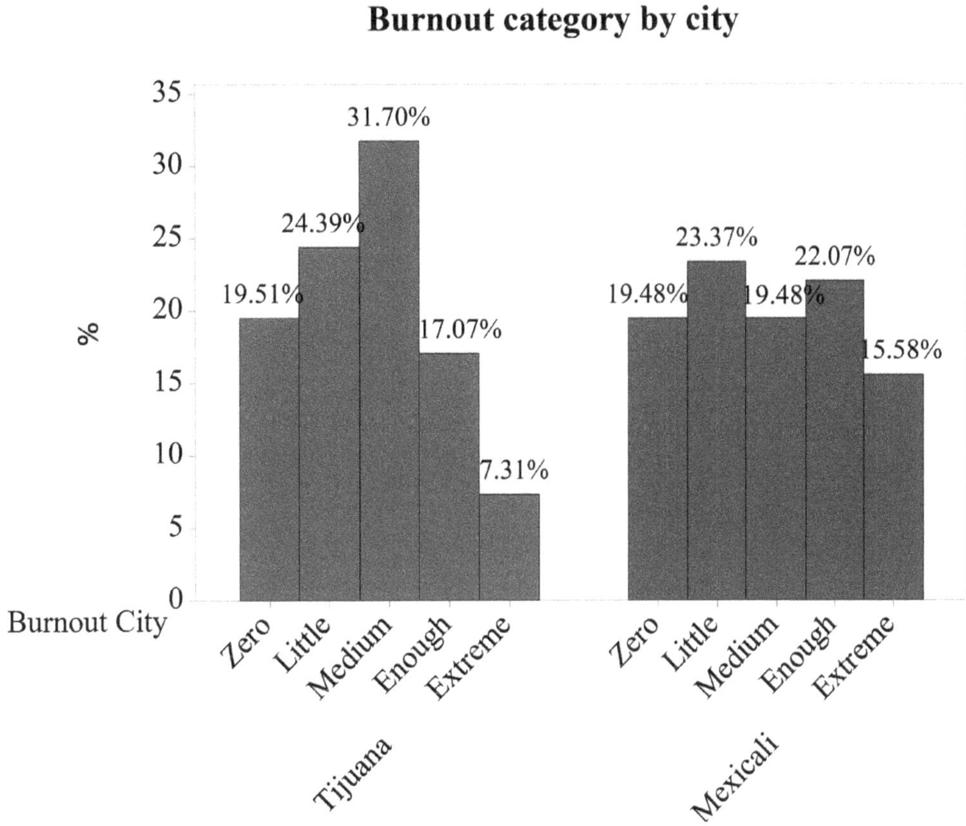

Phase Four: Difference of Means in Relation to Weight

The procedure of difference of means was carried out in order to know whether there were differences between the weight men and women, who occupy middle and upper management positions in the manufacturing industry of the cities of Tijuana and Mexicali. For this analysis, the data had to comply with the assumption of normality (Pearson, 1931; Barlett, 1935; Geary, 1947) and equality of variances, which becomes critical when the sample sizes are noticeably different in each group (Welch, 1938). The analysis was conducted from the 118 pieces of data collected in the two cities.

First, the validity of the assumption of normality for weight in the city of Tijuana was establlished; the test used was that of Ryan Joiner with n = 41; this test is similar to the Shapiro Wilk test, suitable for sample sizes <50. The results showed that the data follows a normal distribution with a p-value of 0.100 (Figure 9).

Normality in the weights in the city of Mexicali was determined with the Kolmogorov-Smirnov test considering n = 77. The results showed that it is possible to accept the null hypothesis with a confidence interval of 95% with a p-value of 0.150 shown in Figure 10; confirming the normality of the data.

For the assumption of equality of variances the results show that there is no significant difference between the groups because in the test through the Bonett's method there is a p-value of 0.060, and through the Levene's method the p-value is 0.069; both values are greater than the significance level

Figure 9. Normality test for Tijuana data (Elaborated by the authors)

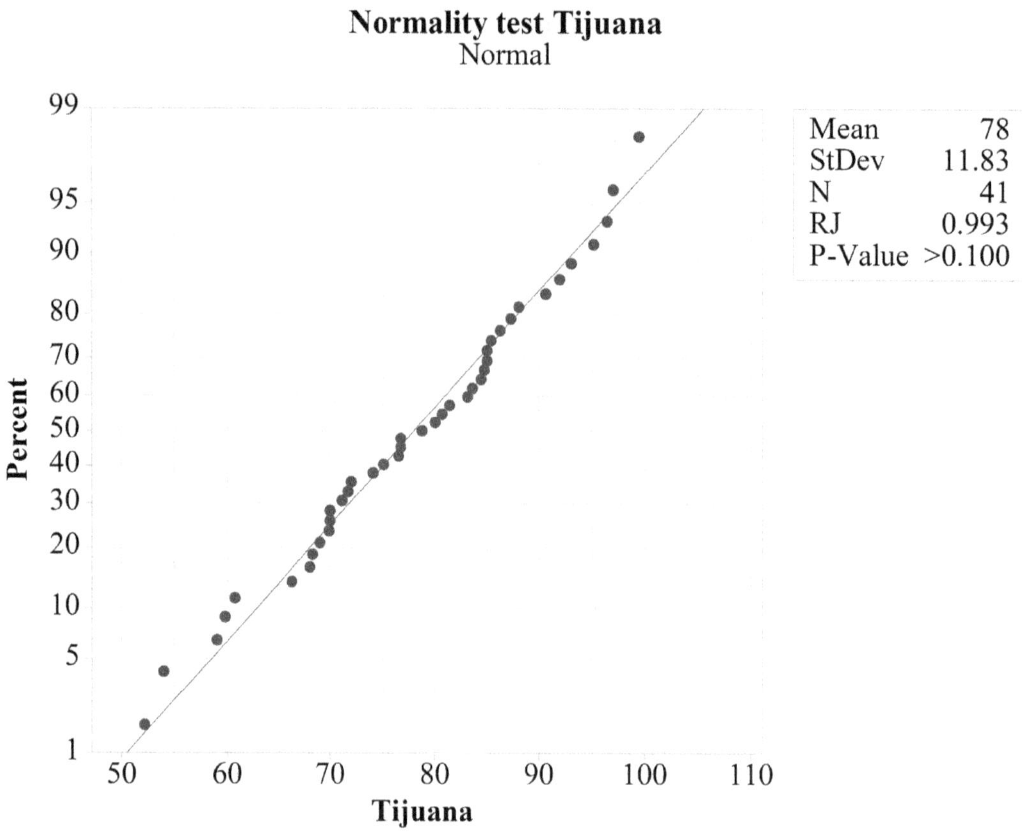

Figure 10. Normality test for Mexicali data (Elaborated by the authors)

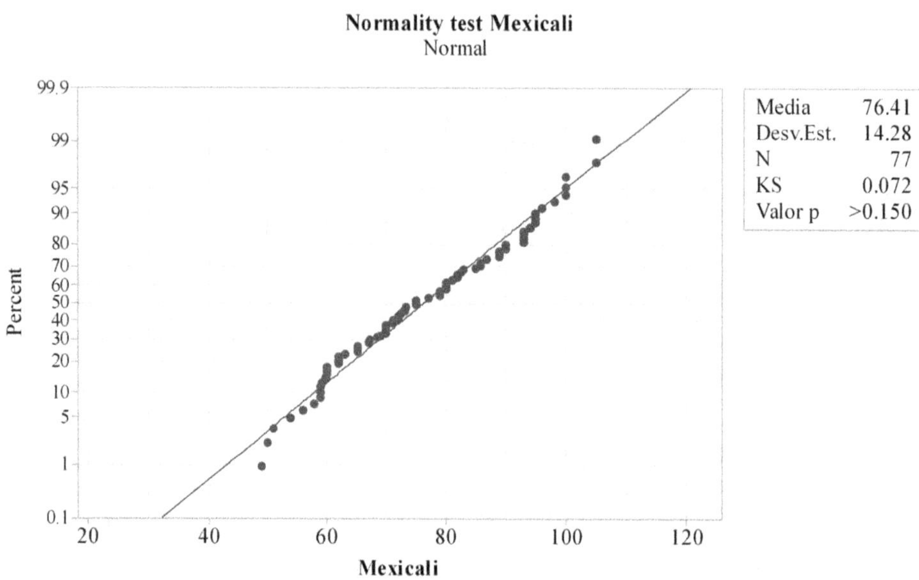

of 0.05; In addition, the group intervals overlap, with values of 9.914 to 14.828 for the group of people from Tijuana; and from 13.219 to 16.761 for the people in Mexicali, thus confirming the assumption of equality of variances for both groups. The pieces of data used in latter analyses were 118.

A comparison was drawn between the weight of the women, who occupy middle and upper management positionsin the manufacturing industry, of the two selected cities, with a confidence interval of 95%. The results are shown in Table 2 and Figure 11. The p-value in this case was of 0.935, so it is possible to accept the null hypothesis and consider that there are no differences in women's weights due to their working in different cities.

In the case of men occupying these positions, the same comparison was made. The results show that the difference in means oscillates between -8.60 and 1.81, with a confidence of 95%, see Table 3 and Figure 12. In this case there is no statistical evidence to reject the null hypothesis, with a p-value of 0.198. It is considered that there is no difference between the weight of men in the city of Mexicali and the weight of men in the city of Tijuana.

Table 2. Difference of means por the weight of women

Weight of Women	N	Mean	StDev	SE Mean
Tijuana	9	67.0	10.2	3.4
Mexicali	35	66.7	10.8	1.8
Difference = μ (Tijuana) – μ (Mexicali)				
Estimate for difference: 0.32				
95% CI for difference: (-8.11, 8.75)				
T-test of difference = 0(vs≠): T-Value = 0.08 P-Value=0.935 DF=12				

Figure 11. Boxplot of the women's weight in middle and upper management (Elaborated by the authors)

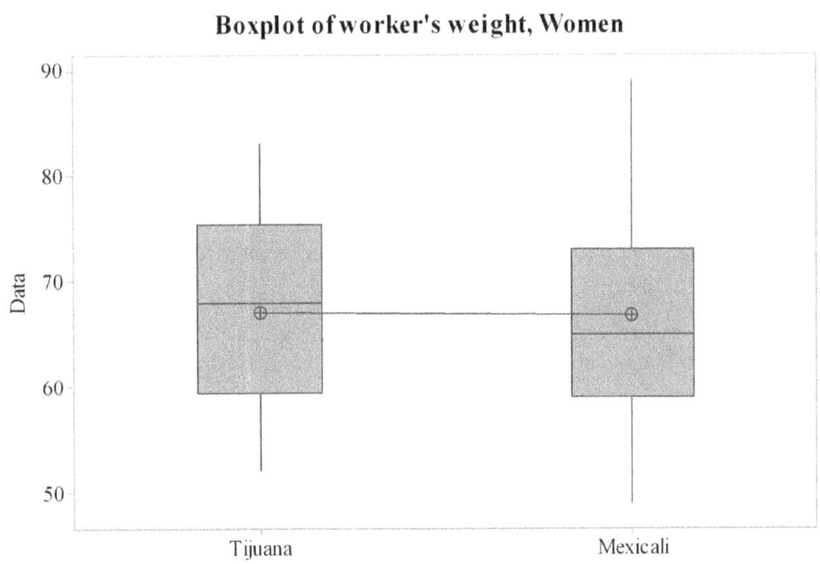

Table 3. Difference of means por the weight of men

Weight of Men	N	Mean	StDev	SE Mean
Tijuana	32	81.1	10.4	1.8
Mexicali	42	84.5	11.6	1.8
Difference = μ (Tijuana) – μ (Mexicali)				
Estimate for difference: -3.39				
95% CI for difference: (-8.60, 1.81)				
T-test of difference = 0(vs≠): T-Value = -1.30 P-Value=0.198 DF=72				

Figure 12. Boxplot of the men´s weight in middle and upper management (Elaborated by the authors)

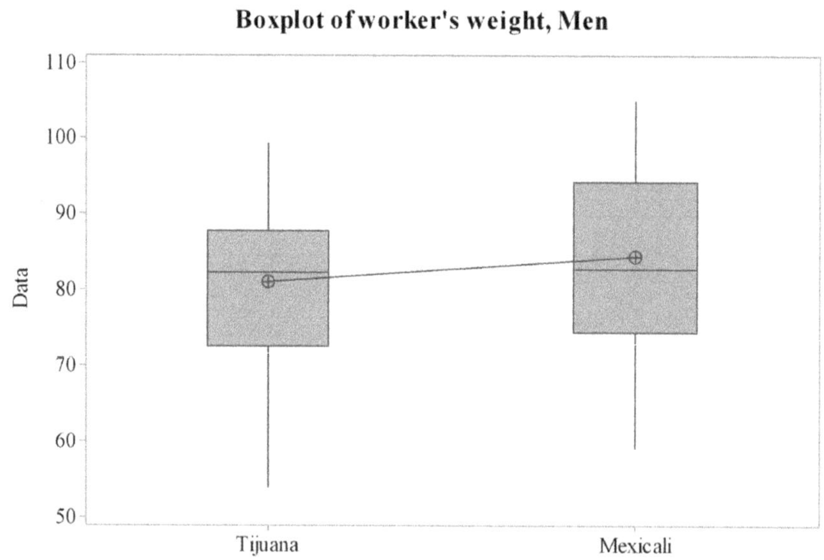

Phase Five: Analysis of Association of Variables

With the aim of analyzing whether there is a relationship between the weight of the personnel that occupies middle and high level management positions and the level of stress they experience, an ordinal regression analysis was carried out based on the information obtained in 74 men and 44 women. For this, the software minitab 17 was used and the model adjustment was first made considering the information referring to men. The results obtained are shown in Table 4.

Additionally, a p-value = 0.078 was obtained for the test that all the coefficients are equal to zero. With the above information it can be said that the weight of men in this positions is related to the level of stress they experience, for a confidence level a little higher than 90%.

In relation to women, the same adjustment was made and the results obtained are shown in Table 5.

For these data, a p-value = 0.496 was obtained for the test that all the coefficients are equal to zero. With the above information it can be said that in the case of women, the weight is not related to the level of stress they experience.

Table 4. Parameters obtained in the analysis of ordinal regression for men

Predictor	Coeficient	P - Value
Constante (1)	-12.199	0.029
Constante (2)	-11.041	0.047
Constante (3)	-9.913	0.074
Constante (4)	-8.805	0.110
Weight	0.250	0.055
Weight * Weight	-0.001	0.050

Table 5. Parameters obtained in the ordinal regression analysis for women

Predictor	Coeficient	P – Value
Constante (1)	-5.605	0.591
Constante (2)	-4.461	0.669
Constante (3)	-3.734	0.720
Constante (4)	-2.229	0.830
Weight	0.091	0.766
Weight * Weight	-0.000	0.842

SOLUTIONS AND RECOMMENDATIONS

Based on the results, it is possible to conclude that at least in a preliminary way that there is a significant relationship between the weight of the personnel with middle and high level management positions and the level of stress they experience. In the case of women, this relationship was not present; however, the authors consider that this was due to the sample size that is not sufficient to detect the relationship. Therefore, it is proposed to explore this relationship, but using larger sample sizes in both cases, either to discard or confirm this relationship.

CONCLUSION AND FUTURE RESEARCH DIRECTIONS

According to the World Health Organization (2016), Mexico is in the first places, along with China and the United States, with the largest number of people stressed worldwide. The performance of the workforce is impaired due to inadequate workspaces and lack of work benefits that encourage frequent medical review, which is why most of the chronic diseases that Mexican society suffers today are triggered by pressure and the work stress that is exposed for years.

Some of the illnesses unleashed by stress that most afflict the health and economy of Mexicans are: depression and anxiety, gastrointestinal diseases, chronic fatigue, obesity and overweight, the latter caused by sedentary lifestyle and poor diet, which together they affect 71.3% of the adult population, which positions Mexico as the first country in the world on this issue, according to the Food and Agriculture Organization of the United Nations (FAO). According to the sample selected for the study presented, the middle and upper management of the manufacturing industry of Baja California, maintain these percentages, having 73.17% of people with this condition, while in Mexicali it is 62.32.

In the sample, obesity was higher in men (22.21%) than in women (17.38%), as well as in overweight, where there was a prevalence of 55.55% in men, while in women there was a prevalence of 30.43%, below what is reported in adults of the Americas region, considered as the highest population in these conditions, which presents 62% of overweight and 26% of obesity in both sexes.

Even though the results show no difference between the weights of those in the middle and upper management of both cities, it is necessary to point out that women who occupy these positions perceive a higher and more extreme level of stress (41.49%) than men (27.77%).

For the above, it is considered of utmost importance to continue with the study to generate a model that explains the relationship that exists between the variables, which will allow the development of strategies to improve the health of employees by reducing work stress and its consequences, what will positively impact the quality of life of people and the productivity of companies.

REFERENCES

Alaminos, A., & Castejón, J. L. (2006). *Elaboración, análisis e interpretación de encuestas, cuestionarios y escalas de opinión.* Universidad de Alicante. Retrieved January 15, 2018 from https://rua.ua.es/dspace/bitstream/10045/20331/1/Elaboraci%C3%B3n,%20an%C3%A1lisis%20e%20interpretaci%C3%B3n.pdf

American Institute of Stress (AIS). (2016). *Workplace Stress.* Retrieved December 20, 2017 from http://www.stress.org/workplace-stress/

Aranda, C., Pando, M., & Salazar, J. G. (2015). Síndrome de burnout en trabajadores de diversas actividades económicas en México. *Revista iberoamericana de psicología: ciencia y tecnología, 8*(2), 23-28.

Automotive Industry Action Group (AIAG). (2010). *Measurement Systems Analysis Reference Manual* (4th ed.). Chrysler Ford, General Motors Supplier Quality Requirements Task Force.

Bartlett, M. (1935). The Effect of Non-Normality on the t Distribution. *Mathematical Proceedings of the Cambridge Philosophical Society, 31*(2), 223–231. doi:10.1017/S0305004100013311

Bittel, L., & Ramsey, J. (1988). Enciclopedia del Management: Organización y Administración de Empresas (Vol. 4). Madrid: Academic Press.

Consejo Nacional de la Industria Maquiladora y Manufacturera de Exportación Index. (2016). *¿Cómo nace la Industria Maquiladora de exportación?* Retrieved from: http://www.index.org.mx/IMMEX/antecedentes.php

Cook, D. A., & Beckman, T. J. (2006). Current Concepts in Validity and Reliability for Psychometric Instruments: Theory and Application. *The American Journal of Medicine, 119*(2), 166.e7–166.e16. doi:10.1016/j.amjmed.2005.10.036 PMID:16443422

Gamero, C. (2010). Evaluación del coste por pérdida de jornadas laborales asociado al estrés laboral: Propuesta para España. *Estudios de Economía Aplicada, 28*(3), 1–20.

Geary, R. C. (1947). *Testing for Normality. Biometrika.* Retrieved from https://pdfs.semanticscholar.org/2114/43e0d4bb0bcb6b200c263e250f5591a18458.pdf

Gil Monte, P. (2008). El Síndrome de quemarse por el trabajo (burnout) cómo fenómeno transcultural. *Informació Psicològica*, *4*(11), 91–92. Retrieved from https://dialnet.unirioja.es/servlet/articulo?codigo=2551783

Gil-Monte, P., & Peiro, J. (1999). Perspectivas teóricas y modelos interpretativos para el estudio del síndrome de quemarse por el trabajo. *Anuales de Psicología, 15*(2), 261-268. Retrieved from http://www.um.es/analesps/v15/v15_2pdf/12v98_05Llag2.PDF

Gonzalez, M., & Landero, R. (2008). Confirmación de un modelo explicativo del estrés y de los síntomas psicosomáticos mediante ecuaciones estructurales. *Pan Am J Public Health*, *23*(1), 7–18. doi:10.1590/S1020-49892008000100002

Guzmán, M. E., Del Castillo, A., & García, M. (2010). Factores psicosociales asociados al paciente con obesidad. In *Universidad Autónoma del Estado de Hidalgo. Obesidad: Un enfoque multidisciplinario.* Madrid: Colección Real, Museo Nacional del Prado.

Huh, J., Delorme, D. E., & Reid, L. N. (2006). Perceived third-person effects and consumer attitudes on prevetting and banning DTC advesiting. *The Journal of Consumer Affairs*, *40*(1), 90–116. doi:10.1111/j.1745-6606.2006.00047.x

IMSS advierte sobre los riesgos del estrés laboral. (2015). *Noticias Tijuana.* Retrieved from http://tijuananoticias.info/bc/imss-advierte-sobre-los-riesgos-del-estres-laboral/

Instituto Nacional de Estadística y Geografía (INEGI). (2016). *Banco de información económica, Manufacturas.* Retrieved from http://www.inegi.org.mx/sistemas/bie/

Kang, M., Koh, S., Cha, B. S., Park, J. K., Woo, J. M., & Chang, S. J. (2004). Association between job stress on heart rate variability and metabolic syndrome in shipyard male workers. *Yonsei Medical Journal*, *45*(5), 838–846. doi:10.3349/ymj.2004.45.5.838 PMID:15515194

Ksiazek, I., Stefaniak, T. J., Stadnyk, M., & Ksiazek, J. (2011). Burnout syndrome in surgical oncology and general surgery nurses: A cross-sectional study. *European Journal of Oncology Nursing*, *15*(4), 347–350. doi:10.1016/j.ejon.2010.09.002 PMID:20951089

Lazarus, R., & Folkman, S. (1984). *Stress, Appraissal and Coping.* New York: Springer. Retrieved from https://books.google.com.mx/books?hl=es&lr=&id=i-ySQQuUpr8C&oi=fnd&pg=PR5&dq=Lazarus+R.S.+%26+Folkman,+S.+(1984).+Stress,+Appraisal+and+Coping.+cita&ots=DeJQjrciQf&sig=VDJ2kDOeTPHWbRlELA9nClz36oU#v=onepage&q&f=false

Leka, S., Griffiths, A., & Cox, T. (2004). Estrategias sistemáticas de solución de problemas para empleadores, personal directivo y representantes sindicales. Institute of Work, Health & Organisations, (3), 5.

Maslach, C. (2003). Burnout the cost of caring. Malor Books.

México, Instituto Nacional de Salud Pública. (2016). *Encuesta Nacional de Salud y Nutrición de Medio Camino (ENSANUT MC) Informe final de resultados.* Retrieved from https://www.gob.mx/cms/uploads/attachment/file/209093/ENSANUT.pdf

México, primer lugar en estrés laboral: OMS. (2015). *Diario el Independiente.* Retrieved from https://www.diarioelindependiente.mx/2015/05/mexico-primer-lugar-en-estres-laboral-oms

Moreno Jiménez, B., Rodríguez Carvajal, R., & Escobar Redonda, E. (2001). La evaluación del burnout profesional. Factorialización del MBI-GS. Un análisis preliminar. *Ansiedad y Estrés*, *7*(1), 69–78. Retrieved from https://www.researchgate.net/publication/230596331_La_Evaluacion_del_Burnout_profesional_Factorializacion_del_MBI-GS_Un_analisis_preliminar

Nakai, M., Chen, D., Nishimura, K., & Miyamoto, Y. (2014). Comparative study of four methods in missing value imputations under missing completely at random mechanism. *Open Journal of Statistics*, *4*(01), 27–37. doi:10.4236/ojs.2014.41004

National Health and Nutrition Survey. (2012). *Resultados Nacionales.* Cuernavaca, Mexico: Instituto Nacional de Salud Pública. Retrieved from http://ensanut.insp.mx/informes/BajaCalifornia-OCT.pdf

Oficial Mexicana, N. O. R. M. A. (2012). *Servicios básicos de salud. Promoción y educación para la salud en materia alimentaria. Criterios para brindar orientación* (NOM-043-SSA2-2012). Ciudad de México. Retrieved from http://dof.gob.mx/nota_detalle.php?codigo=5285372&fecha=22/01/2013

Organización Internacional del Trabajo (OIT). (2016). *Estrés en el trabajo, Día mundial de la seguridad y la salud en el trabajo 28 de abril.* Retrieved from file:///C:/Users/end%20user/Desktop/Papers%20readed/Estres%20en%20el%20trabajo%20OIT.pdf

Pearson, E. S. (1931). The Analysis of variance in case of non-normal variation. *Biometrika*, *23*(1-2), 114–133. doi:10.1093/biomet/23.1-2.114

Portuondo, Y., & Portuondo, J. (2011). La repetibilidad y reproducibilidad en el aseguramiento de la calidad de los procesos de medición. *Tecnología Química, 30*(2), 117-121. Retrieved from http://www.redalyc.org/articulo.oa?id=445543770014

Salgado, P., & Mejía, R. (2008). Estrés en ejecutivos de medianas y grandes empresas mexicanas: Un enfoque de desarrollo humano organizacional. *Estudios Gerenciales*, *24*(108), 15–36. doi:10.1016/S0123-5923(08)70042-1

Secretaria de salud (SS). (2013). *Servicios básicos de salud. Promoción y educación para la salud en materia alimentaria. Criterios para brindar orientación* (NOM-043-SSA2-2012). Diario Oficial de la Federación.

Welborn, T. A., Dhaliwal, S. S., & Bennett, S. A. (2003). Waist-hip ratio is the dominant risk factor predicting cardiovascular death in Australia. *The Medical Journal of Australia*, *179*, 580–585. PMID:14636121

Welch, B. L. (1938). The significance of the difference between two means when the population variances are unequal. *Biometrika*, *29*(3-4), 350–362. doi:10.1093/biomet/29.3-4.350

World Health Organization (WHO). (2006). *Prevención del suicidio. Un instrumento en el trabajo. Departamento de Salud Mental y Abuso de Sustancias*. Manejo de Trastornos Mentales y Cerebrales.

World Health Organization (WHO). (2012). *Estadísticas sanitarias mundiales*. Retrieved from http://www.who.int/gho/publications/world_health_statistics/2012/es/

World Health Organization (WHO). (2016). *Obesidad y Sobrepeso, datos y cifras*. Retrieved from http://www.who.int/mediacentre/factsheets/fs311/es/

Chapter 9
A Framework Designed for Macro-Ergonomical Analysis of Indian Farmers:
Assessment and Analysis of Occupational Injuries of Agricultural Farmers of South Odisha in India

Debesh Mishra
KIIT University, India

Suchismita Satapathy
KIIT University, India

ABSTRACT

Farming provides food, which is the primary need of each and every person, and also provides employment to farmers. Still it is a non-profitable and neglected occupation. Maximum health injuries are observed during field work due to extreme climates and dusty atmosphere. Use of old hand tools and no availability of modernized tools are the cause of 70% of injuries. Although the farming sector plays a vital role on Indian economy, it remains undeveloped and unattractive. The Indian government is also framing many policies for the farmers (i.e., small/nominal farmers), but farming in states like Odisha remains undeveloped. So, in this chapter, a survey is conducted to find the health-related injuries of farmers of Odisha and then a framework is designed by QFD (quality function deployment) to suggest how to avoid injuries and provide occupational safety measures for farmers.

DOI: 10.4018/978-1-5225-7192-6.ch009

INTRODUCTION

Farming sector is a very important but neglected sector all over the world. Day by day man prefers mechanized and comfortable life. So human-civilization is more and more attracted towards the industrial sector. Farming business is an ancient traditional business, but still it is a non-profitable business sector. Occupational safety is a big issue of discussion for agricultural workers. The methods of working in field in extreme climate (heat, rain), the contact with the chemicals (pesticides, fertilizers), the exposure to soil, dust, the contamination due to bacteria, the exposure to animals, cattle's, injury due to hand tools and musculoskeletal lateral disorders are the most important injuries faced by all agri-workers. Agricultural workers need sufficient precaution and safety measures at the time of field and machine work, such that no physical damage occurs to them. The fatality rate in agriculture is far higher than any other economic sector. A large proportion of all fatal workplace accidents occur in agriculture, even though a small proportion of the workforce is employed in farming. The level of farm accidents is not decreasing. Similar accidents occur each year (Hope, Kelleher, Holmes, & Hennessy, 1999). Most of the agricultural injuries are resulted from the improper selection and use of hand tools. In agricultural sector the traditional hand tools play a major role in performing the farming activities. The conventional hand tools like spade/hoe, sickle, hammer, shovel, knife among others have been used since the ancient though some modifications are found nowadays. As most of the farmers in Odisha, India, are from poor economic background, they usually prefer the conventional hand tools in farming instead of using the developed power operated machineries. The hand tools are mostly used in all farming activities like land preparation, weeding, harvesting of crops for mentioning some. The lack of ergonomic considerations in the design of hand tools and equipment, lack of expertise, lack of the knowledge of safety measures and negligence of farmers and adverse environment may lead to agricultural accidents or injuries (Mishra & Satapathy, 2018). Murphy (1992) has defined farming as the occupation which depends on the farming awareness skill of individual and the capability to do complicated and repetitive jobs. Voaklander et al. (2006) have suggested having all the skills as defined by Murphy (1992), enhance the ability and performance of individuals and also the safety level in the working place. Further it is stated that absence of any of the skill can result in farming injury. Karthikeyan, Veeraragavathatham, Karpagam, and Firdouse (2009) have reported that most of the conventional tools and equipment used in the agricultural sector were manufactured locally by the available materials such as wood, iron or stone, respectively. Moreover, after these nominal farmers of Odisha, the large agri investors also faces several types of respirotary problems due to gas of large machinaries. So India's Government should frame policies against accidents, frame traing programme and take sufficient safety measures such that injuries of farming sector can be avoided. Advancement of technology to modify tool design can also play an important role to avoid occupational injury for farmers. From the literature review, it was found that several studies have been carried out about the farmers' injuries across different regions in the world; however still no such study has been done in the state of Odisha particularly. Hence this research presents an attempt that was made to study, collect information and analyze the farmers' injuries in the South Odisha region in India.

AGRICULTURAL INJURIES AND ACCIDENTS IN INDIA

The literature on agricultural injuries and accidents in India is limited to a few. Several authors and researchers have reported the rate of agricultural accidents as higher than the industrial sectors. The lack of infrastructural facility, medical facility, training and rare use & maintenance of machines as well as the lack of defining the work according to gender or age are the causes of agricultural injuries (Knapp, 1965, 1966). Mohan and Patel (1992) have reported 576 agricultural related injuries in 1 year in 9 villages of Haryana in India. 87% injuries were reported as minor, 11% as moderate and remaining as severe. It is found that most of the injuries were due to hand tools, i.e. 24% by spade, 23% by sickles, 6% by bullock carts, 6% by manually operated fodder cutters, 5% each by power operated fodder cutters, diesel engines and tractors. Zhou and Roseman,(1994) found both of the limbs as the most injured body part for 1000 farmers in Alabama by agricultural accidents. Mittal, Bhatia, and Ahuja (1996) have reported 36 agricultural injuries, of which 8.3% as fatal and 91.7% as non-fatal, in 12 villages of Punjab in India during 1 year. The injury rate per thousand machines per year was found to be 23.7 for tractors, 15.5 for sprayers, 7.1 for electric motors, 5.7 for threshers and 2.2 for fodder cutters. DeMuri and Purschwitz (2000) have reported that the agricultural injuries are due to the lack of proper supervision, irrational expectations, economic difficulty and the lack of proper safety device. Xiang et al. (2000) have considered 1500 Chinese farmers from 14 villages and conducted face to face interview of 1358 farmers between July 1997 and September 1997, to evaluate the agricultural related injuries. Rautiainen and Reynolds (2002) have carried out agricultural survey in the United States of America (USA or simply US) and they reported a high fatality rate of 0.22 per 1000 workers per year and the injuries rate as 5 to 166 per 1000 workers per year. A survey was conducted in selected villages of four states viz. Madhya Pradesh, Tamil Nadu, Orissa and Punjab on accidents happened during the year 1995-99 (Gite & Kot, 2003). The limited data collected indicated that the fatalities due to agricultural accidents were 21.2 per 100,000 workers per year. The major source of accidents were farm machines namely tractors, threshers, chaff cutters, cane crushers, sprayers and electric motors, and other sources namely snake bites, drowning in wells/ponds and lightening. Tiwari, Gite, Dubey, and Kot (2002) have found the agricultural accident rate as 1.25 per thousand workers per year in Madhya Pradesh district in India. It is reported that 77.6% of agricultural accidents were because of farming machines, 11.8% because of hand tools and 10.6% due to others. Helmkamp and Lundstrom (2002) have reported the injuries of farmers to be higher than industrial workers. In India, agricultural workers constitute as one of the important sources of farm power. Besides, they also operate animal drawn equipment, tractors, power tillers, self-propelled and power operated machines. Table 1 gives the population dynamics of Indian agricultural workers which shows that by 2020 the population of agricultural workers in the country will be about 242 million of which 50% will be the female workers. Thus, there is going to be a significant role of farm workers in country's agriculture and due attention needs to be given to their safety and occupational health issues so as to have higher productivity, less accidents, and minimum occupational health problems.

Kumar, Singh, Mohan, and Varghese (2008) have found 576 agricultural related injuries with 332 tools such as 58% hand tool related, from 9 villages with 19,723 persons in the first phase. Further, in the second phase with 21 more villages of 78,890 persons, it was reported that 54 hand tools i.e.19% of hand tool was related out of 282 injuries. It was also recommended to have interventions development and training at block levels about the safety measures of equipments. Kumar and Dewangan (2009) have investigated the agricultural accident for six years i.e. between years 2000-2005 of 42 villages of 4 districts in Arunachal Pradesh in India. It was reported to have the accident rate as 6.39 per thousand workers per year with 40% farm implement related injury.

Patel, Varma, and Kumar (2010) have reported the agricultural accident rate as 0.8 per 1000 workers per year in Etawah district of Uttar Pradesh in India. Also, it is reported the lack of study in agricultural injuries in developing countries due to non-availability of compiled information. Nilsson, Pinzke, and

Table 1. Population dynamics of Indian agricultural workers

Serial Number	Particulars	2001	2007	2012	2020
		Number in Millions			
1	Country's population	1029	1130	1210	1330
2	Number of workers as percentage (%) of population	39	41	42	45
3	Total number of workers	402	463	508	605
4	Percentage (%) of agricultural workers to the total workers	58.2	52.0	47.5	40.0
5	Number of agricultural workers	234	241	241	242
6	Percentage (%) of female agricultural workers	39	42	45	50
	a) Number of male agricultural workers	143	140	132	121
	b) Number of female agricultural workers	91	101	109	121

Source: (Banthia, 2004; GOI, 2002, 2006)

Lundqvist (2010) have analyzed the responses from 223 injured farmers collected by the Swedish Farm Registry as part of a survey sent to 7,000 farms by the Swedish University of Agricultural Sciences and Statistics Sweden in 2004. These data showed that there were no significant differences in injuries incurred between the age groups, but that senior farmers seemed to suffer longer from their injuries. This study highlighted the importance of advising senior farmers to bear in mind that their bodies are no longer as young and strong as before. All age groups of farmers should, of course, be careful and consider the risks involved in their work, but since aging bodies need longer to heal, senior farmers probably need to be even more careful and review their work situation and work environment in order to avoid injuries during the final years of their working life. Then, it is recommended to educate senior farmers about the risks of injuries causing increasing damage due to their age. Copuroglu et al. (2012) have studied the agricultural injuries of 41 patients for a period of 3 years and found hand as the most commonly injured part compared to the lower limb and foot. Wibowo & Soni, (2016) have studied the farmers of East Java and Indonesia. Most of the farmer's injury was found in hand and the fatigue was reported as 92.8% in upper back, 93.6% in mid back and 91.8% in lower back. Safe, good and fit in hand were suggested as design criteria for hand tools. The recommended tool handles length was 12.4 cm and the diameter length was 3cm, respectively. The workers are not injured purposefully, but the accident occurs due to the negligence of safety precautions in operating the machines or by the adverse environment. Although agricultural injury does not have any standard definition (A. Kumar, Varghese, & Mohan, 2000), some authors have proposed any ones. For instance, Cooper et al. (2006) have defined the agricultural injury as the injury that occurs while performing farm work or going to or from work. Nag and Nag (2004) have differentiated farming accident as farm implement related agricultural accidents and nonfarm implement related agricultural accidents. Accidents with the use of hand tools or farming machineries are included in farm implement related agricultural accidents, whereas accidents without the use of hand tools or farming machineries are included in nonfarm implement related agricultural accidents i.e. snake biting, hazardous fuels & gases, airborne irritant, noises, vibrations, zoo noses, dusts, chemicals, fungal, end toxins, carrying heavy loads, exposure to heat, falls from height and electrical hazards. This definition was being used in the present study to analyze the farmer's injuries while operating with the hand tools only.

QUALITY FUNCTION DEPLOYMENT (QFD) ANALYSIS

Quality function deployment (QFD), originated in Japan firms to improve the quality. "Deployment" has a much broader meaning than its English translation. In Japan "deployment" refers to an extension of activities. Therefore, "quality function deployment" means that responsibilities for producing a quality item must be assigned to all parts of a corporation (Kogure & Akao, 1983).

QFD a planning tool that uses matrices to show the relationship between two or more sets of concepts and facilitates a customer focused product and process design by making explicit the relationship between design characteristics and customer requirements (Hauser & Clausing, 1988). Akao (1990) has defined QFD as a strategic management tool that provides a structured way for service providers to assure quality and customer satisfaction while maintaining a sustainable competitive advantage and focuses on delivering "value" by seeking out both spoken and unspoken customers' needs, translating them into actionable service features involving all members of the supplier organization. Also, Chen and Weng, (2006) have defined QFD as a systematic method for translating the voice of customers into a final product through various product planning, engineering and manufacturing stages in order to achieve higher customer satisfaction.

QFD History

QFD was developed in the late of 1960's and early 1970's in Japan by Professors Yoji Akao, Shigeru Mizuno and other quality experts as they wanted to develop a quality assurance method that considers customer satisfaction of a product before it was manufactured at the time that quality control methods were primarily aimed at fixing a problem during or after manufacturing (Akao, 1997). This technique took more than ten years to reach the United States of America (Guinta & Praizler, 1993). Many companies have used QFD in all fields and realized significant benefits, and the tool continues to grow in popularity (Griffin & Hauser, 1992). According to Fisher and Schutta (2003), QFD is a system for translating consumer requirements into appropriate company requirements at each stage, from research and product development to engineering and manufacturing to marketing/sales and distribution. QFD is a planning process that translates customer needs into appropriate organizational requirements (Pawitra & Tan, 2003). Maritan and Panizzolo (2009) proposed that when used in the strategic planning process, QFD maintains the integrity of the voice of the customer (VOC) and generates innovative strategies to achieve an organization's vision. They also argue that it leads directly to policy deployment for implementation and performance management.

Functional Fields of QFD

QFD has been introduced to the service sector such as government, banking, healthcare, education and research. Later, QFD's functions had been expanded to wider fields such as design, planning, decision-making and costing. Essentially, there is no definite boundary for QFD's potential fields of applications. Now it is hardly to find an industry to which QFD has not yet been applied (Chan & Wu, 2002).

QFD Phases

Cohen (1996) named the four phases of QFD as product planning, part deployment, process planning and production planning. Cristiano, Liker, and White (2000) conducted a survey that compare between QFD phases in U.S and Japan companies as shown in Figure 1.

Figure 2 illustrates the 4-phase model of QFD, which is accomplished by using a series of matrices (Chan & Wu, 2002).

House of Quality (HOQ)

The House of Quality (HOQ) is a 'voice of customer' analysis tool and the key element of the Quality Functional Deployment (QFD) technique. It starts with the voice of the customer to interpret the customer's requirements in terms of engineering design values by developing a relationship matrix. These techniques help to understand and interpret the customer needs and help to provide a product of superior value. QFD is a planned approach to define the customer needs/ requirements and to translate them into specific plans to achieve the needs. The term, "voice of customers" is used to describe the stated and un-stated customer needs. In the present study the voice of the customers is the needs or requirement of farmers, which was collected by direct discussion and survey. The other methods that may be used are focusing on groups, observations, warranty data, customer specifications, field reports, etc. The customer needs are summarized by using a 'product planning matrix' or 'house of quality', to represent the product requirements/technical characteristics i.e. "how's" based on the needs i.e. "what's".

The structure of QFD can be thought of as a house (so-called 'House of Quality') and shown in Figure 3. The parts of the house of quality are described as follows. The outside walls of the house are shown as the customer requirements and their priorities.

On the left side is a listing of VOC. On the right side is the prioritized customer requirement, which is derived from customer survey. The ceilings of the house contain the technical descriptors or requirements with expert's priorities. The central or interior walls of the house are the relationships between customer requirements and technical requirements. Customer voices (customer requirements) are translated into

Figure 1. Percentage of publications in functional fields of QFD
Source: Chan and Wu, 2002)

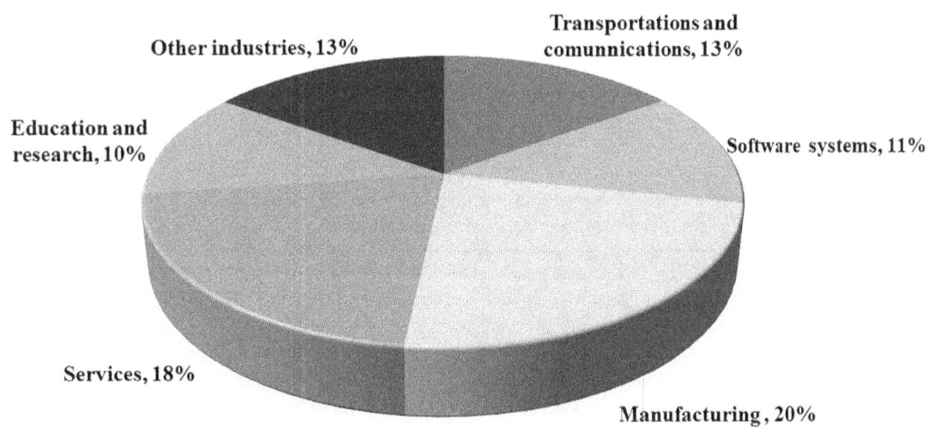

Figure 2. Advanced Phases of QFD Used in Japan and U.S.A
Source: Cristiano et al., 2000)

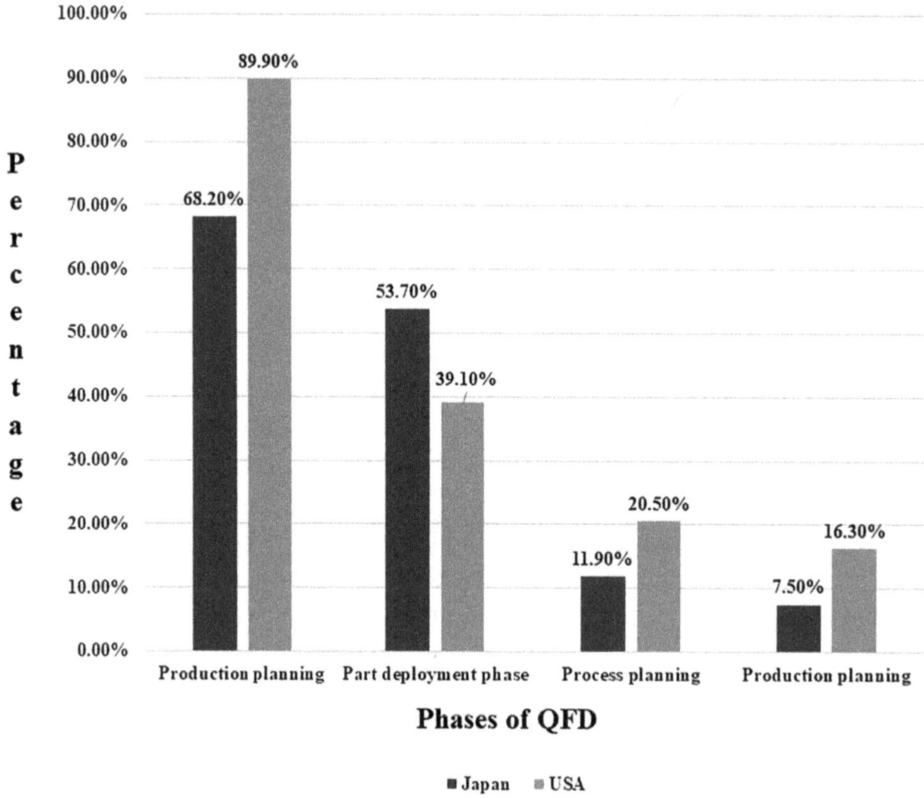

Figure 3. House of Quality

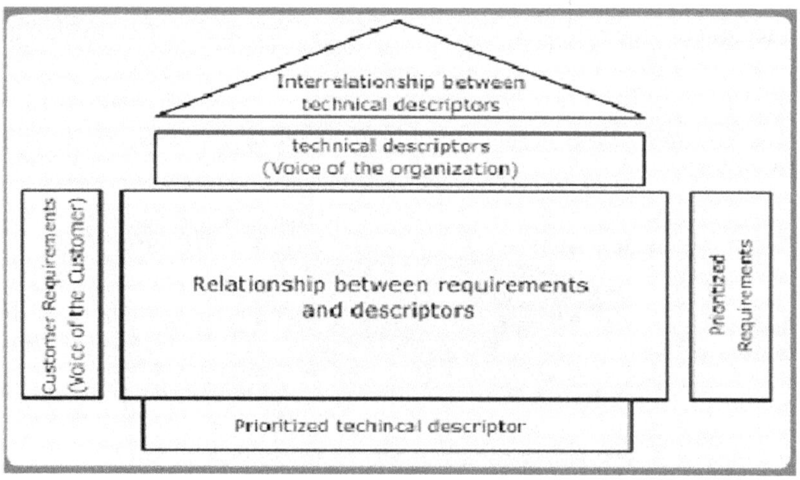

engineering requirements (technical descriptors). The Voice of the Customer is defined as the identification, structuring, and prioritization of customer needs (Griffin & Hauser, 1993). Customer needs are measured in terms of consequences, which are determined by asking customers directly what they are looking for in a product or service. The VOC is obtained primarily by two methods, interviews or focus groups. Griffin and Hauser (1993) have suggested that interviews with 20-30 customers should identify 90% or more of the customer needs in a relatively homogeneous customer segment. Multiple analysts (4-6) should review the transcripts of the focus groups to identify group synergies. Product concepts are then created based on customer priorities.

The roof of the house is the interrelationship between independent technical requirements. Here the trade-offs between similar and/or conflicting technical requirements are identified. The aim of the house is to determine prioritized technical requirement. Technical benchmarking, reverse engineering, tradeoff, and target value comparison are mostly used to determine technical bounds. Ideally in the QFD analysis, no more than 50% of the relationship matrix should be filled, and a random pattern should result (Fisher & Schutta, 2003). The correlation matrix defines the relationships among technical attributes as represented by the roof of the HOQ. The bottom of the house evaluates the competition in terms of service characteristics and target values are defined in this matrix (Tan & Pawitra, 2001).

This is the basic structure for the house of quality. However, based on this format varied QFD matrices are proposed. Quality function deployment starts with a list of goals/objectives. This list is often referred as the WHATs that a customer needs or expects in a particular product. This list of primary customer requirements is usually vague and very general in nature. Further definition is accomplished by defining a new, more detailed list of secondary customer requirements needed to support the primary customer requirements. In other words, a primary customer requirement may encompass numerous secondary customer requirements. As the customer needs and expectations are expressed in terms of customer requirements, the QFD team needs to come up with engineering characteristics (HOW's) that will affect one or more of the customer requirements. Each engineering characteristic must directly affect a customer perception (VOC) and be expressed in measurable terms.

This method helps the development personnel to understand the essential requirements, internal capabilities & constraints and helps to design the product more effectively to meet the customer needs. It helps the development personnel to continue on accurate requirements and reduces the misinterpreting of customer needs/requirements. Thus, QFD is considered as an effective communicating and quality planning tool. Satapathy (2014) has used the artificial neural network (ANN), QFD and the interpretive structural modelling (ISM) approach for the electric utilities services in India for the consumer's satisfaction and discussed the interrelationship between various design requirements. Lombardi and Fargnoli (2018) have used QFD method for assessing and prioritizing the risks and hazards in a small company that produces wheat using a wheeled cabin tractor and other agricultural equipments such as trailers, ploughs, harrows, etc. The primary tool of QFD is the house of quality which is a useful tool for arranging facts (Hauser & Clausing, 1988). Figure 4 illustrates the elements of the HOQ (Erdogan, 2003).

Chan and Wu (2005) reported that a house of quality contains the following six steps:

1. **Customer Requirements (WHATs):** QFD starts with the customers (Griffin & Hauser, 1992). After the collection of all customer requirements, similar requirements are grouped into categories and written into affinity diagram.

2. **Customer Importance Weight:** In order to reveal the importance weight of the customer requirements and identify it the customers are asked to give the importance ratings for each WHAT using rating scales.
3. **Design Requirements (HOWs):** The design requirements (HOWs) are identified by QFD team from literature review, customer/farmer requirements and interviews with the experts.
4. **Relationship Matrix Between WHATs and HOWs:** In this study the assigned numbers and symbols for QFD are: 8 for Strong [O], 6 for Moderate [Θ], 4 for Week [Δ], 2 for Very week [●].
5. **Target Values for HOWs:** Target values represent "How Much" for design requirements. At this step, the target value for each design requirement was established from literature review and interviews with farmers.
6. **Individual Rating of Design Requirements:** Finally, the individual ratings of design requirements of educational carpentry workshop were calculated by using the equation (1):

$$\textit{Individual Rating of Design Requirements} = \sum_{j}^{n} A_{ij} X_j \quad (1)$$

Where A_{ij} and X_j denote the relative importance of the ith characteristics with respect to the jth customer need in the relationship matrix and the importance of jth customer needs (customer ratings), respectively. Moreover, n = number of customer needs.

RESEARCH METHODOLOGY

Odisha state, in India, is one of the ancient states known for its agricultural sectors. Khorda is one of the districts in Odisha, where most of the populations are farmers by their profession. In the Khorda district, 5 villages are selected which are within 10 km radius distance from the great lake Chilka. The villages are so selected where more than 60% population are agricultural farmers in each village.

Figure 4. HOQ Elements
Source: Erdogan, 2003

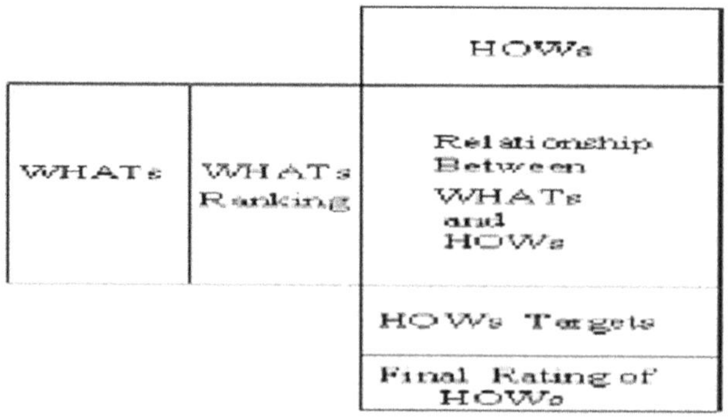

Materials

The following materials were used to perform the study.

- Video/photo camera
- Job Content Survey
- Questionnaire
- Laptop Dell
- Software Excel, Minitab 18®

Methods

The most commonly used hand tools in farming by the farmers such as knife, kudali (pick axe), spade, khurpi (weeding fork), plain edge sickle, serrated sickle and shovel, are only considered in the present study. The type of injury by using the hand tools are ranked in a 5-point scale of (1 = nil, 2 = cut, 3 = sharp force injury, 4 = blunt force trauma, 5 = fracture). The injury level for different hand tools is measured with a 5-point rating scale as (1 = minor, 2 = moderate, 3 = nil, 4 = serious, 5 = severe).

Before collecting the information and data, a team was formulated consisting of the author, co-author and one member from one of the selected villages. Permission was taken from the village head/leader and all 3 members visited the farmer's home randomly in the evening time during their leisure time. The information for the consequent visits to the farmers and other villages were obtained from the previous visits and talking to the farmers. The same procedure was being followed in other villages under consideration. The farmer's injury by using hand tools irrespective of age and gender, were collected for 4 years, i.e. from 2014 to 2017. From all the 5 villages, a total 145 farmers were selected for the study and their provided information was collected in a standard questionnaire containing their name, address, gender, age, the type of injury by using the hand tools and the injury level for different hand tools, respectively (See Appendix , Table 6 and Table 7). The design requirements which are the safety precautions recommended to be followed by farmers to avoid accidents or injuries are found by expert analysis, considering the opinions and suggestions from 3 specialized experts in the academic field of agriculture. Finally, the recommended design requirements are ranked accordingly by using the Quality Function Deployment (QFD) technique. Various steps followed in the present study are illustrated in Figure 5.

RESULTS AND DISCUSSION

In the present study the total number of agricultural farmers selected was 145. Out of the total number of farmers, 96 were males and 49 were females. The farmer's age distribution in the present study is shown in Table 2 below. It is observed that maximum % of female farmers was found between the age groups of 31-45, which is 59.18%. Respect the maximum % of male farmers it was in the same age group of 31-45 as female farmers. In the other age group, i.e. 18-30 & >45, it can be seen that were more compared to the female farmers. Out of 145 farmers, 30 male farmers and 13 female farmers were reported to be injured during the four years period from 2014 to 2017. The farmer's injuries distribution, along with their age groups, is shown in Table 3. From Table 1 and Table 2, it can be seen that maximum number

Figure 5. Steps followed in the present study
Source: The authors

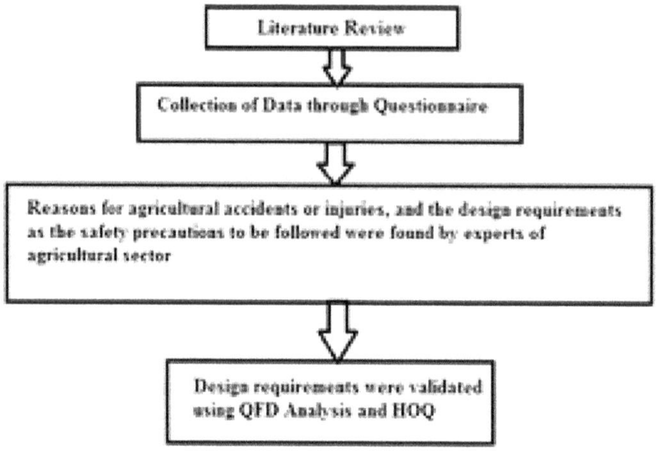

of male and female farmers who were victims of agricultural injury, were in the age group of 31 to 45. While minimum numbers of injured farmers were found in the age group of 18 to 30.

Through the personal interaction and questionnaire administration with the farmers in the present study, it was observed that 43 farmers out of 145 farmers were the victims of agricultural injuries by the use of some selected agricultural hand tools, such as knife, kudali (pick axe), khurpi (weeding fork), spade, plain edge sickle, serrated sickle and shovel. The total number and percentage of accidents due to each of these tools considered separately is illustrated in Table 4. It can be observed that the total number of accidents or injuries are 79 and the higher number of accidents was 14 (17.72%) in each hand tools like knife, kudali (pick axe) and khurpi (weeding fork). The number of accidents by other hand tools such as spade, plain edge sickle, serrated sickle and shovel are 13 (16.45%), 06 (7.59%), 11 (13.92%) and 07 (8.86%), respectively.

Model Development

Based on the interaction with the farmers through questionnaires, some important reasons for agricultural accidents or injuries caused by the hand tools are depicted below,

1. Lack of expertise in the farming process.
2. Lack of knowledge about the use of tools.
3. Negligence of inspecting and ensuring the tools before use.
4. Muskulateral disorder due to hard and repetitive work.
5. Contact with animals and plants bring exposure to bites, infections, allergies and other health problems.
6. Exposure to pesticides and other agrochemicals constitutes another major occupational risk causing poisoning and death.
7. Emergency services are often delayed in time of accidents due to the remoteness of the work sites.
8. Respiratory problem due to exposure to dust and diesel smell.

Table 2. Farmers age distribution in present study

Age in Years	No. of Male Farmers	% of Male Farmers	No. of Female Farmers	% of Female Farmers	Total no. of Farmers (Male and Female)	Total % of Farmers (Male and Female)
18-30	10	10.41	06	12.24	16	11.03
31-45	54	56.25	29	59.18	83	57.24
>45	32	33.33	14	28.57	46	31.72
Total	96	100	49	100	145	100

Source: The authors

Table 3. Farmer's injuries distribution in present study

Age in Years	No. of Male Farmers	% of Male Farmers	No. of Female Farmers	% of Female Farmers	Total no. of Farmers (Male and Female)	Total % of Farmers (Male and Female)
18-30	03	10	01	07.69	04	09.30
31-45	17	56.66	10	76.92	27	62.79
>45	10	33.33	02	15.38	12	27.90
Total	30	100	13	100	43	100

Source: The authors

Table 4. Hand Tool related injuries of farmers

Serial Number	Type of Hand Tool	No. of Accidents	% of Accidents
1.	knife	14	17.72
2.	Kudali (pick axe)	14	17.72
3.	Spade	13	16.45
4.	Khurpi (weeding fork)	14	17.72
5.	Plain edge sickle	06	7.59
6.	Serrated sickle	11	13.92
7.	shovel	07	8.86
Total	--	79	100

Source: The authors

9. Geographical region (hilly wet dry land).
10. Natural disaster flood, draught, cyclone related unsafe working condition.
11. Due to noise hearing problem of large machinaries.
12. Diseases due to extreme weather like heat, rain and cold
13. Work place violence.
14. Mental and financial pressure due to loss of vegetation.

To satisfy the occupational safety requirement of farmers, the below 15 design requirements as the safety precautions to be followed are found by experts of agricultural sector, opinion of farmers and literature review.

1. Training must be provided.
2. Protecting arms, hands and legs during the repetitive work.
3. The tools should be properly and firmly positioned to its respective handles.
4. Sound less and dust less equipments must be used .
5. Wearing of protective shields like gloves, shoes during operational work.
6. Mask and shield must be weared.
7. Organic farming must be done.
8. Advance technology and tools used to avoid repetitive work and lenghty work.
9. Increased frequency of breaks.
10. Safety monitring technologies must be followed.
11. Modified work day schedule.
12. Emergency services must be provided by Govt or stake holders.
13. Work-safety pracices related training must be given.
14. Analysis of working conditions by regular farm visits and assessing and monitoring of work-related health hazards.
15. Regular check up equipments and land and new equipment and procedure for farming.

The HOQ (matrix) is the acknowledged form of QFD. HOQ is constructed from these major components as explained below.

- **Customer Needs (What's):** A structured list of requirements derived from expert's feedback.
- **Design Requirements (How's):** A structured set of relevant and measurable services/characteristics which are required for fulfilling what's.
- **Planning Matrix (Left Matrix):** Gives customer/expert perceptions observed in surveys. It includes the relative importance of requirements.
- **Interrelationship Matrix (Centre Matrix):** Gives the expert's perceptions of interrelationships between design requirements and customer needs. An appropriate scale is applied and illustrated using symbols or figures. Filling this portion of the matrix involves discussions.
- **Design Correlation (Top) Matrix:** Used to identify where design requirements support or impede each other in the system or product design.

The various steps involved in the present QFD methodology:

Step 1: This step identifies fourteen items as illustrated in reasons for agricultural accidents or injuries above i.e. customer needs (WHATs). The items are entered into HOQ in the room called voice of customers.

Step 2: The customer needs are prioritized using a number that reflects the importance of each requirement of farmers.

Step 3: An exhaustive analysis was done to assess the relationship between each item of the customer needs. The interrelations are typically defined as 8 for Strong [O], 6 for Moderate [Θ], 4 for Week [Δ], 2 for Very week [●]. The interrelationship matrix is attached to left side of the HOQ.

Step 4: Keeping in view of stated items of reasons for agricultural accidents or injuries, the fifteen system design requirements are considered through literature reviews and discussions with experts from the industry.

Step 5: A correlation matrix is formed, which indicates how the system designs are related to one another. The scale, used in step 3, is used here for describing the same and is entered on the top of HOQ which forms the top matrix of the house. It is called the interrelation between HOWs.

Step 6: The relationship between WHATs and HOWs are determined and the matrix is constructed using the intersection of each row (what) with each column (how), which represents the strength of relationship between each what and each how.

Step 7: A revised rating is determined from the left matrix of Figure 5 using equation (2) and is placed at right side of HOQ.

$$Rating_i = Z_i + \frac{1}{n-1}\sum_{j=1}^{n} B_{ij} Z_{ij} \tag{2}$$

where Z_i is the initial customer rating for customer need i and B_{ij} denotes the interrelationship between customers' needs i and j.

Step 8: The individual rating for each design requirement is calculated using equation (3) and placed at the bottom row 1 of the HOQ.

$$Individual\ Rating_i = \sum_{j}^{n} A_{ij} X_j \tag{3}$$

where A_{ij} and X_j denote the relative importance of the ith characteristic with respect to the jth customer need in the relationship matrix and the importance of the jth customer need perceived by the customer i.e. the customer rating and n is the number of customer needs.

Step 9: A revised rating for each design requirement is determined as for requirement (customer) needs using equation (1) and is entered in row 2 at bottom of HOQ.

Step 10: The final ratings of design requirements are normalized by dividing each rating with maximum available rating. The final ratings are tabulated in row 3 at bottom of HOQ. Using final ratings, the design requirements are prioritized as per their importance.

The revised rating for each design requirement is calculated for the customer needs using equation (2). The final ratings of design requirements are normalized by dividing each rating with the maximum available rating. The ranking is done considering the values of normalized rating. The item no.8 (i.e. advance technology and tools used to avoid repetitive work and lenghty work) is ranked first in the design requirement as its normalized rating value is 1, and item no.4 (i.e. sound less and dust less equipments

Table 5. Quality Function Deployment (QFD)

Reasons for agricultural injuries	Customer Rating	1	2	3	4	5	6	7	8	9	10	11	12	13	14	15	Customer Need improvement
1	5	●	●	O	●	●	●	●	●	Δ	Δ	Θ	Δ	Δ	●	Θ	5.8769
2	2	Δ	●	●	Δ	Θ	Δ	Δ	Δ	Δ	Θ	Δ	Δ	Δ	Θ	Δ	3.1076
3	3	●	O	Δ	Θ	Δ	Δ	Δ	Δ	O	O	●	●	Δ	Δ	Δ	4.1692
4	2	Θ	O	O	O	●	O	●	●	Δ	Θ	Δ	O	O	●	●	2.9384
5	1	Δ	O	O	●	●	Δ	Θ	Δ	Δ	Δ	Δ	Δ	Θ	Δ	Δ	3.3538
6	3	●	O	O	O	●	Δ	●	Δ	Δ	Θ	●	Δ	Θ	Δ	Δ	4.1538
7	3	Δ	Δ	Θ	Δ	Δ	Δ	Θ	Δ	Δ	Δ	Δ	O	O	●	●	4.3076
8	2	Δ	Δ	O	O	●	●	Δ	Δ	Δ	●	Δ	Δ	Θ	Δ	Δ	2.8461
9	4	●	●	●	●	Δ	Δ	Θ	Δ	●	Δ	Δ	Δ	Δ	Δ	Δ	5.2153
10	5	Δ	Θ	Δ	Δ	Θ	●	Δ	Δ	Θ	Δ	Δ	Δ	Θ	Δ	Δ	6.1692
11	4	Δ	Δ	Δ	Δ	Δ	Δ	Θ	Δ	Δ	Δ	Δ	O	O	●	●	5.3384
12	3	Δ	Θ	Δ	Δ	Δ	Δ	Δ	Θ	●	Δ	Δ	●	●	●	●	4.2769
13	2	Δ	Θ	Δ	Δ	Δ	●	●	Δ	Δ	●	●	Δ	Δ	Δ	Θ	3.2153
14	1	Θ	O	O	●	●	●	●	Θ	Δ	●	●	Δ	Δ	Δ	Δ	1.8461
Initial rating design requirement		1.67	1.96	2.16	2.27	1.45	1.42	1.58	1.66	1.69	1.73	1.51	1.86	2.09	1.34	1.55	
Revised Rating		3.46	3.94	4.16	4.52	2.75	2.97	3.11	4.61	3.16	3.41	3.04	3.43	4.16	2.55	2.87	
Normalised Rating		0.75	0.85	0.90	0.98	0.59	0.64	0.67	1	0.68	0.74	0.65	0.74	0.90	0.55	0.62	
Rank		6	5	4	2	13	11	10	1	9	7	14	8	3	15	12	

must be used)' is ranked second as its normalized rating value is 0.98. Similarly, all 15 items are ranked accordingly.

CONCLUSION

For farmers from the low economic sections, the traditional tools and techniques are the only options to carry out the farming activities. Hence, respective competent authorities must have a high vision for the farmers to provide them with the latest tools and techniques, training about the operation and safety measures to consider, adequate medical facilities and regular inspection of respective activities. The most important effort to protect the farmers from injuries should focus on avoiding the repetitive and complex tasks through 'Job Design', which includes simplifying the tasks or mechanization, job enlargement, job rotation and job enrichment / teamwork. In the cases where elimination of the repetition of work is not feasible, the prevention strategies should be followed such as improvement of the tool & equipment designs, improvement of the workplace layout and work-practices.

ACKNOWLEDGMENT

We would like to convey our sincere thanks to the farmers participated in the present survey and to KIIT (Deemed to be University) for continuous encouragement in carrying out such research work. This research received no specific grant from any funding agency in the public, commercial, or not-for-profit sectors.

REFERENCES

Akao, Y. (1990). *Quality function deployment: Integrating customer requirements into product design.* Cambridge, MA: Journal of Productivity Press.

Akao, Y. (1997). QFD: Past, Present, and Future. *Proceedings of the International Symposium on QFD.*

Banthia, J. K. (2004). *Census of India 2001 - Primary Census Abstracts.* Registrar General & Census Commissioner, Govt. of India.

Chan, L. K., & Wu, M. L. (2002). Quality function deployment: A literature review. *European Journal of Operational Research, 143*(3), 463–497. doi:10.1016/S0377-2217(02)00178-9

Chan, L. K., & Wu, M. L. (2005). A systematic approach to quality function deployment with a full illustrative example. *International Journal of Management Sciences, 33*, 119–139.

Chen, L. H., & Weng, M. C. (2006). An evaluation approach to engineering design in QFD processes using fuzzy goal programming models. *European Journal of Operational Research, 172*(1), 230–248. doi:10.1016/j.ejor.2004.10.004

Cohen, L. (1995). *Quality function deployment: How to make QFD work for You.* Reading, MA: Addison- Wesley.

Cooper, S. P., Burau, K. E., Frankowski, R., Shipp, E., Deljunco, D., Whitworth, R., ... Hanis, C. (2006). A cohort study of injuries in migrant farm worker families in South Texas. *Annals of Epidemiology*, *16*(4), 313–320. doi:10.1016/j.annepidem.2005.04.004 PMID:15994097

Copuroglu, C., Heybeli, N., Ozcan, M., Yilmaz, B., Ciftdemir, M., and Copuroglu, E. (2012). Major extremity injuries associated with farmyards accidents. *The Scientific World Journal*. doi:10.1100/2012/314038

Cristiano, J. J., Liker, J. K., & White, Ch. C. (2000). Customer-driven product development through quality function deployment in the U.S. and Japan. *Journal of Product Innovation Management*, *17*(4), 286–308. doi:10.1016/S0737-6782(00)00047-3

DeMuri, G. P., & Purschwitz, M. A. (2000). Farm injuries in children: A review. *Wisconsin Medical Journal*, *99*(9), 51–55. PMID:11220197

Erdogan, B. (2003). *The extent of information visualization in Turkish construction industry, A QFD approach*. The Middle East Technical University.

Fisher, C., & Schutta, J. T. (2003). *Developing New Service- Incorporating the Voice of the Customer into Strategic Service Development*. Milwaukee, WI: ASQ Quality Press.

Gite, L. P., & Kot, L. S. (2003). *Accidents in Indian Agriculture. Technical Bulletin No. CIAE/2003/103*. Coordinating Cell, All India Coordinated Research Project on Human Engineering and Safety in Agriculture, Central Institute of Agricultural Engineering, Nabibagh, Bhopal.

GOI. (2002). *India Vision 2020. Planning commission*. New Delhi: Govt. of India.

GOI. (2006). *Population projections for India and States 2001-2006, Report of the technical group on population projections constituted by the National Commission on Population*. New Delhi: Govt. of India.

Griffin, A., & Hauser, J. R. (1992). Patterns of communications among marketing, engineering and manufacturing. *Journal of Management and Science*, *38*(3), 360–373.

Griffin, A., & Hauser, R. J. (1991). The Voice of the Customer. *Marketing Science*, *12*(1), 1–27. doi:10.1287/mksc.12.1.1

Guinta, L. R., & Praizler, N. C. (1993). *The QFD Book, The team approaches to solving problems and satisfying customers through quality function deployment*. New York: American Management Association.

Hauser, J., & Clausing, D. (1989). The house of quality. *Harvard Business Review*, *66*(3), 63–73. PMID:10303477

Helkamp, J., & Lundstrom, W. (2002). Tractor Related Deaths Among West Virginia Farmers. *Annals of Epidemiology*, *12*(7), 510. doi:10.1016/S1047-2797(02)00344-7

Hope, A., Kelleher, C., Holmes, L., & Hennessy, T. (1999). Health and safety practices among farmers and other workers: A needs assessment. *Occupational Medicine (Philadelphia, Pa.)*, *49*, 231–235. doi:10.1093/occmed/49.4.231 PMID:10474914

Huiyun, X., Zengzhen, W., Lorann, S., Thomas, J. K., Xuzhen, H., & Xianghua, F. (2000). Agricultural work-related injuries among farmers in Hubei, People's Republic of China. *American Journal of Public Health*, *90*(8), 1269–1276. doi:10.2105/AJPH.90.8.1269 PMID:10937008

Karthikeyan, C., Veeraragavathatham, D., Karpagam, D., & Ayisha, F. S. (2009). Traditional tools in agricultural practices. *Indian Journal of Traditional Knowledge*, *8*(2), 212–217.

Knapp, L. W. Jr. (1965). Agricultural Injuries Prevention. *Journal of Occupational Medicine.*, *7*(11), 553–745. doi:10.1097/00043764-196511000-00001 PMID:5831719

Knapp, L. W. (1966). Occupational and Rural Accidents. *Archives of Environmental Health*, *13*(4), 501–506. doi:10.1080/00039896.1966.10664604 PMID:5922020

Kogure, M., & Akao, Y. (1983). Quality function deployment and CWQC in Japan. *Journal of Quality and Progress*, *16*(10), 25–29.

Kumar, A., Singh, J. K., Mohan, D., & Varghese, M. (2008). Farm hand tools injuries: A case study from northern India. *Safety Science*, *46*(1), 54–65. doi:10.1016/j.ssci.2007.03.003

Kumar, A., Varghese, M., & Mohan, D. (2000). Equipment-related injuries in agriculture: An international perspective. *Injury Control and Safety Promotion*, *7*(3), 1–12. doi:10.1076/1566-0974(200009)7:3;1-N;FT175

Lombardi, M., & Fargnoli, M. (2018). Prioritization of hazards by means of a QFD- based procedure. *International Journal of Safety and Security Engineering*, *8*(2), 342–353. doi:10.2495/SAFE-V8-N2-342-353

Maritan, D., & Panizzolo, R. (2009). Identifying business priorities through quality function deployment. *Marketing Intelligence & Planning*, *27*(5), 714–728. doi:10.1108/02634500910977917

Mishra, D., & Satapathy, S. (2018). Drudgery Reduction of Farm Women of Odisha by Improved Serrated Sickle. *International Journal of Mechanical Engineering and Technology*, *9*(2), 53–61.

Mittal, V. K., Bhatia, B. S., & Ahuja, S. S. (1996). *A Study of the Magnitude, Causes, and Profile of Victims of Injuries with Selected Farm Machines in Punjab. Final Report of ICAR adhoc Research Project.* Ludhiana, IN: Department of Farm Machinery and Power Engineering, Punjab Agricultural University.

Mohan, D., & Patel, R. (1992). Design of safer agricultural equipment: Application of ergonomics and epidemiology. *International Journal of Industrial Ergonomics*, *10*(4), 301–309. doi:10.1016/0169-8141(92)90097-J

Murphy, D. J. (1992). *Safety and Health for Production Agriculture*. St. Joseph, MI: American Society of Agricultural Engineers.

Nag, P. K., & Nag, A. (2004). Drudgery, accidents and injuries in Indian agriculture. *Industrial Health*, *42*(4), 149–162. doi:10.2486/indhealth.42.149 PMID:15128164

Nilsson, K. S., Pinzke, S. P., & Lundqvist, P. (2010). Occupational Injuries to Senior Farmers in Sweden. *Journal of Agricultural Safety and Health*, *16*(1), 19–29. doi:10.13031/2013.29246 PMID:20222268

Patel, S. K., Varma, M. R., & Kumar, A. (2010). Agricultural injuries in Etawah district of Uttar Pradesh in India. *Safety Science*, *48*(2), 222–229. doi:10.1016/j.ssci.2009.08.003

Pawitra, T. A., & Tan, K. C. (2003). Tourist satisfaction in Singapore-a perspective from Indonesian tourists. *Managing Service Quality*, *13*(5), 339–411. doi:10.1108/09604520310495868

Prasanna Kumar, G. V., & Dewangan, K. N. (2009). Agricultural accidents in north eastern region of India. *Safety Science*, *47*(2), 199–205. doi:10.1016/j.ssci.2008.03.007

Rautiainen, R. H., & Reynolds, S. J. (2002). Mortality and morbidity in agriculture in the United States. *Journal of Agricultural Safety and Health*, *8*(3), 259–276. doi:10.13031/2013.9054 PMID:12363178

Satapathy, S. (2014). ANN, QFD and ISM approach for framing electricity utility service in India for consumer satisfaction. *International Journal of Services and Operations Management*, *18*(4), 404–428. doi:10.1504/IJSOM.2014.063243

Tan, K. C., & Pawitra, T. A. (2001). Integrating SERVQUAL and Kana's model into QFD for service excellence development. *Managing Service Quality*, *11*(6), 418–430. doi:10.1108/EUM0000000006520

Tiwari, P. S., Gite, L. P., Dubey, A. K., & Kot, L. S. (2002). Agricultural injuries in Central India: Nature, magnitude and economic impact. *Journal of Agricultural Safety and Health*, *8*(1), 95–111. doi:10.13031/2013.7221 PMID:12002378

Voaklander, K. D., Kelly, D. C., Rowe, B. H., Schopflocher, D. P., Svenson, L., Yiannakoulias, N., & Pickett, W. (2006). Pain, medication, and injury in older farmers. *American Journal of Industrial Medicine*, *49*(5), 374–382. doi:10.1002/ajim.20292 PMID:16526061

Xiang, H., Wang, Z., Stallones, L., Keefe, T. J., Huang, X., & Fu, X. (2000). Agricultural work-related injuries among farmers in Hubei, People's Republic of China. *American Journal of Public Health*, *90*(8), 1269–1276. Retrieved from http://www.ncbi.nlm.nih.gov/pubmed/10937008

Zhou, C., & Roseman, J. M. (1994). Agricultural injuries among a population-based sample of farm operators in Alabama. *American Journal of Industrial Medicine*, *25*(3), 385–402. doi:10.1002/ajim.4700250307 PMID:8160657

KEY TERMS AND DEFINITIONS

Agricultural Injury: It is the injury that occurs while performing farm work.

Ergonomics: It is the process of designing or arranging workplaces, products, and systems so that they fit the people who use them.

Farming: Farming is a part of agriculture. It is the growing of crops or keeping animals by people for food and raw materials.

House of Quality (HOQ): The house of quality is a kind of conceptual map that provides the means for interfunctional planning and communications.

Injury: Injury is damage to the body caused by external force. This may be caused by accidents, falls, hits, weapons, and other causes.

Quality Function Deployment (QFD): It is a structured approach to defining customer needs or requirements and translating them into specific plans to produce products to meet those needs.

Voice of Customer: It is the requirements/feedback from the customer (internal or external) to provide the customers with the best in class service/product quality.

APPENDIX

Questionnaire for Farmers' Injury

Tables 6 and 7.

Table 6. Farmers type of injury

Farming Accidents		Scale: (1=Nil, 2=Cut, 3=Sharp Force Injury, 4= Blunt Force Trauma, 5=Fracture)							
Type of Injury									
Serial Number of Farmer	Gender-Male(M)/ Female(F)	Age	Knife	Kudali (Pick Axe)	Spde	Khurpi (Weeding Fork)	Plain Edge Sickle	Serrated Sickle	Shovel
P1									
P2									
P3									
P4									
P5									
P6									
P7									
P8									
P9									
P10									
P11									
P12									
P13									
P14									
P15									
P16									
P17									
P18									
P19									
P20									
P21									
P22									
P23									
P24									
P25									
P26									
P27									
P28									

continued on following page

Table 6. Continued

Farming Accidents		Scale: (1=Nil, 2=Cut, 3=Sharp Force Injury, 4= Blunt Force Trauma, 5=Fracture)							
Type of Injury									
Serial Number of Farmer	Gender-Male(M)/ Female(F)	Age	Knife	Kudali (Pick Axe)	Spde	Khurpi (Weeding Fork)	Plain Edge Sickle	Serrated Sickle	Shovel
P29									
P30									
P31									
P32									
P33									
P34									
P35									
P36									
P37									
P38									
P39									
P40									
P41									
P42									
P43									
P44									
P45									

Table 7. Farmers level of injury

Farming Accidents		Scale: (1=Minor, 2= Moderate, 3= Nil, 4=Serious, 5=Severe)							
Level of injury									
Serial Number of Farmer	Gender-Male(M)/ Female(F)	Age	Knife	Kudali (Pick Axe)	Spade	Khurpi (Weeding Fork)	Plain edge Sickle	Serrated Sickle	Shovel
P1									
P2									
P3									
P4									
P5									
P6									
P7									
P8									
P9									
P10									

continued on following page

Table 7. Continued

Farming Accidents		Scale: (1=Minor, 2= Moderate, 3= Nil, 4=Serious, 5=Severe)							
Level of injury									
Serial Number of Farmer	Gender-Male(M)/ Female(F)	Age	Knife	Kudali (Pick Axe)	Spade	Khurpi (Weeding Fork)	Plain edge Sickle	Serrated Sickle	Shovel
P11									
P12									
P13									
P14									
P15									
P16									
P17									
P18									
P19									
P20									
P21									
P22									
P23									
P24									
P25									
P26									
P27									
P28									
P29									
P30									
P31									
P32									
P33									
P34									
P35									
P36									
P37									
P38									
P39									
P40									
P41									
P42									
P43									
P44									
P45									

Chapter 10
The Contribution of Neuroscience and Health Psychology to Macroergonomics:
Focusing on Workers as Active Agents

Miguel Angel Serrano-Rosa
Universidad de Valencia, Spain

Francisco Molins
Universidad de Valencia, Spain

ABSTRACT

When a work system is considered we must focus on people that are part of the system as an active agent that can participate and interact all along the working process. In this process, stress contributes to increasing the probability of diseases and accidents. Therefore, one of the main objectives, from a preventive perspective, is to reduce stress levels. Related to this, the characteristics of healthy organizations will be described as a model to promote health. The objective of this chapter is to offer a broad and multidisciplinary perspective of prevention risk in the workplace centered in the worker, taking the view of neuroscience and health psychology. For that purpose, stress interventions in the organizations and in the individuals will be outlined. Finally, this chapter would finish proposing a preventive intervention to improve healthier work environments, taking into account neurosciences and health psychology. This proposal will be based on improving lifestyle, education for health, and self-respect.

DOI: 10.4018/978-1-5225-7192-6.ch010

John Henry he hammered in the mountains
His hammer was striking fire
But he worked so hard;
it broke his heart
John Henry laid down his hammer
and died, Lord, Lord
John Henry (trad. Adapted by Pete Seeger)

GENERAL PERSPECTIVE

John Henry is a traditional character from the history of the United States of America that represents the fight between the man and the machine. The man, John Henry, fighting against the machine to prove that men can defeat the machine; at the end, John Henry wins but finally, due to the great effort, dies. From a humanistic point of view, the need to be superior to the machine is a good view to preserve the supremacy of men in front of the machines; but, this "romantic" view finishes with John Henry's dead. Is it worth it? It is worthwhile dying to demonstrate that one can be better in his job? Is it worthwhile dying/getting sick for a job? That's the point. Fatal injuries related to work are usual and it is necessary to take actions to reduce these fatal consequences.

Considering that macroergonomics is a subdiscipline of Ergonomics which is concerned with the analysis, design, and evaluation of work systems, it is necessary to consider that a new vision of work systems, analyses and evaluation should be done within a wide perspective, trying to avoid mistakes and difficulties from the "classic ergonomics". Recently, Thatcher, Waterson, Todd and Moray (2017) have asserted that "ergonomics should move beyond a Westernised view of worker-organization-technology fit, taking a multidisciplinary approach which engages with other social and biological sciences". From neuroscience, when a work system is considered researchers must focus on people that are part of the system, but not as a mere "mechanical and passive elements" but as an active agent that can participate and interact all along the working process. In this sense, when the authors of this chapter talk about healthy workplaces, they refer to these people that participate of an interactive system. Moreover, every day is increasingly the necessity of person-to-person interaction to increase productivity, due to an important part of the jobs are based on social relationships. Therefore, workers should be the focus to have healthy workplaces. Thus, from a macroergonomic perspective, environmental elements have been related to psychological characteristics and performance (Realyvásquez, Maldonado-Macías, García-Alcaraz, Cortés-Robles & Blanco-Fernández, 2016). Therefore, a new perspective seems necessary to have a more comprenhensive vision of sociotechnical systems and its relation to performance and health.

The main aim of Ergonomics is to find the harmony between efficiency and workers' well-being (health, security and satisfaction) by means of changing or improving working conditions. Considering that Ergonomics is a multidisciplinary discipline that gathers from other disciplines to address more efficiently the problems in workplace, it is surprising that despite taking knowledge from other disciplines there is a lack of addressing more comprehensive of psychosocial factors, especially stress (Serrano & Costa, 2018). Thus, it is considered that broaden the ergonomic perspective would help to cope with workplace issues. Therefore, a prospective view should be defended, not to beat the classical ergonomic perspective, but to contribute with a modern perspective gathering updated knowledge from the world of biopsychosocial sciences. It is well known that scientific knowledge and organizational practices are not close as they should (Caetano & Santos, 2017). In this sense, if human being is the central point of workplace more attention must receive neurosciences that dedicate their efforts to investigate human beings' processes.

On the other hand, from a Social Psychology perspective, organizations are complex constructions of the human being. These complex organizations are created by humans, beyond the individual. Thus, the consideration of context emerges as an increasingly exciting dimension; that is, "the substrates of cognitive functions cannot be studied in isolation but that insight into their mechanisms and consequent outcomes can be gained only from the full contextual setting in which they develop and operate" (Aminoff et al., 2009). In this line, social neuroscience seeks to specify the neural, hormonal and even, genetic mechanisms underlying social behavior. All behaviors under work contexts are social behaviors that have consequences at individual, group and the organizational level. In fact, social neuroscience is trying to understand the associations between social behaviors and biological changes and *viceversa* (Cacioppo, Bernston & Decety, 2010). In this sense, one the major concerns in workplace is the healthy factor; that is, the amount of diseases and injuries that occur in workers. Additionally, as it is stated by the World Health Organization (WHO) health is a complex concept that includes psychological and social health. Therefore, neuroscience comprises an important conceptual approach to understand global mental health (Stein et al., 2017). Complementary, a good performance is necessary for an organization; in this regard, self-regulation and goal pursuit should be taken into account due to be an important part of a good performance (Gallo, Cohen, Gollwitzer & Oettingen, 2013).

But what is needed to prevent from diseases or accidents? A great part of the diseases/accidents come from the high levels of experienced stress (Adler, 2007). One important issue about stress is that people are not good to recognize the own stress state. Usually, stress is recognized by people after months or even years of suffering, when the consequences are present. Associated with stress, there is an increase of mental load that is another important risk in work environments (Serrano & Costa, 2018). Therefore, one of the main objectives, from a preventive perspective, is to reduce stress levels and make workers conscious about their levels of stress. In this regard, from a psychosocial perspective, jobs require abilities (i.e. communication, social skills, conflict management, decision making, relaxation, social abilities, self-control, and others) that are necessary to cope with stress, mental load and the diseases related to (cardiovascular diseases, injuries, stress, burnout, among others).

Considering all above, the objective of this chapter is to offer a broad and multidisciplinary perspective of prevention risk in the workplace centered in the worker, taking the view of social neuroscience (including social psychology) as a basis that, to the knowledge, has not been employed in ergonomics, in spite that is a comprehensive sight of human beings in social contexts. Finally, from the Psychology point of view, individual differences will be emphasized as an important factor that should be considered when human behavior is studied. All of this will be considered, with the assumption that productivity should be one of the main objectives (together with the health and satisfaction of workers). In fact, taking care of workers increases *per se* productivity. To carry out this aim, this chapter will dedicate the first part to explain and show what is needed for a healthy workplace, from the point of view of Positive Organizational Psychology. In the final part, a broad conception of prevention risk factors will be addressed integrating neuroscience and health psychology into ergonomics, to offer a long-term prevention based on the worker responsibility and the ability to detecting psychosocial risks.

HEALTHY ORGANIZATIONS

Trying to optimize the functioning of an organization is not new. In the nineteenth century, with the Second Industrial Revolution, with the rise of companies and new forms of work, it was considered essential to use scientific knowledge to study and optimize each position, so that workers could reach their maximum performance and satisfaction. This idea came up with the engineer Frederick Winslow Taylor and the concept "scientific management" (1984), who firmly believed in the existence of "one best way"; that is, to analyze the characteristics of the work to assign it to the worker that best fits in it. However, it was considered a reductionist proposal focused on mechanical aspects, such as the division of tasks, which forgot intrinsic characteristics of human being, such as motivations, goals or needs (Waring, 2016).

Human factors began to gain importance after the studies of the psychologist Elton Mayo at the beginning of the 20th century (Hawthorne Plant experiments). These studies began with the aim of studying what working conditions favored performance: schedules, luminosity, temperature, ...; however, to the surprise of Elton Mayo, the performance increased in any of the conditions that were tested; then, improvement was interpreted to the fact that all workers were being supervised by the researchers (Sedgwick & Greenwood, 2015). Thus, if the mere fact that a worker felt "special" served to increase their performance, it was evident that Taylor's rational approach, where the worker worked solely for money, lacked complexity.

Therefore, and despite its limitations and methodological criticism (Peiró, 1983; 1984, Veen and Korver, 1998), Hawthorne's studies served as a starting point to studying human nature in its total complexity within the organizational context, something that, would lead to the development of the Psychology of Organizations and Human Resources. The European Network of Organizational Psychologists (ENOP) defines the Psychology of Organizations (or Organizational Psychology) as a subdiscipline of psychology that is responsible for the study of collective behavior in relation to the sociotechnical system called "organization" (Martínez, Ramos and Moliner, 2015). This discipline analyzes and studies all the components of the organization, both individually and collectively, on the basis that workers are people that have personality traits, values, attitudes, cognitive limitations, objectives, etc.

On the other hand, thanks to the contribution of Martin Seligman with the development of Positive Psychology or the "scientific study of the optimal functioning of the human being" (Seligman, 2000), Positive Organizational Psychology has also begun to be discussed (Negruti, Hristova, Larsen, Krumov, 2015). This discipline goes beyond the weaknesses of the organization, also influencing its potential to promote its optimal development and achieve a healthy organization (Salanova, Martínez & Llorens, 2014; 2005).

How to Get Healthy Organizations?

As defined by Salanova (2008), healthy organizations are all those organizations that "make systematic, planned and proactive efforts to improve the health of employees through good practices related to the improvement of tasks (i.e. designing and redesigning of positions), the social environment (i.e. improving communication) and the organization (work / private life conciliation strategies)". But why is it convenient to get a healthy organization? Some results have been associated with these organizations at different levels (Martínez, et al., 2015):

- Individual: the presence of resources in the workplace is associated with employee welfare, such as high levels of engagement, job satisfaction and performance. Other results also mentioned in the literature are the commitment of the worker and the low turnover and intention of abandonment.
- Group: organizational resources and engagement (measured at the team level) are related to collective outcomes such as the service climate and collective performance.
- Organizational: a key result of healthy organizations is the satisfaction and loyalty of clients, social responsibility and economic-financial results.

Based on these results, it is essential to promote healthy organizations. However, this goal is not simple. There are many variables to consider and it seems complicated to be able to address them all. Therefore, in order to simplify the task and based on the Empirical-Comprehensive Healthy Work Organization (ECOLS) model (Wilson, DeJoy, Vandenberg, Richardson & McGrath, 2004), the authors of this chapter will offer a perspective to detect the main variables and explain what strategies to use to address them.

Job Design

Job design involves, on the one hand, (1) the characteristics of the task; that is, its complexity, duration, physical conditions in which it will develop (temperature, luminosity, noise, etc.) and (2) the personal characteristics of the worker.

Today, it is known that human beings show a limited rationality (Thaler, 2016; Tversky and Kahneman, 1974, Simon, 2000, Sunstein and Thaler, 2003) and are influenced by their emotions, values, attitudes, personality and experiences when interpreting their work context. After this conceptual evolution, it is understood that, in reality, the objectivity of the characteristics of work is not as important as the worker perception. In addition, not all workers have the same motivations. While some people may aim for a higher salary (extrinsic motivation), there will be others who value more a job that allows self-actualization (Maslow, 1954; 2016), be creative, have autonomy, assume responsibilities and feel that their work is being productive (intrinsic motivation). Therefore, if an optimal work design is intended, the objective characteristics of the position and the personal characteristics of the worker must be analyzed in interaction. All of this point to the need to individually address each specific situation, taking into accounts both dimensions to achieve the greatest possible adjustment.

Organizational Climate

By organizational climate is understood the individual perceptions that members of an organization have regarding their social environment (Martínez, et al., 2015); that is, their personal experience, although to some extent shared with some or all of the other workers, regarding to the work environment. Many studies relate different perceptions of climate to work performance, the involvement of workers, satisfaction (eg Patterson et al., 2005, Parker et al., 2003, Ostroff & Bowen, 2000, Reichers & Schneider, 1990; Vaca, Vaca & Quintero, 2015; Zambrano, Véliz & Barzola, 2017) and, ultimately, with its importance to maintain a healthy organization (Prado, 2014; Vega, Gálvez & Santamaría, 2017). Work climate is related to three relevant constructs: social support, communication and participation and involvement.

Social support is "the feeling of being appreciated and valued by other people and of belonging to a social network" (Barra, 2004), and it has been frequently linked to maintaining health and overcoming the disease (Almagiá, 2014; Hakulinen, et al., 2016; Holt-Lunstad & Uchino, 2015; Brodman, 1993;

Kaplan, Sallis & Patterson, 1993). The lack of social support within an organization is an important risk factor in the emergence of Burnout (Aranda, et al., 2004). Therefore, it is a key to favor a climate of warmth and companionship that allows employees to feel integrated into the social network of the organization. For this purpose, it is proposed to increase accessibility, for example, through group leisure activities or self-help groups within the work, which favor the link between employees.

On the other hand, communication is the process by which members of an organization are interrelated by exchanging information, establishing links and commitments, delegating, supervising and assuming functions (Moriano-León, 2005). Optimal communication is necessary to maintain a good organizational climate and employee's satisfaction (Pellegrin & Curry, 2011). Finally, participation consists in an active role in the activities and decisions that are made within the work environment. There is a clear relationship between those employees who participate and autonomy and the increase in motivation (Martínez, et al., 2015). Being motivated is the key to developing engagement, "a psychological state characterized by vigor, dedication and absorption or concentration at work." (Schaufeli, Salanova, González-Roma and Bakker, 2002; Bakker & Albrecth, 2018). Those people who experience engagement, are more committed and involved and strive more in their work, establishing a link with it and valuating more their role and tasks (Norton, Mochon & Ariely, 2010). In addition, engagement would be related to higher performance, well-being and the decrease in absenteeism (Schaufeli, Salanova, González-Roma & Bakker, 2002). Therefore, it seems appropriate to promote the delegation of responsibilities and the decentralization of authority so that workers participate in the organizational decisions in order to experience greater autonomy, involvement and motivation.

Labor Perspectives

Labor perspective refers to the perception that workers have about the stability and security in their job, as well as the possibility of advancing in their professional career and feeling satisfied. In relation to this construct, in this chapter the authors will refer to the perception of justice and self-realization.

The perception of justice refers to the "personal evaluation of the ethical and moral standards that characterize the organization" (Omar, 2015), and will depend on organizational practices and policies, mainly focused on leaders (Peña-Ochoa & Durán, 2015). Many authors (i.e. Omar, 2015; Cheng, 2014; Moon, Hur, Ko, Kim, & Yoon, 2014; Bal, de Lange, Ybema, Jansen, & Van der Velde, 2011) point to a clear relationship between the perception of justice and a greater commitment and involvement with the organization. In addition, perceiving a climate of justice maintains the confidence of workers, an aspect that, according to Wei-Yuan, Shang-Ping, Chwei-Jen, & ChinFang (2013), links organizational justice with commitment. Omar (2015), also highlights, that justice is directly related to an increase in performance and job satisfaction. Some practices that contribute to developing the perception of justice would be transparency, explanation and justification in decision-making, attention to employees, or the exercise of voting to promote an organizational democracy.

On the other hand, self-actualization would be the need that, according to Maslow (1954; 2016), many individuals express that, far from wanting a higher salary, they prefer to transcend, develop their skills, be creative, and leave their mark. It is necessary, therefore, to analyze the real motivations of workers to identify which elements is an incentive for them and to improve their satisfaction and performance. Otherwise, wages can be raised and economic bonuses can be provided that do not produce beneficial effects, or even harm (Ariely, Gneezy, Loewenstein, & Mazar, 2009; Chib, De Martino, Shimojo, &

O'Doherty, 2012), when it would be better to promote tasks according to this intrinsic motivation of self-actualization.

Central Organizational Attributes

This domain is mainly related to the attitudes, values and policies adopted by an organization (Martínez, et al., 2015), that deals with the culture and organizational practices in more detail.

Pellegrin & Currey (2011) defined the organizational culture as values and beliefs that guide behavior in the workplace, whose strength and stability is related to the performance of the organization (Salas-Arbeláez, Solarte & Vargas, 2017; Sorensen, 2002). Culture goes beyond people (Ashkanasy, Wilderom & Peterson, 2011), so, unlike climate, it is a much more stable element that complicates its approach to the point of suggesting a preventive proposal: people who want to access to an organization should know their culture and if it is related to their beliefs to decide if it is convenient to enter, before being immersed in it. In the case of organizacional culture, each intervention must be developed *ad hoc*, depending on the particular needs that require modification. For example, to flexibilize a highly controlled and hierarchical organization, it would be useful to apply group dynamics that help establish links and for supervisors to trust subordinates more, granting them greater autonomy and flexibility. Or perhaps, it would be useful to provide formation that develop the ability to manage stress and pressure so that employees can better adapt to a culture of innovation that constantly faces moments of uncertainty and risk decisions.

On the other hand, organizational practices refer to all those strategies that are exercised in the workplace by the leaders and/or supervisors that influence the employees. Throughout the chapter the authors have already mentioned several practices that would favor the development of healthy organizations, for that reason, here, besides inviting to review those already commented, the authors will limit themselves to mention the distinction of organizational practices proposed by Salanova (2009): organizational (selection and socialization of new employees, work-life balance strategies, employability development), social (creation of open channels for organizational communication, promote healthy leadership styles, foster a culture of quality customer service and to the citizen) and of task (to combine tasks to avoid boredom, development of feedback channels, propose challenging tasks). It should be noted, however, that there are no ideal practices and that, as stated by Salanova (2009), "each organization should develop its own menu of good practices taking into account its own culture and values".

All the above aspects are basic when it comes to understanding the welfare of organizations and workers, and they entail the need for a psychosocial adjustment between the worker and his work. In this sense, psychological adjustment can be reach with a balance between (1) the needs and competencies of the employee and (2) the demands of the work environment (Cullen, 1995). This relationship necessarily reminds the demands-control model (Karasek & Theorell, 1990). Demands-control model (also called the "job strain model" or DC) emphasizes different aspects in the relation between jobs and health. Work stress and its effects on physical and mental health are not only the result of aspects of the job, but rather of the joint action between the demands ("job demands") and the degree of perceived control ("decision latitude"). Thus, the model identifies two basic aspects within the work situation: work demands and the possibility of control. According to this model, job stress appears when a situation is characterized by high demand and low latitude decision (reviewed in Serrano & Costa, 2018). When people talk about the possibility of control, in reality they are talking about the subjective perception of control, that is, it is the worker who perceives that there are insufficient resources to face demands. Lazarus and Folkman (1984), affirmed that an event would only be a stressor (or will be perceived as a demand) when people perceive it, and this evaluation "will depend on the coping strategies, the individual and their previous experience". In this sense, stress, although it is an adaptive response of the organism (can cope with and solve a situation), can become chronic and harmful if it is frequent; that is if the working conditions are not adequate or the demands are always perceived as very high. Thus, chronic stress is related to various pathologies and negative moods that affect the health and performance of the worker (Serrano & Costa, 2018). For example, many studies point to its immunosuppressive effect (i.e. Borrás, 1995, Gómez-González & Escobar, 2006; Pellicer, Salvador & Bénet, 2002; Selye, 1936, 1950), facilitating the appearance of diseases and difficulting their cure; others relate stress to the greater risk of the appearance of cardiovascular pathologies (i.e. Cohen, Edmondson & Kronish, 2015; Kivimäki & Kawachi, 2015; Lagraauw, Kuiper & Bot, 2015). In addition, stress has also been linked to the feeling of demotivation and dissatisfaction, lack of involvement with work, poor performance and the desire to leave (i.e. Barkhuizen, Rothmann & van de Vijver, 2014; Gracia, 2015; Padula, Chiavegato, Cabral, Almeid, Ortiz, & Carregaro, 2012; Sonnentag, Mojza, Demerouti & Bakker, 2012).

However, it is important to remember that a stress must be perceived by the worker and this will depend on their individual characteristics. Therefore, the employees should be empowered; that is, provide coping strategies and improve their abilities, so that they feel more effective to cope with stressors (Grau, Salanova & Peiró, 2012) or even prevent their appearance.

A NEW PREVENTION PROPOSAL TO BENEFIT OF HEALTHIER WORK ENVIRONMENTS FOR HUMAN WELL-BEING

Previous section has finished emphasizing the need to provide empowerment and coping strategies that reduce the likelihood of stress and, as a consequence, of accidents, errors and reduction of workers' performance. For this, there are numerous effective interventions. However, accidents and workplace deaths continue, which makes believe that a broader approach is necessary. Thus, as stated at the beginning, the purpose of this chapter is to optimize organizations to be healthy places where workers do their jobs efficiently without compromising their health, and even improving it. Only in that case, companies will get healthy employees. Thus, healthy employees are characterized by high self-efficacy, optimism, resilience, energy, motivation, confidence and engagement. Therefore, promote these types of employees seems the smartest option to increase the success of the company. However, instead of worrying about promoting healthy workers, organizations and workers are usually limited to trying to reduce stress through organizational or individual interventions (Colligan & Higgins, 2005). The following section is dedicated to show strategies to reduce stress.

Strategies to Reduce Stress

Although stress is a problem that is perceived individually, when talking about work stress, the authors of this chapter mean that a significant number of workers are submitted to stress and that it is due to working conditions. Therefore, it is assumed that these are not optimal and are susceptible to intervention. In this sense, interventions are usually carried out to solve a problem already generated. On the other hand, work stress can be difficult to specify in an organizational intervention, as well as assessing whether this has been successful, and this generates distrust when investing in interventions that improve the management of the problem. However, when the problems begin to translate into negative consequences regarding productivity and benefits, the need for intervention begins to be recognized. In fact, the new trends in business management go in the direction of emphasizing and protecting "human resources".

The organizational strategies of intervention can be classified based on different aspects (Michie, 2002): worker control, information and communication systems, management styles, task interest, conflict regulation, socialization process and training.

- **Worker Control:** One of the measures that can be taken to reduce the risk or impact of work stress is to increase the degree of control. This will positively influence health and satisfaction. The degree of control should be about what the employee should do (his / her competences and functions, results, among others) and how he / she does it. In addition, the worker should be able to regulate their work and rest times, as well as participate in aspects related to their work. Obviously, this is the ideal and always taking into account that the worker meets the objectives assigned to him. Control involves good training and responsibility, so the worker has to be well trained.
- **Information and Communication Systems:** The information systems of companies must be precise as they affect all workers. Thus, the information that workers have must be precise (this does not necessarily mean giving all the details), consistent with other information, as well as with the decision making of the organization. The language must be adapted to the receiver (so the level of complexity must be varied depending on the worker). On the other hand, the treatment and transmission of information must be adequate and accessible. In this regard, an information system must allow the worker to know what is expected from him (tasks or objectives to be met) and get feedback on the work done. In this sense, workers and managers must be trained in communication so that exactly what is transmitted is conveyed and not what is understood. Therefore, messages must be clear in a context where there is no noise and the worker is receptive to that message. All this will greatly reduce the risk of work stress due to communication mistakes.
- **Management Styles:** In an organization an appropriate design of the hierarchy is required and a normative and regulatory system that is accepted by the workers. For this purpose, it is recommended to have an effective style of leadership focused on the democratic and participative direction. Therefore, the manager must allow the worker's effective participation making him see that he is part of the organization. On the other hand, the regulations within an organization must be affordable and logical and understood by all employees. In this line, working with clear objectives is of great acceptance since it allows to reach the objectives giving a certain degree of freedom to the worker.

In short, the most effective supervisory styles are those that focus on employees considering their characteristics. For this, managers should be trained in human resource management skills, and specifically in social skills, to avoid management styles that may be negative to subordinates. All of this ends up impacting on a greater identification of the worker with the company, increasing their sense of responsibility, as well as increasing the degree of autonomy, which will result in greater control and in a reduction of stress (provided that the work is well planned and workers have the necessary skills to perform the tasks).

- **Interest of the Task:** One aspect that greatly increases the feeling of stress is the monotony and boredom at work. Thus, in those jobs that have a high degree of monotonous or tedious tasks it is recommended the rotation of positions (that the workers do not do the same task throughout the day), modifying the way of doing the work or increasing the degree of complexity of the work. In this sense, involving the worker in decisions about their tasks will increase the feeling of belonging and will give a greater degree of satisfaction (reducing the degree of stress).

- **Conflicts:** A basic aspect to reduce work stress is to reduce the number and degree of labor conflicts. Any conflict will cause an increase in the stress response and a reduction in well-being. Therefore, proper management of the conflict is necessary. For this, the adequate training of the workers, as well as the clarity in the definition of positions, will allow the reduction of labor conflicts. On the other hand, the implementation of mediation protocols will allow conflicts to be resolved based on rules.
- **Socialization and Training Processes:** In general, all the above strategies are aimed at the organization. However, there are other interventions focused on the individual that are also very effective. In this sense, the processes of socialization of the new workers are basic actions for the integration and knowledge of the new incorporations. This greatly reduces the stress of the "new" worker by having contacts (including a "guide") that will make it easier to know the company. On the other hand, giving the right training for the position that is going to be covered is fundamental for the reduction of stress. Training refers to training in job-specific skills and training in social skills and management of the position itself. In addition to these "formal" skills, knowing the ins and outs of the organization will also allow the worker to predict possible changes in the organization. In short, for the reduction of stress the worker must have skills that allow a better adaptation to the position and the company.

In summary, increasing the degree of control, reducing conflicts, knowing how to manage people and adequate training are factors that help reduce the level of stress in organizations. However, all this does not guarantee success. For an intervention to be effective there are a number of aspects that will increase the probability of success. In particular, any intervention must be ordered and reflected, in addition to having the medium / long term support of the management staff. On the other hand, an adequate risk analysis must be carried out, identifying them adequately, analyzing them and assessing the causes. With this, the intervention proposal must be designed in detail: knowing how to do it; who and where it will take place; when; what stages it will take; to which groups it will be directed, etc. Then carry out the intervention as planned and later evaluate if the intervention has been successful. Finally, it is essential that there is a follow-up and subsequent control to avoid that the risk factor turn again. All this must be complemented with a specificity of solutions regarding the context and using professionals with experience in the field of organizational intervention. Last but not least, there should be worker participation allowing their collaboration in the intervention. In short, a global and integrated intervention within the organization would be the most appropriate to prevent or solve the problems of stress in companies.

Measures of a global, organizational and collective nature are the measures that are usually recommended in organizational environments. However, there is another type of individual intervention that would consist of providing workers with strategies to adapt to the work environment. All individuals are exposed to stress and its consequences. Therefore, stress prevention is the optimal option. But, what are some of the preventive strategies that can serve as protective factors? Here there are some of the stress prevention strategies that, from Psychology, are recommended most frequently:

- Having social support: as mentioned previously, social support buffer stress.

- Training of social skills: being able to relate more efficiently, be assertive and properly handle criticism would be protective factors against stress.
- Change negative thoughts into positive ones: negative thoughts have a direct influence on how people value their self-concept and self-esteem, reducing them and believing themselves less able to cope with stress, with its consequent harmful consequences.
- Interpret properly difficult events: assess them as a challenge and not as a threat.
- Time management: a good distribution of tasks helps to keep calm and reduces anxiety and stress.
- Time for leisure: leisure is necessary to break the monotony, which also generates stress, and it is important to free oneself from the burdens for a while.
- Maintain an active style of coping: it is possible to request help and even receive training to know how to perform a task better, or solve a problem, but it is convenient to face it and not procrastinate, otherwise, it will last over time and its effects will be worse.

Both preventive and intervention strategies on stress are intended to increase control in workers. However, although many companies know and follow the guidelines mentioned above and aim to achieve such objectives that would result in health, employees are at the same time facing with contradictory results: in the 21st century a real pandemic of stress is raging and causing harm. These facts lead researchers to think that, perhaps, due to many efforts to intervene in organizations, the focus should be placed on the knowledge provided by the behavioral and neurosciences as key disciplines to understand the human being in interaction with their psychosocial and organizational environment.

Beyond the Interventions That Reduce Stress

From Psychology and Neuroscience many and very diverse techniques are known that people can use to learn how to manage and reduce stress. However, generally within organizations, this knowledge is not usually taken into account or is unknown. These facts lead to think that, perhaps, in spite to many efforts made to intervene in organizations, the focus should be placed on the individual, going beyond a much more preventive level.

From the psychology point of view, there are several stress interventions programs that are effective to reduce the consequences of stress or to cope with it. All of them are adequate to teach workers in order to reduce the negative effects of stress. Briefly, stress interventions can be classified depending on the principal effects: physiological effects (reducing the levels of physiological arousal produced by the stress; i.e relaxation or breathing techniques) or cognitive effects (modify the coping of people in stressful situations; i.e. inoculation of stress, mindfulness, problem-solving, among others). All these techniques are adequate to reduce the negative effects of stress or improve coping. However, these techniques are relegated to the individual. Thus, it is found that work stress generates many negative consequences such as stress-related diseases (cardiovascular, metabolic), performance reduction and job dissatisfaction. In the case of diseases related to stress, science is emphasizing the importance of encouraging healthy life habits to avoid or reduce diseases such as cardiovascular, metabolic or psychological that are directly related to stress. In this sense, the promotion of a healthy lifestyle entails the realization of physical activity and a healthy diet. It has been demonstrated that the performance of moderate and regular physical activity prevents the appearance of cardiovascular and psychological diseases (Mikkelsen, Stojanovska, Polenakovic, Bosevksi & Apostolopoulos, 2017), as well as a healthy diet reduces the risks of cardiovascular and metabolic diseases (Dawson, Dash & Jacka, 2016; Hoerr, Fogel & Van Voorhees, 2017).

Therefore, in addition to the actions that can be taken by organizations to reduce stress for workers, certain behaviors or a healthy lifestyle can reduce the likelihood of stress-related illnesses or allow a better coping with stress (Adler, 2007). In this regard, from the perspective of the authors of this chapter, it is considered that stress reduction or prevention programs should be complemented with distal prevention programs from the work situation itself; that is, the proposal of this chapter would be that, given that a healthy lifestyle prevents the appearance of some diseases related to stress and that there are psychological techniques that improve stress behaviors, prevention should be focused on anti-stress lifestyles, that at least, reduce the probability of the appearance of the negative consequences of stress. We consider that these healthy behaviors should be part of the lifestyle of people, so it should be taught from school, fostering in children physical activity and healthy eating as basic elements of health in general, as well as teaching of psychological techniques (for example, relaxation, problem solving or mindfulness techniques, among others) that would be useful to shape a healthy psychological lifestyle. All this would result in a better coping with work environments, based on a lifestyle that would be contrary to stress and therefore its negative consequences. In this sense, Neuro-education is a promising new perspective (Monfardini, Reynaud, Prado & Meunier, 2017) that links neuroscience and education in order to "to create a better understanding of how people learn and how this information can be used to create more effective teaching methods, curricula, and educational policy" (Carew & Magsamen, 2010).

Therefore, the proposal of this chapter aims at a distal prevention, educating people from the school in healthy lifestyle (physically, psychologically and socially), using the knowledge of neurosciences, who will later face their jobs and sources of stress in a more efficient way and who will also have an organism more prepared to suffer the negative consequences of stress. On the other hand, and associated with the above, it is also considered that it would be necessary to train in a "universal" occupational risk prevention from school in order to teach children their labor rights and to avoid, for example, the abuses / excesses that may be committed in the work context, as well as, his obligations and responsibilities. This section emphasizes the need to train workers in the limits of what is healthy and what is not; from the experience of the authors of this chapter, they have detected workers who consider normal tolerate high levels of stress or labor conflicts, which gives them to understand that workers assume certain working conditions as normal, when they may not be. That is why, in such risk prevention training it would be necessary to emphasize self-awareness about the protection of one's health; that is, working people must know discern when working conditions can cause health problems.

In short, the proposal of this chapter proposal aims to promote healthy lifestyles to have workers who know to manage their work so that reduce the negative consequences. In this way, employees would have committed, and they would know their rights and that would make them more active in their jobs and in the long run they would get sick less, producing, ultimately, greater benefits.

CONCLUSION

In conclusion, this proposal is based on improving life-style, education for health and self-respect, considering that a long-term prevention will be more effective than a short-term preventive approach. The reach that are expected from this proposal is not only prevent psychosocial risk factors (stress or burnout) but physical risks, taking into account that the proposal is focused on the individual as an active agent that is worried (and responsible) about his health; that is, the most important thing is that workers must take care of their own. This perspective goes beyond organizations but governments, trying to capture the attention of politicians to propose a complementary model of education useful to learn behaviors about healthy work behaviors and, in consequence, to have healthy workplaces.

In short, the most effective is to make global interventions that reduce psychosocial risk factors but that take into account the worker in terms of their abilities to cope with stress. Consequently, this comprehensive intervention will improve the functioning of stressed people making it more efficient in the environments as well as improving the organization, in terms of productivity and quality.

REFERENCES

Adler, N. E. (2007). *Reaching for a Healthier Life: Facts on Socioeconomic Status and Health in the U.S.* J.D. and C.T. MacArthur Foundation Research Network on Socioeconomic Status and health.

Almagiá, E. B. (2014). Apoyo social, estrés y salud. *Psicología y salud, 14*(2), 237-243

Aminoff, E. M. (2009). The landscape of cognitive neuroscience: challenges, rewards, and new perspectives. In M. S. Gazaniga (Ed.), *The cognitive Neurosciences* (4th ed.; pp. 1255–1262). MIT.

Aranda, C., Pando, M., Salazar, J. G., Torres, T. M., Aldrete, M. G., & Pérez, M. B. (2004). Factores psicosociales laborales y síndrome de burnout en médicos del primer nivel de atención. *Investigación en salud, 6*(1).

Ariely, D., Gneezy, U., Loewenstein, G., & Mazar, N. (2009). Large Stakes and Big Mistakes. *The Review of Economic Studies, 76*(2), 451–469. doi:10.1111/j.1467-937X.2009.00534.x

Ashkanasy, N. M., Wilderom, C. P. M., & Peterson, M. F. (2011). *The Handbook of Organizational Culture and Climate*. Sage.

Bakker, A., & Albrecht, S. (2018). Work engagement: Current trends. *Career Development International, 23*(1), 4–11. doi:10.1108/CDI-11-2017-0207

Bal, P. M., De Lange, A. H., Ybema, J. F., Jansen, P. G., & Van der Velde, M. E. (2011). Age and trust as moderators in the relation between procedural justice and turnover: A large-scale longitudinal study. *Applied Psychology, 60*(1), 66–86. doi:10.1111/j.1464-0597.2010.00427.x

Barkhuizen, N., Rothmann, S., & Vijver, F. J. (2014). Burnout and work engagement of academics in higher education institutions: Effects of dispositional optimism. *Stress and Health, 30*(4), 322–332. doi:10.1002mi.2520 PMID:23949954

Barra, E. (2004). Apoyo social, estrés y salud. *Psicología y Salud, 238*(2), 237-243.

Borrás, F. X. (1995). Psiconeuroinmunología: Efectos del estrés psicológico sobre la función inmune en humanos sanos. *Ansiedad y Estrés, 1*(1), 21–35.

Brodman, C. (1993). Social relationships and health-related behavior. *Journal of Behavioral Medicine, 16*(4), 335–351. doi:10.1007/BF00844776 PMID:8411141

Cacioppo, J. T., Bernston, G. G., & Decety, J. (2010). Social neuroscience and its relationship to social psychology. *Social Cognition, 28*(6), 675–685. doi:10.1521oco.2010.28.6.675 PMID:24409007

Caetano, A., & Santos, S. C. (2017). The Gap Between Research and Professional Practice in Work and Organizational Psychology: Tensions, Beliefs, and Options. In E. R. Neiva, C. Vaz, & H. Mendonça (Eds.), *Organizational Psychology and Evidence-Based Management*. Springer. doi:10.1007/978-3-319-64304-5_1

Carew, T. J., & Magsamen, S. H. (2010). Neuroscience and Education: An ideal partnership for producing evidence-based solutions to guide 21st century learning. *Neuron*, *65*(5), 685–688. doi:10.1016/j.neuron.2010.08.028 PMID:20826300

Cheng, S. (2014). The mediating role of organizational justice on the relationship between administrative performance appraisal practices and organizational commitment. *International Journal of Human Resource Management*, *25*(8), 1131–1148. doi:10.1080/09585192.2013.816864

Chib, V. S., De Martino, B., Shimojo, S., & O'Doherty, J. P. (2012). Neural Mechanisms Underlying Paradoxical Performance for Monetary Incentives Are Driven by Loss Aversion. *Neuron*, *74*(3), 582–594. doi:10.1016/j.neuron.2012.02.038 PMID:22578508

Cohen, B. E., Edmondson, D., & Kronish, I. M. (2015). State of the art review: Depression, stress, anxiety, and cardiovascular disease. *American Journal of Hypertension*, *28*(11), 1295–1302. doi:10.1093/ajh/hpv047 PMID:25911639

Colligan, T. W., & Higgins, E. M. (2005). Workplace stress: Etiology and consequences. *Journal of Workplace Behavioral Health*, *21*(2), 89–97. doi:10.1300/J490v21n02_07

Cullen, A. (1995). CE Credit: Burnout: Why Do We Blame the Nurse? *The American Journal of Nursing*, *95*(11), 23–28. doi:10.1097/00000446-199511000-00017 PMID:7485275

Dawson, S. L., Dash, S. R., & Jacka, F. N. (2016). The importance of diet and gut health to the treatment and prevention of mental disorders. *International Review of Neurobiology*, *131*, 325–346. doi:10.1016/bs.irn.2016.08.009 PMID:27793225

Gallo, I. S., Cohen, A. L., Gollwitzer, P. M., & Oettingen, G. (2013). Neurophysiological correlates of the self-regulation of goal pursuit. In P. A. Hall (Ed.), *Social Neuroscience and Public Health* (pp. 19–34). New York: Springer. doi:10.1007/978-1-4614-6852-3_2

Gómez-González, B., & Escobar, A. (2006). Estrés y sistema inmune. *Rev Mex Neuroci*, *7*(1), 30–38.

Gracia, E. (2015). Stress, coping, and work engagement within the specific job context: Comment on Kaiseler, et al. (2014). *Psychological Reports*, *116*(2), 377–380. doi:10.2466/01.20.PR0.116k22w1 PMID:25799121

Grau, R., Salanova, M., & Peiró, J. M. (2012). Efectos moduladores de la autoeficacia en el estrés laboral. *Apuntes de Psicología*, *30*(1-3), 311–321.

Hakulinen, C., Pulkki-Råback, L., Jokela, M., Ferrie, J. E., Aalto, A. M., Virtanen, M., ... Elovainio, M. (2016). Structural and functional aspects of social support as predictors of mental and physical health trajectories: Whitehall II cohort study. *Journal of Epidemiology and Community Health*, *70*(7), 710–715. doi:10.1136/jech-2015-206165 PMID:26767407

Hoerr, J., Fogel, J., & Van Voorhees, B. (2017). Ecological correlations of dietary food intake and mental health disorders. *Journal of Epidemiology and Global Health*, *7*(1), 81–89. doi:10.1016/j.jegh.2016.12.001 PMID:28003094

Holt-Lunstad, J., & Uchino, B. (2015). Social support and health. *Health behavior: Theory, research and practice*, 183-204.

Kaplan, R. M., Sallis, J. F., & Patterson, T. L. (1993). *Health and human behavior*. New York: Mcgrath-Hill.

Karasek, R., & Theorell, T. (1990). *Healthy work. Stress, productivity, and the reconstruction of working life*. New York: Basic Books.

Kivimäki, M., & Kawachi, I. (2015). Work stress as a risk factor for cardiovascular disease. *Current Cardiology Reports*, *17*(9), 74. doi:10.100711886-015-0630-8 PMID:26238744

Lagraauw, H. M., Kuiper, J., & Bot, I. (2015). Acute and chronic psychological stress as risk factors for cardiovascular disease: Insights gained from epidemiological, clinical and experimental studies. *Brain, Behavior, and Immunity*, *50*, 18–30. doi:10.1016/j.bbi.2015.08.007 PMID:26256574

Lazarus, R. S., & Folkman, S. (1984). *Stress, Appraisal and Coping*. New York: Springer.

Martínez, V., Ramos, J., & Moliner, C. (2015). *Psicología de las organizaciones*. Madrid: Síntesis.

Maslow, A. (1954). *Motivation and Personality*. Harper.

Maslow, A. (2016). *El hombre autorrealizado: hacia una psicología del ser*. Editorial Kairós.

Michie, S. (2002). Causes and management of stress at work. *Occupational and Environmental Medicine*, *59*(1), 67–72. doi:10.1136/oem.59.1.67 PMID:11836475

Mikkelsen, K., Stojanovska, L., Polenakovic, M., Bosevski, M., & Apostolopoulos, V. (2017). Exercise and mental health. *Maturitas*, *106*, 48–56. doi:10.1016/j.maturitas.2017.09.003 PMID:29150166

Monfardini, E., Reynaud, A. J., Prado, J., & Meunier, M. (2017). Social modulation of cognition: Lessons from Rhesus macaques relevant to education. *Neuroscience and Biobehavioral Reviews*, *82*, 45–57. doi:10.1016/j.neubiorev.2016.12.002 PMID:27923731

Moon, T., Hur, W., Ko, S., Kim, J., & Yoon, S. (2014). Bridging corporate social responsibility and compassion at work: Relations to organizational justice and affective organizational commitment. *Career Development International*, *19*(1), 49–72. doi:10.1108/CDI-05-2013-0060

Moriano-Leó̃n, J. A. (2005). La comunicació̃n organizacional. In Psicología de las organizaciones. Pearson, Prentince Hall.

Negruti, A., Hristova, P., Larsen, K., & Krumov, K. (2015). *Positive Organizational Psychology: Advances in Creating Improved Workplaces and Employee Well-Being*. Kassel University Press GmbH.

Norton, M., Mochon, D., & Ariely, D. (2010). *The IKEA Effect: When labor leads to love*. Universidad de Hardvard.

Omar, A. (2015). Justicia organizacional. In H. F. Littlewoord & S. A. Vega (Eds.), *Psicología Industrial-Organizacional. Una visión latinoamericana*. Instituto Tecnológico y de Estudios Superiores de Monterrey.

Ostroff, C., & Bowen, D. E. (2000). Moving HR to a higher level: HR practices and organizational effectiveness. In K. J. Klein & S. W. J. Kozlowski (Eds.), *Multilevel theory, research, and methods in organizations: Foundations, extensions, and new directions* (pp. 211–266). San Francisco, CA: Jossey-Bass.

Padula, R. S., Chiavegato, L. D., Cabral, C. M. N., Almeid, T., Ortiz, T., & Carregaro, R. L. (2012). Is occupational stress associated with work engagement? *Work (Reading, Mass.)*, *41*(Supplement 1), 2963–2965. PMID:22317170

Parker, C. P., Baltes, B., Young, S., Altmann, R., LaCost, H., Huff, J., & Roberts, J. E. (2003). Relationships between psychological climate perceptions and work outcomes: A meta-analytic review. *Journal of Organizational Behavior*, *24*(4), 389–416. doi:10.1002/job.198

Patterson, M., West, M., Shackleton, V. J., Dawson, J. F., Lawthom, R., Maitlis, S., ... Wallace, A. (2005). Validating the organizational climate measure: Links to managerial practices, productivity and innovation. *Journal of Organizational Behavior*, *26*(4), 379–40. doi:10.1002/job.312

Peiró, J. M. (1983-1984). *Psicología de la organización.* UNED.

Pellegrin, K., & Currey, H. (2011). Demistifyin and improving organizational culture in health-care. In J. Wolf, M. Moir, H. Hanson, L. Friedman, & G. Savage (Eds.), Organization Development in Healthcare: Conversations on Research and Strategies. Emerald Group Publishing.

Pellicer, O., Salvador, A., & Benet, I. A. (2002). Efectos de un estresor académico sobre las respuestas psicológica e inmune en jóvenes. *Psicothema*, *14*(2).

Peña-Ochoa, M., & Durán Palacio, N. M. (2016). Justicia organizacional, desempeño laboral y discapacidad. *Revista Colombiana de Ciencias Sociales*, *7*(1), 201–222. doi:10.21501/22161201.1540

Prado, J. F. U. (2014). *Clima y ambiente organizacional: trabajo, salud y factores psicosociales*. Editorial el manual moderno.

Realyvásquez, A., Maldonado-Macías, A. A., García-Alcaraz, J., Cortés-Robles, G., & Blanco-Fernández, J. (2016). Structural Model for the Effects of Environmental Elements on the Psychological Characteristics and Performance of the Employees of Manufacturing Systems. *International Journal of Environmental Research and Public Health*, *13*(1), 104. doi:10.3390/ijerph13010104 PMID:26742054

Reichers, A. E., & Schneider, B. (1990). Climate and culture: An evolution of constructs. In B. Schneider (Ed.), *Organizational climate and culture* (pp. 5–39). San Francisco, CA: Jossey-Bass.

Salanova, M. (2008). Organizaciones saludables y desarrollo de recursos humanos. *Estudios Financieros*, *303*, 179–214.

Salanova, M. (2009). Prevención de riesgos laborales en tiempos de crisis. *Gestión Práctica de Riesgos Laborales, 58.*

Salanova, M., Martínez, I. M., & Llorens, S. (2005). Psicología Organizacional Positiva. In *Psicología de la Organización* (pp. 349–376). Madrid: Pearson Prentice Hall.

Salanova, M., Martínez, I. M., & Llorens, S. (2014). Una mirada más "positiva" a la salud ocupacional desde la psicología organizacional positiva en tiempos de crisis: Aportaciones desde el equipo de investigación WoNT. *Papeles del Psicólogo*, *35*(1).

Salas-Arbeláez, L., Solarte, M. G., & Vargas, G. M. (2017). Efecto de la cultura organizacional en el rendimiento de las PYMES de Cali. *Suma de Negocios*, *8*(18), 88–95. doi:10.1016/j.sumneg.2017.11.006

Schaufeli, W. B., Salanova, M., González-Romá, V., & Bakker, A. B. (2002). The measurement of engagementand burnout and: A confirmative analytic approach. *Journal of Happiness Studies*, *3*(1), 71–92. doi:10.1023/A:1015630930326

Sedgwick, P., & Greenwood, N. (2015). Understanding the Hawthorne effect. *British Medical Journal*, *351*, h4672. doi:10.1136/bmj.h4672 PMID:26341898

Seligman, M., & Csikszentmihalyi, M. (2000). Positive Psychology: An introduction. *The American Psychologist*, *55*(1), 5–14. doi:10.1037/0003-066X.55.1.5 PMID:11392865

Selye, H. (1936). A síndrome produced by diverse nocouous agents. *Nature*, *138*(3479), 32. doi:10.1038/138032a0

Selye, H. (1950). *Stress: the physiology and pathology of exposure to stress*. Montreal: Acta Inc.

Serrano, M. A., & Costa, R. (2018). *Stressin the stress or the complexity of the human factor:* psychobiological *consequences of distress*. In J. L. García-Alcaraz, G. Alor-Hernández, A. A. Maldonado-Macías, & C. Sánchez-Ramírez (Eds.), *New Perspectives on applied industrial tools and techniques* (pp. 431–447). Springer. doi:10.1007/978-3-319-56871-3_21

Simon, H. A. (2000). Barriers and bounds to Rationality. *Structural Change and Economic Dynamics*, *11*(1-2), 243–253. doi:10.1016/S0954-349X(99)00022-3

Sonnentag, S., Mojza, E. J., Demerouti, E., & Bakker, A. B. (2012). Reciprocal relations between recovery and work engagement: The moderating role of job stressors. *The Journal of Applied Psychology*, *97*(4), 842–853. doi:10.1037/a0028292 PMID:22545619

Sorensen, J. (2002). The strength of corporate culture and the reliability of firm performance. *Administrative Science Quarterly*, *47*(1), 70–91. doi:10.2307/3094891

Stein, D. J., He, Y., Phillips, A., Sahakian, B. J., Williams, J., & Patel, V. (2017). Global mental health and neuroscience: Potential synergies. *Lancet*, *2*(2), 178–185. PMID:26359754

Sunstein, C. R., & Thaler, R. H. (2003). Libertarian paternalism is not an oxymoron. *The University of Chicago Law Review. University of Chicago. Law School*, *70*(4), 1159–1202. doi:10.2307/1600573

Taylor, F. W. (1984). *Management científico*. Madrid: Orbis.

Thaler, R. H. (2016). Behavioral economics: Past, present, and future. *The American Economic Review*, *106*(7), 1577–1600. doi:10.1257/aer.106.7.1577

Thatcher, A., Waterson, P., Todd, A., & Moray, N. (2017). State of Science: Ergonomics and global issues. *Ergonomics*, *61*(2), 197–213. doi:10.1080/00140139.2017.1398845 PMID:29076757

Tversky, A., & Kahneman, D. (1974). Judgment under Uncertainty: Heuristics and Biases. *Science, 185*(4157), 1124–1131. doi:10.1126cience.185.4157.1124 PMID:17835457

Vaca, C. A. M., Vaca, L. O. M., & Quintero, J. N. (2015). El clima organizacional y la satisfacción laboral: Un análisis cuantitativo riguroso de su relación. *AD-Minister*, (26): 5–15.

Veen, P., & Korver, T. (1998). Theories of organization. In Handbook of work and organizational psychology: Vol. 3: Organizational psychology. Psychology Press.

Vega, M. M. C., Gálvez, S. A. H., & Santamaría, J. (2017). Organizational Climate and Psychological Health: An Organizational Duality. *Dimensión empresarial, 15*(1), 63-76.

Waring, S. P. (2016). *Taylorism transformed: Scientific management theory since 1945*. UNC Press Books.

Wei-Yuan, G., Shang-Ping, L., Chwei-Jen, F., & Chin-Fang, Y. (2013). Effects of organizational justice on organizational citizenship behaviors: Mediating effects of institutional trust and affective commitment. *Psychological Reports, 112*(3), 818–834. doi:10.2466/01.21.PR0.112.3.818-834 PMID:24245075

Wilson, M. G., Dejoy, D. M., Vandenberg, R. J., Richardson, H. A., & McGrath, A. L. (2004). Work characteristics and employee health and wellbeing: Test of a model of healthy work organization. *Journal of Occupational and Organizational Psychology, 77*(4), 565–588. doi:10.1348/0963179042596522

Zambrano, I. I. B., Véliz, V. M. B., & Barzola, W. J. F. (2017). Factores del clima laboral que influyen en el rendimiento de los trabajadores del sector público en el Ecuador. *Dominio de las Ciencias, 3*(3), 917–937.

Section 3
Macroergonomic Applications

Chapter 11
Organizational Development in Improving Operations of a Language Center:
Impact on Development of Students

Luz Elena Tarango
Instituto Tecnológico de Ciudad Juarez, Mexico

Manuel Alonso Rodriguez-Morachis
Instituto Tecnologico de Ciudad Juárez, Mexico

Yolanda Frausto
Instituto Tecnológico de Ciudad Juárez, Mexico

Edgardo de Jesus Rojas
Instituto Tecnologico de Ciudad Juarez, Mexico

Marisela Lucero Gaytán
Instituto Tecnologico de Ciudad Juarez, Mexico

ABSTRACT

In the period of 2013 through 2016, several actions were implemented to improve the English communication skills of the students of Instituto Tecnológico de Ciudad Juárez (ITCJ). These actions seek to comply with the expectations defined by employers, then to satisfy the English language proficiency test for the students to be able to graduate from their career and lastly to meet the requirements stablished by Tecnológico Nacional de México (TecNM) contained in the 2013-2018 Institutional Innovation and Development Program (IIDP). All these actions were accomplished applying a systemic approach and organizational development.

DOI: 10.4018/978-1-5225-7192-6.ch011

INTRODUCTION

The Instituto Tecnológico de Ciudad Juárez (ITCJ) is a higher education institution belonging to a national system of technological institutions in Mexico. It is composed of more than 250 institutions throughout the country, with an approximate student population of 500,000. In order to help students reach an appropriate level of English, they have the language centers within the institutions, which contribute to the development of this skill so essential for their development and professional growth, since in the core curriculum is not included. This project was carried out in an institution located on the northern border of Mexico-United States, which has a high demand for its students and graduates in the industrial and commercial sector, many of whom already work in those companies before graduation, but these increasingly require and need them with a high command of English, since most of these are foreign companies. Starting in 1999, Dirección General de Institutos Tecnológicos (DGEST), nowadays Tecnológico Nacional de México (TecNM), started to require all graduates to possess a certain level of English before graduating. As time went by, this requirement became more demanding and measurable. By 2015, it is required for the students to prove a B1 level in the Common European Framework of Reference (CEFR) of a demand. This requirement posed a great challenge to the Language Center, which motivated a change on the requirements causing a great deal of stir among students for they have to invest more time in order to finish their career. In 2013, it was required as part of its strategic planning to expand the Language Center when the number of enrolled students doubled. Also, more groups were opened and more teachers were hired.

In the same year, the demand for ITCJ graduates with a high command of English language increased, this was expressed in meetings of CONREDES (Regional Council for the Development of Education and Sustainability) members. In addition to the provisions of the TecNM requiring students to meet the requirement of English to graduate, students should have at least level B1. This requirement increased the student enrollment in the Language Center by more than 300%. In order to face this situation, immediate containment measures were taken that later concluded in a strategic planning of the change. All these drastic changes required planning, for it involved the administrative staff and teachers to participate in the generation of ideas, development of programs and their implementation, the steps that were used were: assessing the environment, determining performance measurements, diagnosing organizational problems, determining and communicating a vision for the future, developing and implementing an action plan, anticipating the resistance and taking actions to reduce it and, above all, monitoring the changes. All these analyzes were reflected in the changes that were made as long as the internal policy of the institution and the TecNM allowed.

All these changes have benefited the students of the ITCJ, who mostly come from the lower middle class. By optimizing resources, it has been possible to maintain the hour-class cost at only $9 Mexican pesos (approximately half US dollar), this means it is the lowest cost in the entire state of Chihuahua, Mexico. It benefits them when they graduate (or in the development of professional practices), so that they can be inserted more easily in the labor field, which for 2017 had 83% of the graduates working in their areas of competence. In addition, when they handle a second language there are more possibilities for job promotions, salary increases and if they wish, they can aspire to scholarships for postgraduate studies abroad, opportunities that are not taken advantage of when they do not have a second language.

BACKGROUND

The Language Center at ITCJ is an additional service offered to students and the general public. In 2013, the volume of students enrolled in English classes did not reach 700 students per year. It has always been intended for the Language Center to be self-sustainable, that is, with the income received from student registration, office supplies and teachers can be paid. In 2009, internal students (students of the ITCJ) paid only $800 Mexican pesos and nowadays in 2017, enrollment cost has only increased $100 Mexican pesos. This fee still remains sustainable and is the lowest cost throughout the city and the state of Chihuahua, at only $9 Mexican pesos per hour-class.

On the other hand, an Interchange® editorial was managed that has a prestigious content as a textbook, but very high cost for young people. In 2014, the TecNM suggested three different publishers to choose one that was accessible and met the quality standards and that would ensure the B1 level in at least 400 class hours, so they had to review the content by the teachers and choose one that provides the best conditions.

The administrative control carried out in 2010, was based on Excel® spreadsheets that only served to register students in groups, and if needed to obtain statistics, elaborate reports, scorecards or consult the history of any student. It had to be done manually on the printed sheets. In 2010, there were only 12 teachers who were enough to cover the demand, but with little class hours assigned forced some teachers teach English at other institutions. The new provisions for courses offered at the ITCJ have changed the English proficiency requirements to graduate (Tecnológico Nacional de México, 2010):

- For the programs of 2009 and before, it mentions: "Presenting a written exam, where the understanding of technical-scientific articles of his area of studies in the English language is demonstrated" (Tecnológico Nacional de México, 1999).
- For the plans as of 2010, it specifies in the 2010 Authorized Study Plans of those years: *"It must validate the competence of oral and written communication in a foreign language in order to graduate"*.
- For programs starting in 2015, it defines: "The certificate must validate the ability to understand, elaborate and communicate, orally and in writing, standard notions in the personal and professional field, in accordance with CEFR Level B1 for English language (Tecnológico Nacional de México, 2015).

As it can be observed, these modifications are radical and substantial that caused changes to improve the operation of the Language Center, which impacted the students positively and negatively since to fulfill the requirement they would have to invest more time and effort to reach it.

In addition to all the above, the industrial and commercial sector requires more and more practitioners, residents and graduates with a higher level of English, this only to aspire to a position within the workplace. Also, reviewing the answers of the graduates' follow-up, they comment that English is needed to reach promotions, or to get scholarships for postgraduate studies. especially for foreigners. There are also calls for academic exchanges abroad to which the students have not been able to aspire because they are asked to have a B1 of English. All this has led to change the way of operating the language center.

Problem Statement

Under current conditions, students and graduates are not being able to achieve a high command of the English language, since Language Centers do not have the ideal conditions to respond to the demands of the environment, nor is it known if they will comply with the qualification requirements imposed by the TecNM, and it is unknown if the goal established by the ITCJ will be reached in the Institutional Program of Innovation and Development 2013-2018. On the other hand, the current database made in Excel®, for registration, monitoring, control records, student and teacher's records does not support the administration of the service that is being granted nor is it apt to obtain from it statistics.

Justification

According to interviews with some local employers, ITCJ graduates are recognized for having excellent technical training, however, due to the geographical location of Juárez, they must be skillful communicators in English. This is one of the main obstacles graduate students should overcome when looking for a job in their field of study. As it is known, competence on a second language is a necessity for today's professionals, especially when, as in this case, the city borders with the United States. During the analysis conducted using the results of the Industrial Engineering program graduates´ follow up surveys, it was found that their professional promotion or growth within any organization has been limited by the level of English they master (González-Martínez, 2015). This is why it is fully justified to conduct an analysis of the impact of the undertaken actions. To lower resistance to change, it is necessary to consider the implementation of organizational development to help strengthen the operation of the Language Center.

Implementing changes is a complex task involving redesigning structures, processes, and routines within the organization. These modifications must be planned properly and must respond to the proposed goals and objectives. They are adjustments in a progressive way; innovations that are implemented in the organization as a whole; they must be done with a systemic approach.

Objectives

Implement the necessary changes in the operation of the Language Center so that the students who carry out their studies in it, reach the level demanded by the working environment and can meet the qualification requirement and at the same time are ready to access graduate studies.

As part of the continuous improvement effort, all the information related to actions and changes performed must be documented and available. It is essential to know its impact and define a standard to set new goals which allow progress; this project aims to gather evidence and document the implementation of these changes, which occurred in the last four years. All such efforts are performed for improving the teaching and learning of the English language at all levels taught in the Language Center and at the same time to meet the requirements of the TecNM.

Hypotheses

Based on the demands of the environment and the requirements of the TecNM, as well as in the impact on the goals established by the ITCJ in the 2013-2018 Innovation and Development Institutional Program, the following hypothesis is proposed:

The percentage of students enrolled in the Language Center is at least 60% the total number of students enrolled in the ITCJ for the year 2018 and 2019, with respect to the current situation. As the ITCJ has student enrollments twice a year, therefore an average of students per year (μ) was calculated.

Null and alternative hypotheses (H_0 and H_1, respectively) are expressed in Equation (1):

$$H_0 : \mu_{before} \leq \mu_{after} \tag{1}$$

The historical data of the number of students from 2010 to 2017 for both the ITCJ and the students of the Language Center are known, Minitab® was used to elaborate the percentage projection for 2018 and 2019, as well as the population of students using forecast models.

The assumption made is that, with all the actions undertaken, not only the volume of students in the language center will be reached, but that the number of students who reach the CEFR B1 will be increased.

THEORETICAL AND CONTEXTUAL FRAMEWORK

Conceptualization of Organizations

Garbanzo-Vargas (2016) mentions that global trends associated with globalization and the so-called information and knowledge society, manifest the demand for transformations and imposes challenges for organizations, especially educational ones, for their formative role. In this context, organizations must respond with solidity and relevance, ensuring their social viability.

Organizations cannot manifest certain behaviors, these are the behaviors of its individual and group members: they participate and perform in different modalities according to the nature of the same organization. These do not work at random and the success that is achieved is the product of a set of synergies that, conducted in a certain direction, achieve it. Success does not come at random, organizations are social cells that respond to stimuli from the internal and external environment and, according to these, are their products, their behaviors.

Organizations require leadership, an administration to initiate organizational development. In addition, to locate the environmental characteristics, it is pertinent to know the human capital that counts, their talents based on the contribution expected from each of the parties. According to Chiavenato (2009), there are three independent variables that need to be understood in organizations for the achievement of the stated objectives, which are the basis for organizational development: the organizational system, the group level, and the individual level.

Organizational Change

Organizational change is considered by Hellriegel, Jackson and Solcum (2005) as any transformation associated with the design or operation of an organization. Efficient and effective management knows the opportune moment when a change is required and also knows the strategies to lead the organization through the process of change. Changes can be promoted by internal or external forces. Externally, it usually happens as a consequence of the social transformations resulting from the dynamic and complex society, as well as knowing good practices applied in other organizations. Internally, changes are due to the organizations' own needs such as the normal processes of structural adjustments and changes in its objectives.

Internally, due to the need to adopt new processes, objectives, and methods, the organization is likely to require transcendental organizational changes. An example of these necessary changes has been the implementation of technologies in organizations, which has come to represent new methods and processes in their operation. Changes in organizations can occur radically or in a slow way.

Change in organizations can be defined in degrees, as radical or progressive and by chance, as reactive or anticipatory (Hellriegel et al., 2005). The radical changes occur when organizations produce innovations in their methods of doing things, it is the rethinking of the organization, and they are usually difficult to manage. The "slow" or progressive change is a process of continuous evolution over time, where various adjustments are made to a lesser degree to the internal processes of the organization.

Lewin, cited in Hellriegel et al. (2005), refers a series of steps for the radical change process:

- **Unfreezing:** Preparing some management members to be accountable for an important transformation.
- **Transitioning:** Practical response to change.
- **Refreezing:** Consolidating change. At this stage, a constant revision is required so that the members who implement the change do not fall back into the old habits.

Change can occur at two specific moments: when organizations experience a deterioration in their reactive performance or in advance, thanks to the vision of leadership. In the latter case, as there is no obvious crisis, these can be properly planned and executed gradually. The ability to efficiently respond to the different changes that organizations face, determines their sustainability and social positioning. In general, any organization capable to maintain high levels of positioning and sustainability, internally possess a culture of systemic dynamism towards the environment; and - internally - they have proactive, not reactive, vision possessing the ability to identify efficient actions to implement change, as well as to know about the opportune moment to act.

In this regard, organizations can apply different actions for the effective implementation of change to improve it: gradually develop initiatives in the methods used with a constant review in the evolution of these until achieving the projected action, according to the objective to be achieved. Some of the actions that organizations can apply are the following:

- Ability to influence others, projecting and obtaining legitimacy that allows its members to channel their efforts towards the defined direction, implies credibility on the leader of the organization towards the expected change.
- Developing effective skills to obtain from its members the commitment and responsibility required, implies the ability to persuade in search of the projected objectives. Acting consistently in the projected direction is knowing how to deflect the obstacles that usually arise in any process of change and assume a persistent attitude until reaching the objective.

A change requires planning for it to be more orderly and allowing people to prepare and participate in planning through different processes. In this case, planning seeks to provide members of the organization with useful means to address them successfully, despite the demands from both the organization and society; an increasingly critical and informed society about the quality of the services it receives and that are indispensable for their permanence. The processes to consider for the effective planning of change, according to Hellriegel et al. (2005) are:

- Evaluate the environment.
- Determine the performance gap.
- Diagnose organizational issues.
- Define and communicate the vision.
- Develop and implement the action plan.
- Anticipate resistance and take action to reduce it.
- Monitor the changes.

Reorganizing to Implement Change

Garbanzo-Vargas (2016) mentions that the implementation of change is a complex task; since it involves redesigning structures, processes and routines within the organization. These modifications must be properly planned and must respond to the proposed goals and objectives. They are adjustments in a progressive way; innovations that are implemented in the organization as a whole; they must be done with a systemic approach. The trends associated with globalization require solid and pertinent responses to the complex chains of transformations and challenges to which organizations are embedded. In this regard, Hellriegel et al. (2005) warn that in some cases it implies, even, a total redesign or re-engineering of the structure, or partially, which implies reconfiguring the distribution of authority, middle management and departmental controls and job functions; all this to offer society a quality service in accordance with the needs. The organization to implement the planned changes requires adjusting its structures and procedures. The change is not always implemented without changing the organizational structure, which is why the theory of organizational development offers a series of considerations to take into account to apply transformations and lead the organization towards development.

Planning From a Systemic Perspective

Ackoff (1979) mentions that planning is a very effective way to achieve the goals and objectives set by organizations, for this it is necessary to consider the attitudes that managers can present in the process, such as inactive, reactive, active and interactive, which should be seen not as an obstacle but as an opportunity to enrich the process, the proactivists have drawn four planning principles that are:

- Planning for participation: The leader planner must help all the members of the organization to jointly carry out the plans, for this they must be provided with information, instruction and motivation.
- Coordinated planning: All aspects of the organization must be planned simultaneously and interdependently.
- Integrated planning: Refers to the fact that there must be planning at all levels of the organization.
- Continuous planning: As all systems constantly evolve, so should the plans.

Interactive planning is a system of activities, Ackoff (1979) considers the following five interdependent bases in planning:

- **Planning the Purpose:** Determining what is desired, it is necessary to specify goals, objectives and ideals; short, medium and long-term goals.
- **Planning the Means:** Determining how to achieve the objectives and goals, which requires establishing or creating course of action, practices, programs and policies.
- **Planning the Resources:** Defining the kind of resources needed and the amount of such, the way to acquire and generate them and their allocation to the activities once they are available.
- **Planning the Organization:** Determining the requirements in the organization, and if necessary, changes in the administration.
- **Planning the Implementation and Control:** Determining how to implement and control important administrative variables of the system.

Educational Organizations: Organizational Development With Focus on Systems

Organizational development has its origins in the behavioral sciences, and it is defined as a properly planned strategy and future projection, as it serves to understand, modify and develop the staff members to achieve effectiveness (Hellriegel et al., 2005). Organizational development considers the organization as a whole, it is not possible to conceive a change without considering each of its parts, each presenting a strategic function, conceived from the position it occupies: it is the only way to achieve the required changes with greater possibilities of effectiveness (Quiñones, Otarla, Ramos & Aguilera, 2008). It is essential to achieve the changes that organizational development inspires to consider different components: a holistic vision of the organization, systemic methodology, designation of agents of change, real identification of problems, constructivist learning, group processes, feedback, flexibility and contingency, and collaborative work.

In the early 1980's, a special connotation was given by governments, directors, intellectuals, international organizations such as the United Nations Educational, Scientific, and Cultural Organization (UNESCO), the World Bank, and the Inter-American Development Bank, to consider education as the main instrument to pursuit social development, and overcome existing gaps (Brunner, 2000). The understanding of this role within educational organizations is fundamental to trace the desired trajectory and efficiently and effectively respond to social transformations derived from the globalization phenomenon and the information and knowledge society. Moving away from this position poses the risk, to organizations, of responding in the wrong way according to their purposes.

It corresponds to the different educational levels and especially to those of higher education as an instrument that enhances economic, social and cultural changes, to efficiently assume their management processes; fostering education of greater social inclusion and, therefore, aspiring to create a society with greater inclusion, greater culture, higher levels of development, and ethical and moral values. To achieve this social relevance, it is essential to rely on a participatory management, built on a collective basis and with solid livelihoods regarding what happens in the environment. These conditions are the basis for the processes of innovation and change to occur, and to respond to the requirements of a holistically social development.

It is essential that those who have the responsibility to lead these organizations have solid knowledge of the general events at a global, regional and local level, as well as deep knowledge in the specific field of higher education, in order to maintain the social relevance as an institution. Tunnerman (2000) considers that efforts to achieve the relevance of higher education in the knowledge and information society require universities to rethink - creatively - their organizational doctrine, that is, their objectives, mission, and functions. That is, to reinvent them in their operation so that they are in articulation with the demands of the time. This condition also applies to the other educational levels.

Higher education institutions, besides forming competitive, responsible, supportive, pro-active people willing to contribute to the development of a globalized society, have a double moral responsibility: to serve the people and train responsible professionals, with high moral codes and ethics. In this context, social relevance must be the main challenge from the management of higher education, which, according to Tunnerman (2000), includes the commitment of higher education in accordance with the needs that society demands.

2013 - 2018 National Development Plan (2013-2018 NDP)

Published in the Official Journal of the Federation on May 20, 2013, five national goals and three cross-cutting strategies are established. The national goals are: Mexico in peace, an inclusive Mexico; a Mexico with quality education, a prosperous Mexico and a Mexico with global responsibility. For the cross-cutting, enforcement strategies for all agencies and organizations are: democratize productivity, closer and modern government and gender perspective. The achievement of each goal and strategy presupposes its own requirements, thus, the goal of a Mexico with quality education demands the efficient coordination and real fulfillment of the commitment of the different actors involved in it: Union Congress, in the corresponding approval of the necessary reforms, laws and regulations; the educational authorities, federal and state, in the timely allocation of greater resources for all types, levels and modalities of education; the managers, with the efficient and transparent exercise of those resources; teachers, through permanent academic improvement; parents, in their careful and responsible participation in the education of their children, and students, in their efforts to achieve professional, comprehensive and full training, to which they aspire. The Education Sector Program 2013-2018 (Programa Sectorial de Educación, PSE 2013-2018), published in the Official Journal of the Federation on December 13, 2013, establishes six objectives, five of which directly concern higher education: a) quality and relevance, b) coverage, inclusion and equity, c) physical and sports activities, d) art and culture, and e) scientific and technological education. The 2013-2018 PSE determines higher education is to improve in the following indicators: increase from 61.7 to 72% the total number of students enrolled in undergraduate programs recognized by their quality; increase the gross rate of student enrollment in higher education from 29.9% to 40%; increase to 10% the number of students in the National Sports Registry, and increase to 44% the number of students participating in artistic and cultural activities. Also, to increase to 71.6% the number of in-classroom science and technology PhD programs registered in the National Postgraduate Quality Programs (Programa Nacional de Posgrados de Calidad, PNPC in Spanish) from the Science and Technology National Council (Consejo Nacional de Ciencia y Tecnología, CONACyT in Spanish). Taking into account the aforementioned, the ITCJ designs its strategic planning, seeking to contribute significantly to the challenges posed in the 2013- 2018 NDP and the 2013- 2018 PSE.

2013 - 2018 Institutional Innovation and Development Program (IIDP)

The 2013-2018 IIDP for ITCJ was formulated with strict adherence and pursuant the goals established in these documents, and also considering the particular focus of the 2014-2018 Special Program on Science, Technology and Innovation and the commitments on the 2014-2018 National Program of Social

Development (Instituto Tecnológico de Ciudad Juárez, 2015). Likewise, sectoral and special programs whose perspective or channel was related to the work of the Instituto Tecnológico de Ciudad Juárez (ITCJ), aligning goals with public policies of interest to the nation, were considered.

In the IIDP, the main objective establishes strengthen the quality of educational services and the quality of technological higher education taught in ITCJ. This objective is focused on ensuring the relevance of the educational offer, improving the qualifications of faculty by means of training and certifications; to improve their professional development and the recognition of the performance of the teaching and research function, as well as to strengthen the academic capacity and competitiveness indicators and their impact on the quality of educational programs. As well to promote and consolidate the international positioning of the ITCJ.

From this document, ITCJ implements strategy 1.6 from the Annual Institutional Program 2016 (AIP), that mentions "Encouraging the Internationalization of Technological Institutes" and one specific line of action 1.6.3: promoting proficiency in a Second Language, preferably in English, among professors and students at least at level B1 of the CEFR or its equivalent. For this, goals related to this area have been established as it is mentioned that by 2018, it is expected that 60% of the students must be enrolled in a foreign language course.

MATERIALS AND METHODS

A simplified version of the Administrative Audit Methodology (Franklin, 2007) was used, which included questions, direct observation, and documentary review. For the diagnostic stage, the following aspects were considered: planning the vision, mission, objectives, goals, strategies/tactics, processes, policies, procedures, programs, approaches, levels, and horizon. Similarly, for the organization stage the following aspects were considered: organizational structure, dividing and assigning roles, organizational culture, human resources, organizational change, administrative studies, technical support instruments. For the management stage the following elements were considered: leadership, communication, motivation, groups and work teams, stress management, conflict and crisis, information technology, decision-making, creativity, and innovation. Finally, for the control stage authors considered: nature, systems, levels, process, application areas, tools, and quality. All these elements were considered only when they apply to the case in question.

It is necessary to mention that the data collection was through the records provided by the ITCJ. The first author of this chapter was responsible for the area known as Technological Management and Liaison Department that includes the Language Center. The statistics that were used are the basic statistics and are administered and controlled in the area known as Department of Planning of the ITCJ. The previous methodology is complemented with the suggested by Ackoff (1979), for interactive planning, which considers: planning the goals, means, resources, organizational planning and planning implementation and control.

Stage 1: Planning

On October 20, 2015, the vision, mission, values, objectives, and services were established for the Language Center. All of them resulted from the participation of the institution's top management and the participation of members of the Language Center (see APPENDIX 1). On the other hand, there are processes for advertising, registration, English Proficiency tests, payments to teachers and assigning teachers to groups. All these processes were not documented and have been modified to adapt them to the changes that have emerged to improve the Language Center services. This document was named: Operation of the Language Center dated June 21, 2016. Also created a Facebook® account for the Language Center which serves to provide relevant information such as new schedules, notices, registration dates, among others.

Stage 2: Organization

An analysis related to student population increment was conducted and some organizational changes were suggested. These changes included defining and specifying the role of both the academic coordinator and the administrative coordinator, since they were not clearly defined. Teachers were also given the opportunity to improve their work, they were directly involved in the placement interviews, as well as in the application and grammar grading of the English Proficiency tests, thus positively impacting the performance of the students. Also, per teachers' suggestion, the English Proficiency courses were eliminated, since they did not assure with so few hours (100 hours) that the students would reach level B1 of the CEFR. Even though students were asked to take level II, the fail rate was very high, this from January 2016. Following the instructions of TecNM, three proposed bibliographies were analyzed and Traveler®, from MM Publications®, was chosen since it offered training, availability, affordable prices to students, free material for teachers and courtesies for students. The contents of this material are very similar to the previous book, which avoided resistance to change as well as being very didactic and promotes learning acquisition.

Stage 3: Management

The Language Center management adopted an "open doors" policy, although controls were increased, such as teacher time tracking, full compliance with all tasks covered in the textbook, elaborating and submitting the course planning, eliminating classroom parties, among others. Communication is close, direct and cordial. Problems among administrators, students, and teachers are solved with good communication and clear and fair instructions for everyone. TecNM requested all teachers who did not have the TKT (Teaching Knowledge Test) to receive training for them to reach Level B2 of the CEFR. Starting in 2011, the Language Center management attends the new students welcoming meetings to let them know to get early involved in their English language acquisition and not to leave it until the last semesters of their careers, coupled with the fact that they consider it only a requirement to fulfill. Since 2011, the Language Center Management attends the massive new income meetings, inviting young people to prepare themselves better in this area, explaining the necessity to be proficient in the second language for them to able to get a better job, a scholarship and/or stays abroad.

Stage 4: Control

The Information System of the Language Center (SICELE) was designed and implemented to control groups, students, income, payments, academic content, among others, and a website was also created so students could verify their enrollment and which group they were assigned to before starting classes, with the option of consulting their grades and progress during the course. The performance and progress of teachers with surveys among students is also monitored.

Detailed Description of All Actions Undertaken

1. Searching among different publishers (Four Corners®, Interchange®, Passages®, Touchstone®, Top Notch / Summit®, Traveler®) to compare the levels they offer and verifying the contents. This to have a reference point and make decisions on the new material to be considered. Then, the assessment and selection of new bibliographic material that promotes the construction of learning in the classroom and that is within the reach of the student's economic possibilities was carried out. Besides the aforementioned elements, it was given special interest to the student assessment part concerning oral expression including participation in class and project presentation according to the level they are studying. This, in order to gradually increase, is the ability and the perspective required in level IV, where they have to adequately express themselves.
2. It is a graduation requirement for any student to prove he/she has the required English Proficiency level equivalent to level B1 of the CEFR. The parameters of the English Proficiency test were modified to include an oral interview with a weight of 50%, this to ensure graduates can sustain a conversation as it happens when they go through an employment interview. Additionally, it evaluates their reading comprehension by means of a technical-scientific article by a teacher of the academic area, with a weight of 30%, and responding a question in writing with a weight of 20%. The minimum grade to accredit this assessment is 70 on a scale from 0 to 100. The levels (classes) offered by the Language Center and the Traveler® series comparison to CEFR are in Table 1.
3. The offered English Proficiency courses consisted of several chapters of several levels. This in order to offer a quick alternative for those who already had a certain level of knowledge about the language, but they could not meet the CEFR English Proficiency requirement defined by TecNM. Therefore, it was decided to stop offering them, other alternatives to schedules for Level III were offered to make it more accessible to students.
4. Implementing a new MySQL® database and a C-Sharp® application, which has the following advantages: easier information backup and prevents unauthorized access, better portability (easier to change the data platform), and greater data manipulation capacity. This application is one of the most widely used to create Windows® based applications. With this application, it is much easier to relate various variables in the database and easier obtain group lists, report cards, student

Table 1. Levels offered by the Language Center and Traveler® series comparison to CEFR

Class Name	Name of Book Traveler ®	CEFR
Intro	Beginners	-
Level I	Elementary	A1
Level II	Pre-Intermediate	A2
Level III	Intermediate	B1
Level IV	Advanced	B1+
Level V	Advanced	C1

Source: The authors

academic records, as well as facilitate the registration process, since the number of students per group is observed at the moment of entering the student enrollment.

Data Statistical Analysis

Using Minitab® to analyze data, the information related to the ITCJ and the Language Center population for the last eight years was considered to forecast 2018 and 2019. Remember that ITCJ has student enrollments twice a year, therefore an average of students per year was calculated. This information is shown in Figure 1.

Information is available on ITCJ and Language Center student population. This data is considered for the indicator to be measured is the percentage of students enrolled in the Language Center with respect to the global enrollment in ITCJ. This percentage ratio is shown in Table 2.

Data was analyzed using several forecasting models (linear regression, moving average, double moving average, exponential smoothing, double exponential smoothing, and Winter´s). After considering each one of the previously mentioned models, it was decided to use the Double Exponential Smoothing Model for showing the best values of MAD (Mean Absolute Deviation) MAPE (Mean Absolute Percentage Error), and MSD (Mean Squared deviation). Figure 2 shows the graphs of the aforementioned application for the 3 variables in question. Figure 3 presents the data double exponential smoothing model for student population, whereas Figure 4 shows the data double exponential smoothing model for the language center population.

In order to prove the model is statistically adjusted to the real data, its behavior is verified in order to establish the appropriate (parametric or non-parametric) statistic, so, as a first step data normality is tested, both real and model adjusted, to prove the equality of means (ITCJ student population, Language Center population and their ratio). The results are shown in Figure 5. Figure 6 shows probability plot of percentage smoothing normal. Similarly, Figure 7 shows the boxplot of percentage, percentage smoothing. Finally, Figure 8 shows the proof of average equality between percentage and adjusted percentage.

Figure 1. Language Center and ITCJ population
Source: The authors

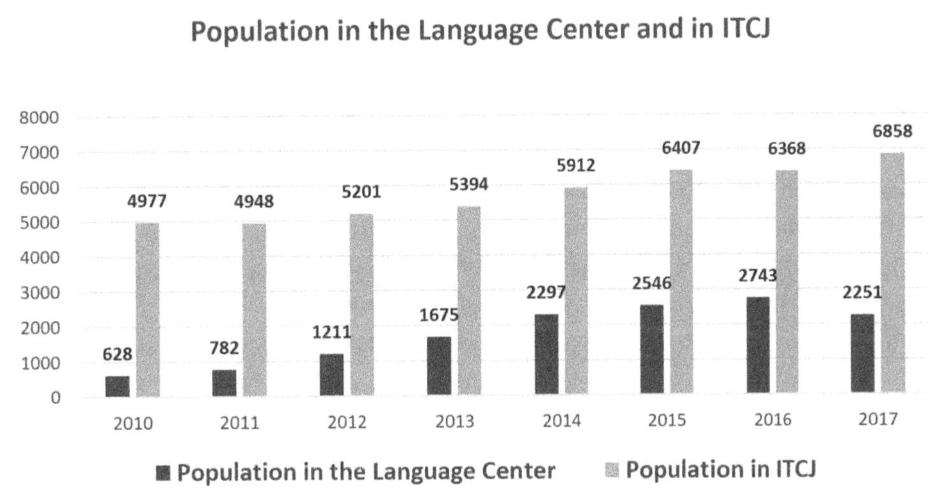

Table 2. Language Center to ITCJ student ratio

Year	Percentage
2010	12
2011	15
2012	23
2013	31
2014	38.8
2015	39.7
2016	43
2017	32

Source: The authors

Figure 2. Data double exponential smoothing model for percentage
Source: The authors

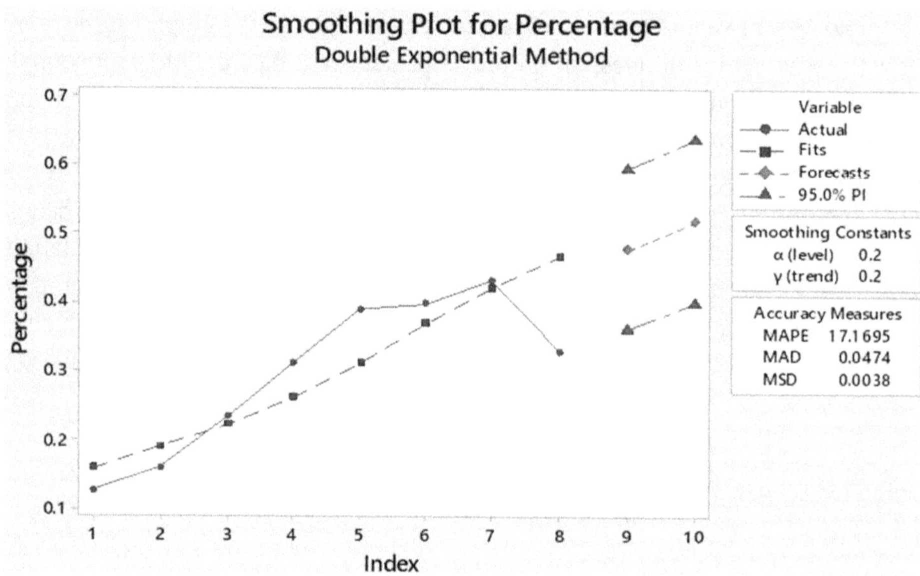

Figure 9, Figure 10 and Figure 11 show the student population normality and student population adjusted by the model chart. Figure 9 specifically shows probability plot of student population smoothing normal. Similarly, Figure 10 shows the probability plot of student population normal, whereas Figure 11 shows the boxplot of student population and student population smoothing. Finally, Figure 12 presents the proof of mean equality between student population and adjusted student population.

Figure 13, Figure 14 and Figure 15 present the student population in the center language normality and student population in the language center adjusted by the model chart. Figure 13 specifically shows probability plot of language center population normal, whereas Figure 14 shows probability plot of language center population smoothing. Similarly, Figure 15 presents the boxplot of language center population, language center population smoothing. Also, Figure 16 presents the proof of mean equality

Figure 3. Data double exponential smoothing model for student population
Source: The authors

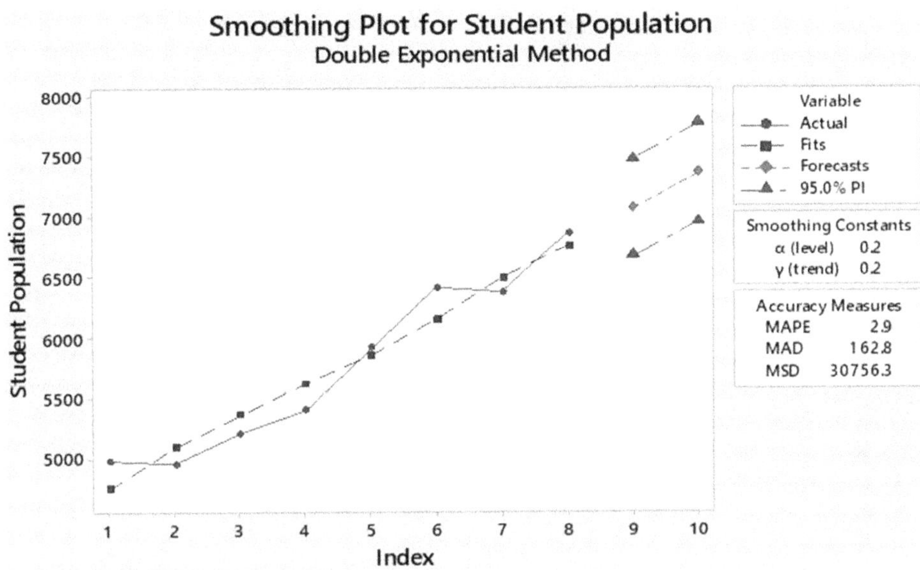

Figure 4. Data double exponential smoothing model for language center population
Source: The authors

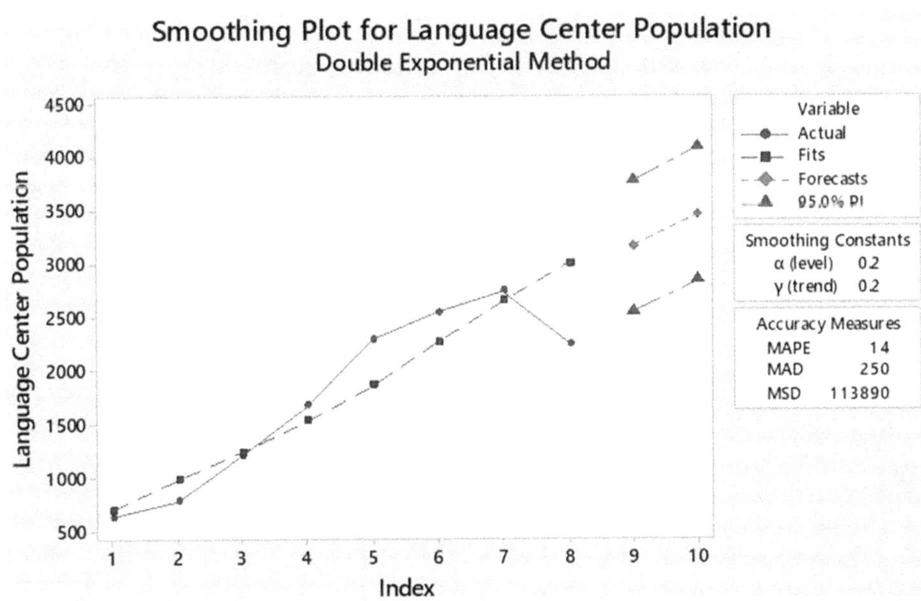

Figure 5. Normality percentage and percentage adjusted by the model chart, normal
Source: The authors

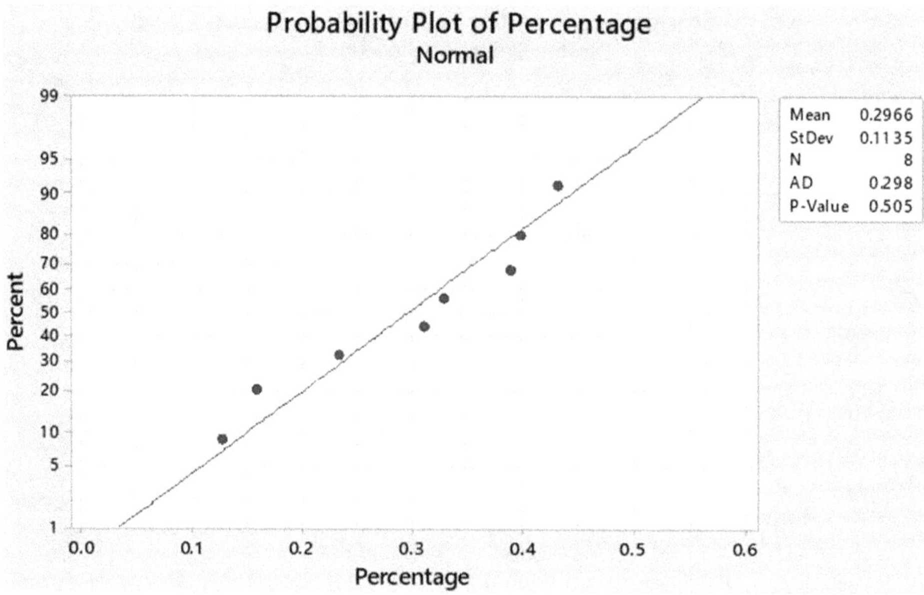

Figure 6. Probability plot of percentage smoothing normal
Source: The authors

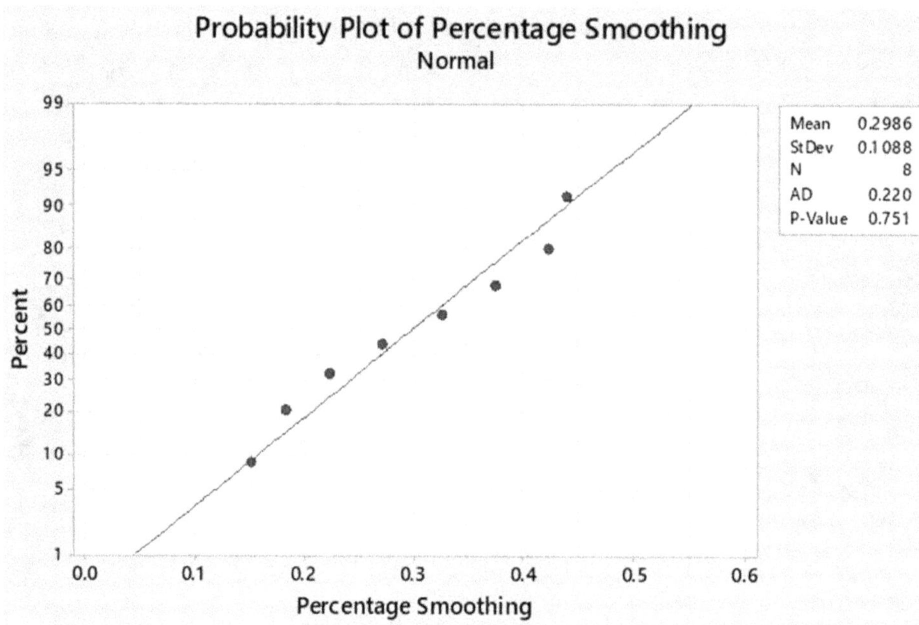

Figure 7. Boxplot of percentage, percentage smoothing
Source: The authors

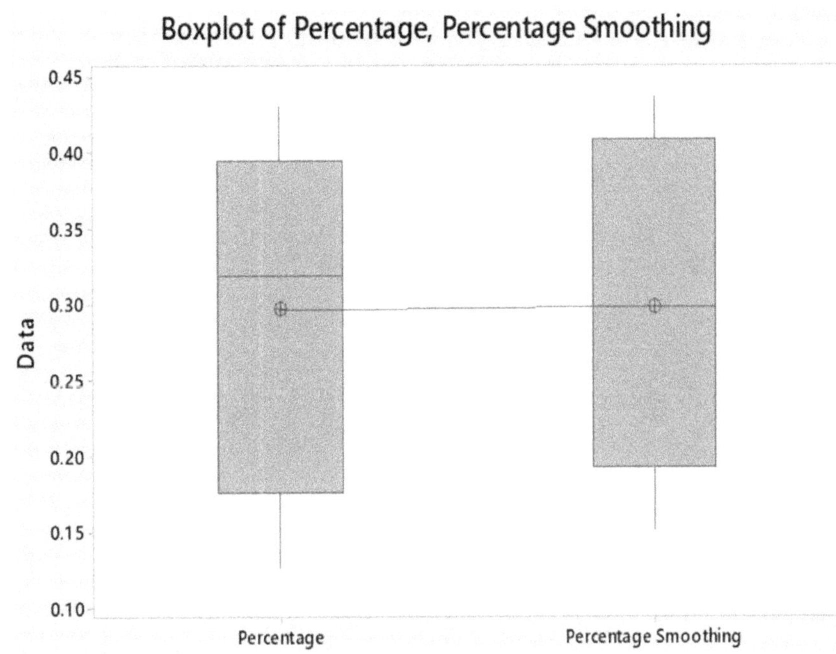

Figure 8. Proof of average equality between percentage and adjusted percentage
Source: The authors

Two-Sample T-Test and CI: Percentage, Percentage Smoothing

Method

μ_1: mean of Percentage

μ_2: mean of Percentage Smoothing

Difference: $\mu_1 - \mu_2$

Equal variances are assumed for this analysis.

Descriptive Statistics

Sample	N	Mean	StDev	SE Mean
Percentage	8	0.297	0.113	0.040
Percentage Smoothing	8	0.299	0.109	0.038

Estimation for Difference

Difference	Pooled StDev	95% CI for Difference
-0.0020	0.1111	(-0.1212, 0.1172)

Test

Null hypothesis	$H_0: \mu_1 - \mu_2 = 0$
Alternative hypothesis	$H_1: \mu_1 - \mu_2 \neq 0$

T-Value	DF	P-Value
-0.04	14	0.971

Figure 9. Probability plot of student population smoothing normal
Source: The authors

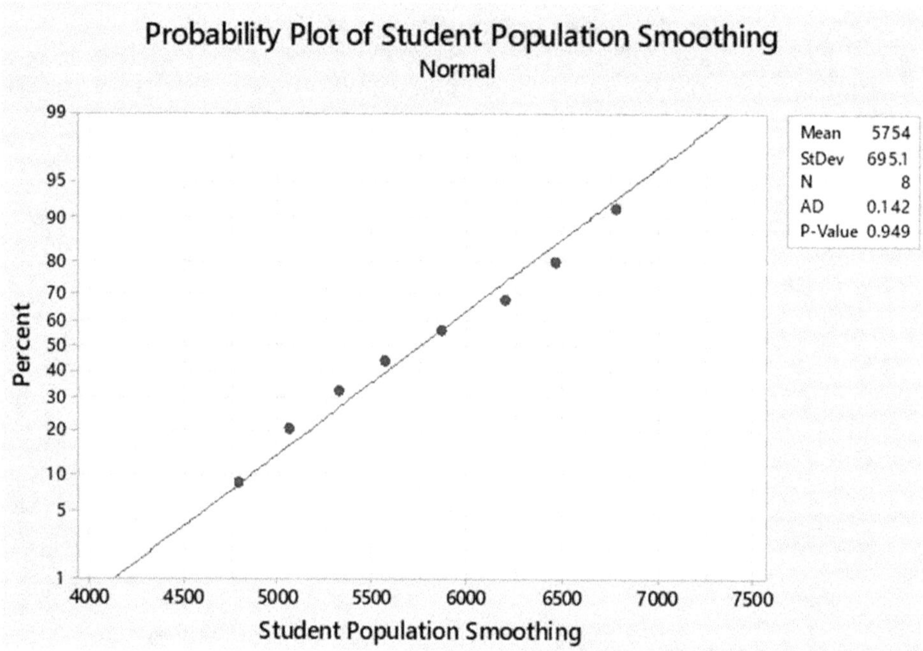

Figure 10. Probability plot of student population normal
Source: The authors

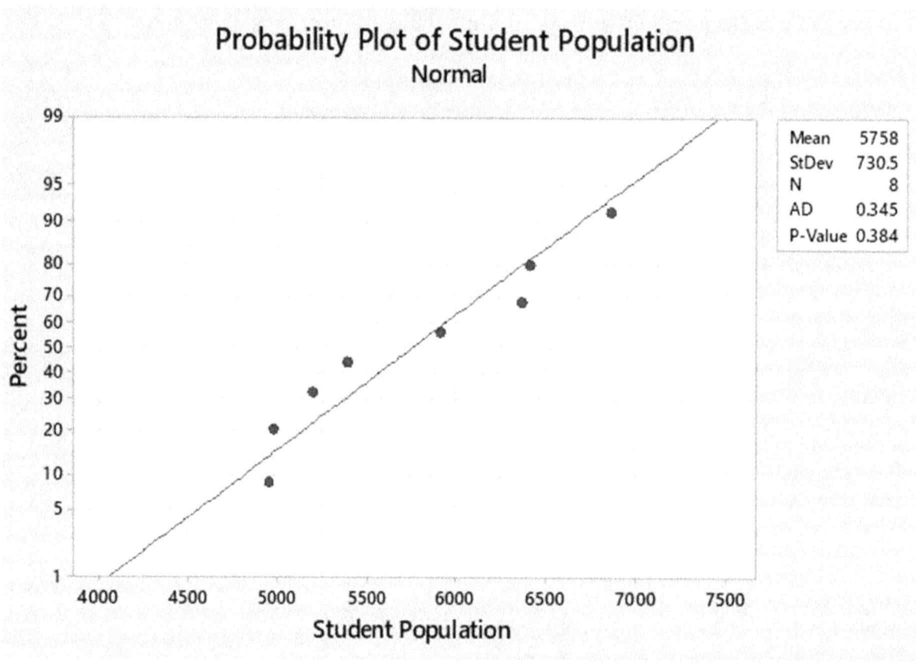

between student population in the Language Center normality and student population in the Language Center adjusted.

Considering a value of $\alpha = 0.05$, the results of the equality of means tests statistically show the model adjusted data is equal to the real data, so, with the Double Exponential Smoothing Model used, it can be obtained a reliable forecast. Table 3 shows the Minitab® predicted results. As it can be observed in the percentage upper limit for 2018 is 59.36% and the one for 2019 is 63.51%. This means that, although a decrease in the population has been seen, it is very probable to reach the expected value of 60%.

RESULTS

As a result of the undertaken actions, the following can be cited:

1. Comparing 2012 to 2017 shows an increment of 160% of students in the Language Center, although this percentage decreased in 2017, possibly for the students enrolling in online English courses. It is not yet known if the students taking these courses achieve the same level of language acquisition. Another observation is that careers in ITCJ such as mechatronics engineering had an 85% increase, which corresponds to the increase in student population for that career, but there are other careers with the same or more population that have not seen an increase in student population in the Language Center. These careers are systems and industrial engineering and the economic-administrative careers.
2. Figure 17 shows the last 4 years comparison between the students that accredited the English Proficiency course and those who accredited levels III, IV and V, since these levels are required to achieve level B1. It is observed that only 120 students reached that level in 2013, the number increased to 302 in 2014, then to 369 in 2015 and to 524 students in 2016, which represents a 19%,

Figure 11. Boxplot of student population and student population smoothing
Source: The authors

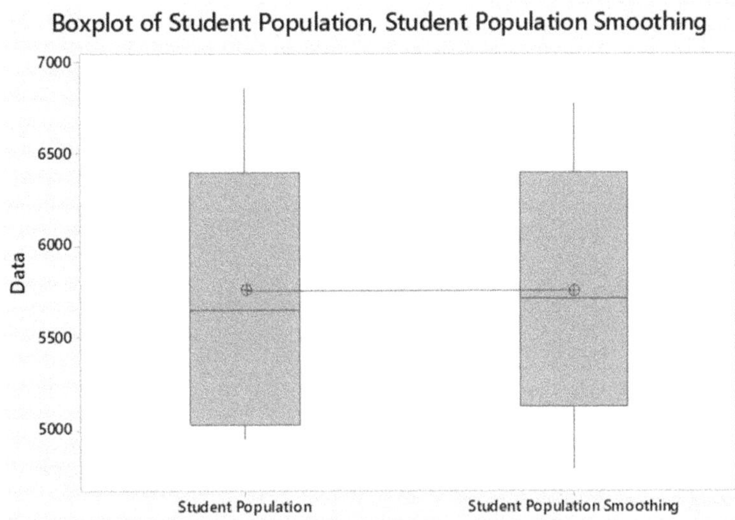

Figure 12. Proof of mean equality between student population and adjusted student population
Source: The authors

Two-Sample T-Test and CI: Student Population, Student ... Smoothing

Method

μ_1: mean of Student Population

μ_2: mean of Student Population Smoothing

Difference: $\mu_1 - \mu_2$

Equal variances are assumed for this analysis.

Descriptive Statistics

Sample	N	Mean	StDev	SE Mean
Student Population	8	5758	731	258
Student Population Smoothing	8	5754	695	246

Estimation for Difference

Difference	Pooled StDev	95% CI for Difference
4	713	(-761, 768)

Test

Null hypothesis H_0: $\mu_1 - \mu_2 = 0$

Alternative hypothesis H_1: $\mu_1 - \mu_2 \neq 0$

T-Value	DF	P-Value
0.01	14	0.992

Figure 13. Probability plot of language center population normal
Source: The authors

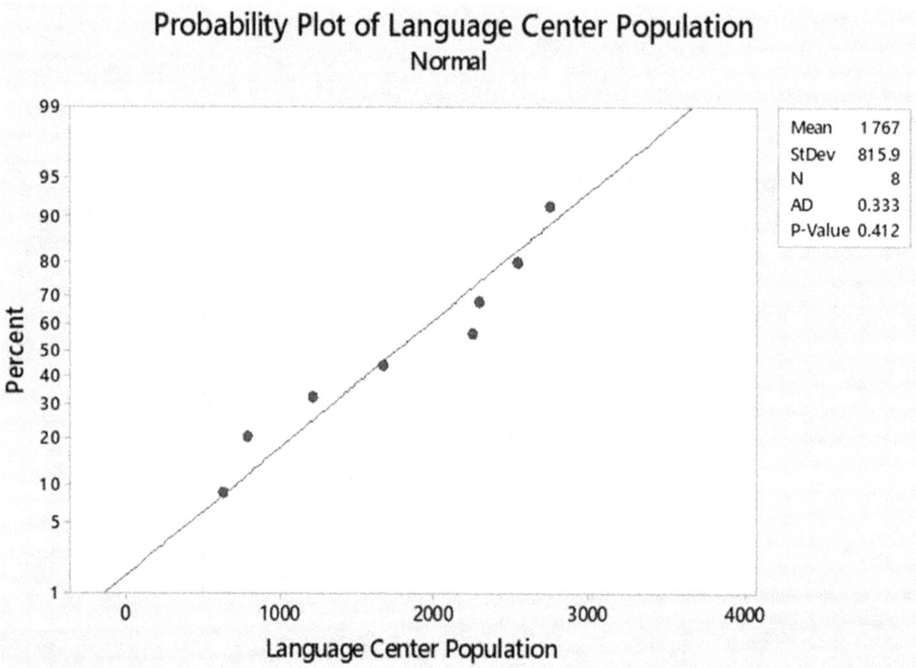

Figure 14. Probability plot of language center population smoothing
Source: The authors

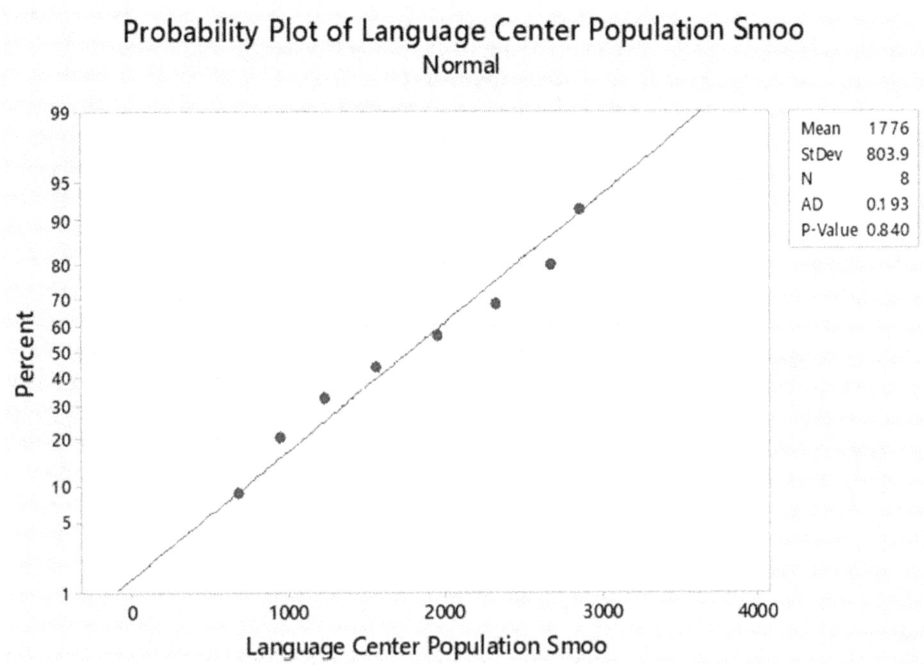

Figure 15. Boxplot of language center population, language center population smoothing
Source: The authors

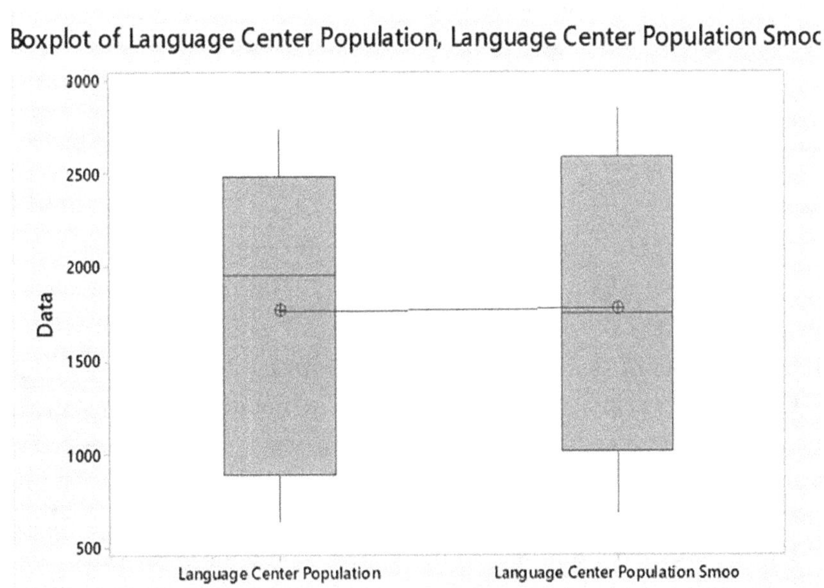

Figure 16. Proof of mean equality between student population in the center language normality and student population in the language center adjusted
Source: The authors

Two-Sample T-Test and CI: Language Center Population, Smoothing Method

μ_1: mean of Language Center Population

μ_2: mean of Language Center Population Smoothing

Difference: $\mu_1 - \mu_2$

Equal variances are assumed for this analysis.

Descriptive Statistics

Sample	N	Mean	StDev	SE Mean
Language Center Population	8	1767	816	288
Language Center Population Smooth	8	1776	804	284

Estimation for Difference

Difference	Pooled StDev	95% CI for Difference
-9	810	(-878, 859)

Test

Null hypothesis H_0: $\mu_1 - \mu_2 = 0$

Alternative hypothesis H_1: $\mu_1 - \mu_2 \neq 0$

T-Value	DF	P-Value
-0.02	14	0.982

Table 3. Forecasted results for 2018 and 2019

Year	Forecasted Percentage	Upper Limit	Lower Limit	Forecasted Population in ITCJ	Upper Limit	Lower Limit	Forecasted Language Center Population	Upper Limit	Lower Limit
2018	47.742%	59.36%	36.117%	7065	7464	6666	3156	3769	2543
2019	51.648%	63.51%	39.782%	7357	7764	6950	3459	4085	2833

Source: The authors

and although the number of students decreased to 356 in 2017, it was known that 1,100 students took online English classes in 2016. These students were not included this statistic, because it had not been measured and checked with the same instrument as the in-classroom students.

3. Another analysis carried out is shown in Figure 18, the ratio of students obtaining level B1 was calculated and compared with all other students who took an accredited some English course, that is, for level B1 the students of the courses III, IV and V, and the other value all the accredited in the other levels were included the levels INTRO, I and II, and proficiency. Here, it can be seen that this ratio increases, not only with more students but with more students who reached level B1. This graph is to be updated in the periods to come and it is expected to increase by the current population of the Language Center of 946 students only in the first period, pending 2 other periods. Before the

end of the year, this trend would present at least 2600 students in 2016; with at least 480 students with level B1 of the CEFR.

Result of Hypothesis Testing

It can be concluded that the proposed H_0 (null hypothesis): $\mu_{before} \leq \mu_{after}$, is accepted for the previous percentage in 2017 is less than or equal to the one in 2018. In 2017 it was 32% as shown in Table 2 and for the periods in 2018 it will be according to the forecast of 59% in the upper limit and 36% in the lower limit, as shown in Table 3.

Figure 17. Comparison of students who took the special degree course against those who reached B1
Source: The authors

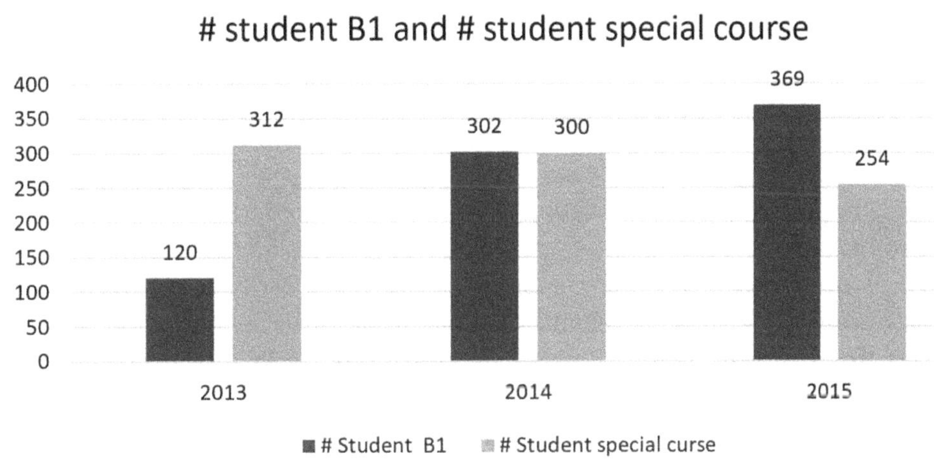

Figure 18. Ratio of B1 level students compared to other levels
Source: The authors

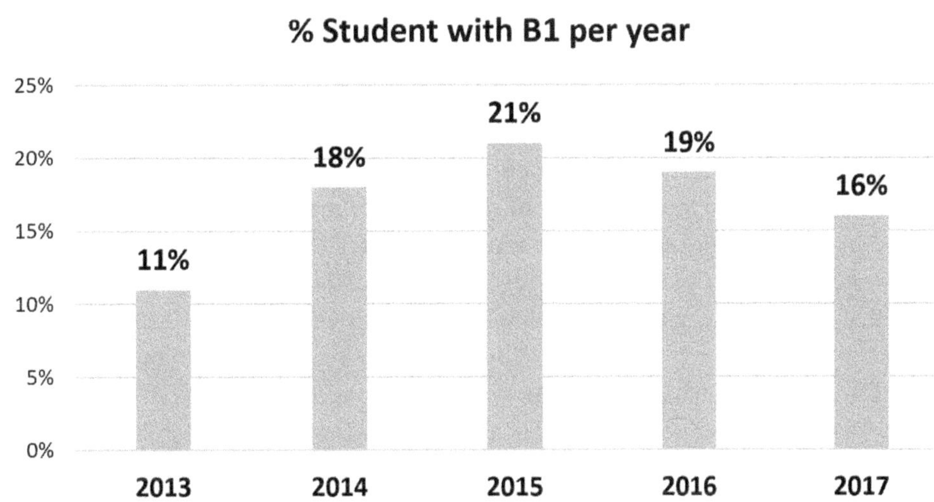

DISCUSSION

In Mexico there are several published papers relating to the subject here presented; amongst them, the one performed by the Universidad Veracruzana in 2009 stands out given the enormous effort made to establish a Development Plan in their Language Center based on their own 2005-2009 Work Programme (Universidad Veracruzana, 2014). A desirable scenario was established, and the whole enterprise was taken on because of the results of a diagnostic of needs, which in turn produced guidelines that were put into place in 2013. The document does not mention if the guidelines were properly followed, or if the desired results were obtained, but these actions were created to improve the service and to answer to the environment, just as the ones performed at the ITCJ Language Center. Another similar work was performed by Vergara-Garibaldi (2012). This work contained a proposal to improve the academic and administrative processes in the Professional Improvement and Language Center in the Technological Institute of La Paz, Baja California Sur, Mexico. In her work, Vergara-Garibaldi performed a diagnostic study regarding the conditions of the service provided at this center, and provided suggestions as to how to improve them. The conditions found in this study were remarkably similar to the ones found at the Language Center in the ITCJ, with the difference that the latter has three times the students. Regretfully, there are no results regarding this study, and it is unknown if the suggestions provided were considered for implementation.

Latorre-Medina and Blanco-Encomienda (2012) performed a study in which they suggested that, in order to face the challenges in an education that used to be rooted in the past and is now fixated in the future, two main improvements have to be made: strategic educational management and effective educational leadership. Although the document is essentially referenced to the teaching and learning process, the focus given to leadership is similar to the one made in this research. In an international level, Fernandez and Rosales (2014) made an interesting proposal in the Ibero-American Cience, Technology, Innovation, and Education Congress regarding educational administration: strategic planning and managerial practices integrating technology, and their impact in education. Given that in this document Fernandez and Rosales explain what educational administration consists of in light of the Total Quality model and also they identify several managerial practices that integrate technology which aid in becoming successful educational administrators, in this research information technology was used to assist in the administration and forecast the future of the Language Center.

On the other hand, in this research work a systems approach was used, just as the work from Puteh, Kaliannan and Alam (2015), who executed a very interesting research with a focus in systems, in which they studied the potential and the challenges of learning and developing amongst work colleagues. They consider this learning as informal, since it is not based on theoretical or formal knowledge. This means that senior co-workers teach or train their juniors. The notion that each and every person in an organization has a special role led to the use of the systems theory to observe how senior colleagues influence or aid in the development of their juniors' competencies. This is owed to the fact that every individual employee possesses a set of knowledge, abilities, and skills that are unique, as well as other characteristics that are useful to achieve or maintain a competitive advantage. The same occurs with the owners and executives. Compared to what was done at the Language Center of the ITCJ, it can be noted that here as well were applied the skills, abilities, and knowledge of certain staff members to train and improve the performance of the newcomers. Thirteen new teachers and three collaborators were added to the staff using the same method in the course of development and documentation of the achievements of the Language Center.

CONCLUSION

Changes tend to cause resistance since comfort and inertia are disturbed, but the world changes and the environment demands increase. These must be considered to grow, to improve and to be able to compete, if this is not done, systems tend to become obsolete and then extinct. For a higher education institution with the prestige and importance of the ITCJ, these changes in the environment cannot and should not go unnoticed, the voices of the employers of the graduates ask that these have more mastery of a second or third language amid other skills as communication, teamwork, leadership, management of new technologies; the ITCJ is already working on it.

The Language Center faces other challenges, such as being certified by the TecNM as a Test of English as a Foreign Language (TOEFL) certification center, obtaining scholarships for low-income students, searching for training courses for teachers, among other challenges. It is intended to equip all classrooms with video projectors to take advantage of the multimedia material offered by the publisher, among others.

Also, all efforts made by the Language Center were documented and sent to take part in the "Recognition to Managerial Improvement" organized by Secretariat of Public Education (Secretaría de Educación Pública, SEP), resulting in 2nd place nationwide, which represents an excellent achievement for all the hard teamwork of all staff up to 2016.

REFERENCES

Brunner, J. J. (2000) *Globalization and the future of education. Latin America and the Caribbean perspective on Education*. Santiago: UNESCO. Retrieved from Internet on January 2018: http://mt.educarchile.cl/archives/Futuro_EDU%25UNESCO-2000.pdf

Chiavenato, I. (2009). *Comportamiento organizacional* (2nd ed.). Mc-Graw Hill.

Fernández, S., & Rosales, M. (2014). Administración educativa: la planificación estratégica y las prácticas gerenciales integrando la tecnología, su impacto en la educación. Congreso Iberoamericano de Ciencia, Tecnología, Innovación y educación. Artículo 1582. Buenos Aires Argentina.

Franklin, F. E. B. (2007). *Auditoría administrative, Strategic management of change* (2nd ed.). Editorial Pearson.

González, M. N. L. (2015). *Employer satisfaction assessment of Industrial Engineering graduates of classes of 2009 - 2014*. Professional Experience Report, ITCJ.

Guiselle, M. G-V. (2016). *Desarrollo organizacional y los procesos de cambio en las instituciones educativas, un reto de la gestión de la educación*. Revista Educación Enero junio 2016, Volumen 40, Num. 1. Recuperado de: 26 enero 2018. https://revistas.ucr.ac.cr/index.php/educacion/article/view/22534

Hellriegel, D., Jackson, S., & Solcum, J. (2005). *Management. A Competency-Based Approach* (10th ed.). McGraw-Hill de México.

Instituto Tecnológico de Ciudad Juárez. (2015). *2013-2018 Institutional Innovation and Development Program*. TNM.

Latorre, M. J. M., & Javier, B.-E. F. (2013). *Strategic Management as Key to Improve the Quality of Education*, 1st World Congress of Administrative & Political Sciences (ADPOL-2012). *Procedia: Social and Behavioral Sciences*, *81*, 270–274. doi:10.1016/j.sbspro.2013.06.426

Puteh, F., Kaliannan, M., & Alam, N. (2015). Learning for professional development via peers: A System Theory approach. *Procedia: Social and Behavioral Sciences*, *172*, 88–95.

Quiñones, V., Otarla, S., Ramos, F., & Aguilera, N. (2008). *Organizational Development*. Retrieved from Internet on January 18, 2018: https://www.bing.com/search?q=Desarrollo+Organizacional.+Blogspot%2C+Grupo15.&PC=U316&FORM=CHROMN

Russell, A. L. (1979). *Redesigning the future* (13th ed.). Editorial Limusa.

Tecnológico Nacional de México. (1999). *Technological Institutes Procedures and Guidelines Manual*. Academic Press.

Tecnológico Nacional de México. (2010). *Authorized Plans and Programs for ITCJ, 2012, 2010 y 2016*. Academic Press.

Tecnológico Nacional de México. (2015). *Academic-administrative Guidelines Manual*. Academic Press.

Tünnerman, C. (2000). Pertinencia social y principios básicos para orientar el diseño de políticas de educación superior. *Educación Superior Sociedad*, 181-196. Recuperado de http://ess.iesalc.unesco.org.ve/index.php/ess/article/viewArticle/364

Universidad Veracruzana. (2014). *Plan de Desarrollo del Centro de Idiomas Poza Rica 2009-2013*. Universidad Veracruzana. Retrieved from Internet on August 13, 2018: http://www.uv.mx/pozarica/ci/files/2012/12/C.-I.-POZA-RICA-CORREGIDO.pd

Vergara Garibaldi María Olivia. (2012), *Plan estratégico para el centro de actualización profesional e idiomas del Instituto Tecnológico de la Paz*. División de Estudios de Posgrado e Investigación, Thesis of Administration degree. México, 2012.

APPENDIX

Language Center at ITCJ

Mission

Our mission is to have the necessary elements such as programs, teachers and high-level infrastructure, to facilitate the student community of ITCJ and society in general, acquire the English language competence, at the level that the work environment demands, at an affordable cost.

Vision

The Language Center at ITCJ, seeks to become a Center recognized by the TNM, to respond to the needs of the Institution and the challenges of satisfying the demands of the community we serve, creating a close link among the educational, social and business sectors.

Values

- Responsibility
- Commitment
- Ethics
- Equity
- Honesty

General Objective

Fulfill the need of ITCJ students and the general public to learn a second language, by offering mainly English language courses and, if in any other language meets the appropriate demand, seek the resources for its delivery.

Specific Objectives

- Support the students in the mastery of the level of English that the working environment demands in their profession.
- Encourage students to obtain a diploma in a foreign language.
- Support students and graduates to meet the requirement of knowing a second language for them to comply with the graduation requirement.
- Provide service to the general public that demands it, at an affordable cost.

Services

- English language courses for beginners, intermediate and advanced.
- Placement exams, so that the applicant knows the possessed degree of English.
- Placement exams for applicants to postgraduate courses and the issuance of the corresponding certificate.
- Motivate students towards obtaining a diploma in a foreign language.
- Promote events for students to practice the second language, such as dead altars, contests, scholarships, among others.
- Support students and graduates to meet the requirement of knowing a second language for them to comply with the graduation requirement.

Requirements

To request a placement test, the interested party must be present ten minutes before the established time for the application, with an official photo identification and the corresponding payment slip.

To enroll in the Beginners initial level, no placement test is required for students only their data and control number and go to the Language Center before making the corresponding payment to check availability of schedules.

For external applicants only give their general information and corresponding payment.

Chapter 12
Distribution of Food in a Specialized Hospital Using Ambient Intelligence to Improve a Model of Macroergonomics

Alberto Ochoa Zezzatti
Universidad Autónoma de Ciudad Juárez, Mexico

Juan Luis Hernandez Arellano
Universidad Autónoma de Ciudad Juárez, Mexico

Gilberto Rivera
Universidad Autónoma de Ciudad Juárez, Mexico

Daniel Azpeitia
Universidad Autónoma de Ciudad Juárez, Mexico

Luis Fernando Maldonado
Universidad Autónoma de Querétaro, Mexico

ABSTRACT

SIDA (Intelligent Food Distribution System, for its acronym in Spanish) is a proposed tool for the distribution of food that can be personalized depending on the medical characteristics of each patient. The target of the tool is to provide foods that contain higher nutrients in the diet set by a hospital. A model of decision trees was based on data from the organization of the United Nations Food and Agriculture Organization (FAO) and used for decision making in the simulated three basic foods based on the diet of Latin American countries typically integrated by rice, potatoes, and lentils from the parameters of fat, energy, and protein, respectively, that contains every type of food.

DOI: 10.4018/978-1-5225-7192-6.ch012

INTRODUCTION

All forms of intelligent life must make decisions that will lead to a victory, a life victory, maybe food, water, or only a place for a rest. The tools like math's help us to choose the better decisions, the decisions tree is another tool to help an industrial, administrative and commercial business. In this case, a new tool for taking decisions about what kind of food is better for the patient using the food that has the hospital will be used. Intelligent Food Distribution System (SIDA, for its acronym in Spanish) is a proposed tool for the distribution of food that can be customized according to the medical characteristics of each patient that provides foods that contain greater nutrients in the diet established by the hospital. *Each hospital kitchen has menus for each diet that will be seen in a diet which food are the ones that benefit patients, in any case, will not eliminate the food or produce only food for each patient.* Since it stands to reason that all foods are required and must be suitably combined, however, it is known that foods more beneficial to a patient, it can serve to take to decide what kind of food decrease portion and benefit of any food its more beneficial for they (see Figure 1).

DECISION TREE USED FOR THE ALLOCATIOIN OF FOOD IN A HOSPITAL

In a visit to the hospital in order to see the way that assigned diet to patient and who and as diets design, basically four types of diets were found; 1) Normal diet where there is no restriction, can consume any food 2) in the soft diet foods must be chewed so that they can consume, however, all liquids or food restricted gelatinous allowed. 3) The low sodium diet does not restrict foods but limits the amount of salt that they possess and 4) low-carb diet limits foods that contain sugar. However, although the four types of constraint are different, in all of them was included rice, lentils, and peeled potatoes to the latter should be prepared as a mash diet soft. Cebola et al. (2016) suggest the need to regularly evaluate the nutritional status which allows to identify and reduce the complications associated with malnutrition.

For this razon, there are several quality systems that ensure care in the handling of food, and in the specific case of hospitals, a rotation of ingredients that help the patient's recovery according to their specific needs. Thus, Caracuel García (2007) have listed the benefits of having to support systems for the management of food in hospitals, among which it stands out; the guarantee that the food that is being

Figure 1. Decision tool attached with SIDA Tool of ambient intelligence

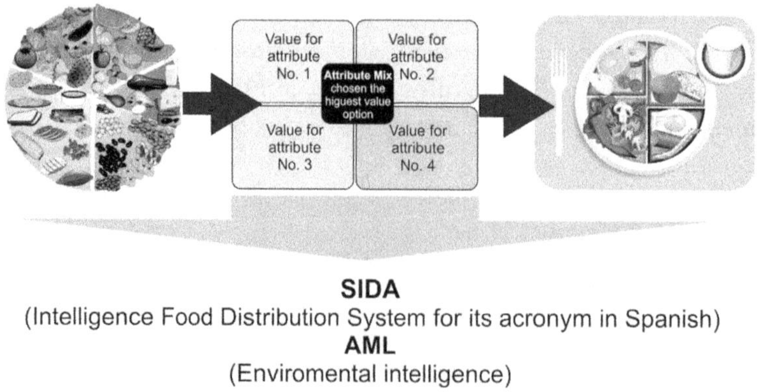

administered to patients is safe and an improvement in the expectation that users (patients) have regarding the hospital. For this, they created an integrated management system where the overall process of the food can be observed, from its reception until the moment, it is in the patient's room (see Figure 2).

Using information from the United Nations Food and Agriculture Organization (FAO) (Foods, 2002) which lists the number of nutrients that have each of these foods (see Table 1). A test was designed to develop a model of decision making based on three meals and three nutritious properties for a decision tree model to choose the food under those parameters is best suited.

Table 1 shows that the three characteristics of each of the selected foods were energy, fat, and protein. Each of the values mentioned is the starting point for feeding the decision tree that was developed using Excel tables. The reason why this tool is used is because more variables can be added easily, for example, age, weight, gender, and travel allowances paid to patients with similar characteristics to create an intelligent system to choose an optimal diet for the patient to help the medical personnel present to be accepted or rejected the proposal. This method fed to SIDA which will be completed as ambient intelligence system once the necessary space for the food logistics automated.

MACROERGONOMICS

Macroergonomics, as a multidiscipline, is a fundamental tool for the design, evaluation, and redesign of the jobs within the organizations. The results of their evaluations allow an improvement of the quality of life of the occupants of these jobs and. In sum, the organization making the socio-technical systems of the organization a healthier and resilient system (Hendrick & Kleiner, 2002). Based on these benefits, we define the ergonomics as the multidiscipline of occupational health that seeks to improve conditions of work and favoring with their practice spaces for the praxis of social responsibility.

Figure 2. Global food process in hospitals
Source: Caracuel-García, 2007

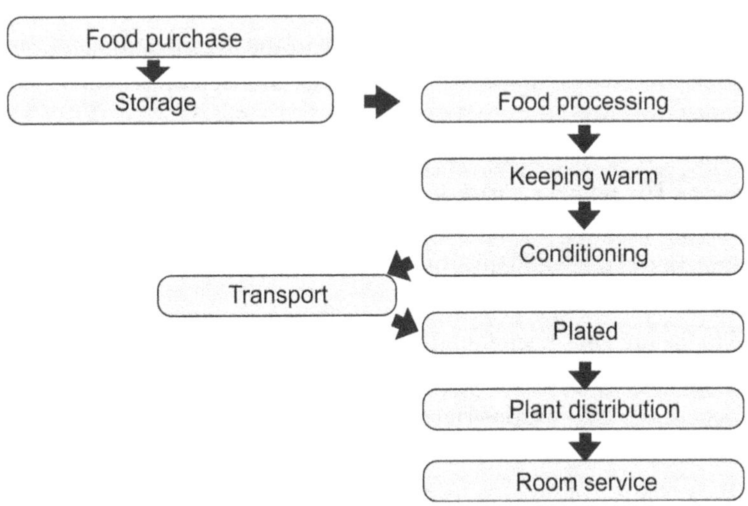

Table 1. Fragment Table FAO: Nutrients in 100 grams of edible portion of food

	Energy	Fat	Proteín
Rice, cooked	364	6,7	1,0
Lentils boiled	116	9,0	0,4
Boiled potatoes skinless	86	1,7	0,1
Chicken	139	19,0	7,00
Beef cattle	115	22,0	1,90

Under this premise, ergonomics promotes adequate spaces for the application of standards, the design of workplaces, improvement of proposals technological, as well as a wide field for scientific-academic construction; but most importantly, the most conducive environment for regulation of a sociotechnical space of the organization, which involves the man and the work, seen the latter as a social fact, which consequently has ethical implications.

One of the aspects that the macro ergonomics focuses on is taking care of the physical, social and psychological aspects of the organizations, for that reason, studies have been found in which the determining factors to measure the quality of working life of people have been identified. These factors are measured and weighted in the organization to know their level of quality of work life in general, in an instance of the worker himself as to determine the level of quality of life he has, work stress, health level, and work-life balance, the perception of ergonomics in their work space, job security and work efficiency (Maldonado & Ochoa, 2016).

Therefore, ergonomics, as a human discipline that studies the relationship between man and the system of work, has the virtue of extending its scope of study towards much more specific disciplines, such as the case of macro ergonomics, to be able to study the relationship man sociotechnical system.

Particularizing in the object of this investigation on the perspective of organizational ergonomics as a tool for proactive attitude and inclusive sensitivity, it can be established that it describes a provision positive towards the improvement of the working conditions of individuals. In the organization, its scope encompasses the general processes of the organization without referring partial views, its motivation is vital because it deals with the survival and sustainability of the organization and, finally, its nature is of foresight-prospective and non-remedial vision of situations and disagreements that arise in the organization.

From the perspective of sensitivity, organizational ergonomics exhibits aspects such as a clear willingness to develop approaches that allow the plurality, flexibility, and opening of new and dissimilar changes in the processes. His interest is part of the valorization of work (add-add), with a clear purpose of generating new positive situations that allow the organization take on new clients or markets, thanks to flexibility and adaptation; and in conclusion the synergistic and cooperative aspects within the organization and its members.

Research related to the area has been found, in which the researchers considered the concept of participative ergonomics, where people from the commercial, logistics, operation, and occupational health areas worked in conjunction with the industrial designers and ergonomists who methodologically led the project. The results of the project translate into procedures for selecting and projecting a new distribution center, that is, the operational process model, a description of ergonomic systems that will enable specific work stations to be designed, and the procedure for adapting existing warehouses.

Strategically, this work helped to optimize the company's processes and ensure that knowledge would be transferred within it. In turn, it became a primary prevention strategy in the field of health, aimed at reducing occupational risks, improving the quality of life at work, something in which it coincides with the present research in which it is intended to improve the system of distribution of food to a socio-technical system to avoid human errors (García Acosta & Lange Morales, 2008).

Studies have also been found in which the creation of improvement committees are sought to implement the policies for improving working conditions, since a change in the organization is required to accept the improvements and through a committee reviewing the process can work, these approaches are adjusted to macro ergonomics, since it integrates problems of strategy, organization, and work. Also, with that, it is possible to analyze the benefits of this approach for companies and provide the conditions for workers to participate effectively in these processes, in which the support and commitment of senior management are essential elements of success to maximize the effectiveness of this organizational approach (Bolis & Sznelwar, 2016).

Based on the above considerations, organizational ergonomics is showing in a clearly proactive-inclusive scenario because none of ethics is possible nor is a social responsibility guaranteed before other more reactive and excluding options, since it would be subject to scrutiny, honest expertise, and especially unfair decisions, as is shown in Figure 3.

Each tray corresponds according to the case, to a room, or to a bed putting the plates with food in tray one for room one, food in tray two for room two and so on counting from the top to the bottom and left to the right, as you can see in Figure 3.

Figure 3. Food distribution cart based on a macroergonomic model, in which meals for different patients are distributed on different floors of the specialty hospital

What is recommended is that starting from the system exit which sends the location of the patient to the cart, and the tray will be configured with a radiofrequency system that indicates that must enter the area of action where the patient is of interest and only in that place. Therefore, the tray of the room one can only enter room one, otherwise, a light is proposed both in the patient's bed and in the food cart that is clicked to indicate that there is an error in the distribution of the diets. As future work, a return of information indicating the number of sometimes a wrong tray is attempted to be delivered to a patient, to identify possible faults that are still operative in the sub-process of food distribution (see Figure 4).

IMPLEMENTATION TOOL

The simulator consists of a simple Excel environment where first priorities to be evaluated and established using a value for each priority are given according to need, for this case, the parameters are evaluated to find the most features in the three foods (see Table 2).

Figure 4. Form in which the food is served that should be distributed to each patient according to their established diet based on their suffering derived from the disease they suffer

Table 2. Priorities to evaluate

Active Males Higher Fat and Protein Energy			
Features of Interest	**Fat**	**Energy**	**Protein**
Fat	1	4	1/6
Energy (kcal)	1/4	1	1/6
Protein	6	6	1
Total score	7	11	1

In order to assign values properly, it must compare between a value and its proportionality of portion and representing a function of this comparison the most important feature for the decision as well. For example, the casc of the protein is more important that the fat because a higher value is given. Subsequently, a table is developed according to each of the characteristics being analyzed. Is very important note that in initial table with 100 features would require one hundred of the tables shown in Table 3.

For Tables 3, 4, and 5 some simple calculations that mix each selected with each of the three variables of interest (fat, protein and energy) to finally get a result that can be seen in Table 6 are provided food. In Table 6, the three-selected food, based on figures provided by the FAO nutrients (Foods, 2002) can be analyzed. In Table 1, we observe that the diversity is very important and from the particular needs of this simulation which is finding between rice, lentils, and potatoes without skin, a balanced and correct specify food as much fat and protein energy that the food complies with greater success, where 1 portion is related with 100% compliant with the requirements and to 0% 0 complies with the requirements associated with the rice.

Table 3. Food about the Feature of Interest (Fat)

Fat			
Selected Foods	**Cooked Rice**	**Boiled Lentils**	**Boiled Potatoes Without Skin**
Cooked rice	1	0	6
Boiled lentils	1/3	1	9
Boiled potatoes without skin	1	0	1
Total score	2	1	16

Table 4. Food about the Feature of Interest (Energy)

Energy			
Selected Foods	**Cooked Rice**	**Boiled Lentils**	**Boiled Potatoes Without Skin**
Cooked rice	1	7	9
Boiled lentils	1/4	1	6
Boiled potatoes without skin	0	1	1
Total score	1	9	16

Table 5. Food about the Feature of Interest (Protein)

Protein			
Selected Foods	**Cooked Rice**	**Boiled Lentils**	**Boiled Potatoes Without Skin**
Cooked rice	1	7	8
Boiled lentils	1/7	1	4
Boiled potatoes without skin	0	1	1
Total score	1	9	13

This is not to say that the best food is rice potatoes or lentils, which means is that, for the particular needs and low perception of a patient's doctor about specific rice is the food most convenient at this time, remember that the Table 2 has to assess priorities with parameters set in person, where each individual gives a score depending on the needs by comparing the data of Table 1, which is the rating you think that you need to evaluate each priority.

In other words, if a second patient performed this same exercise, but considers that the fat is more important than your need protein because so indicates, the result you get may be totally different. Finally, each result is different because the needs to evaluate are dotted subjectively, as is shown in Table 6.

CONCLUSION AND FUTURE RESEARCH

The great fortress of this intelligent tool is also its most formidable weakness, since the results depend heavily on the subjectivity of the score to be given priorities to assess, it serves as a tool for decision making and can be configured according to priorities, but does not replace the decision of the doctor or nutritionist who assigns the data associated with each food. Under no circumstance, the system replace medical personnel arise. However, a tool intends to facilitate your work, streamline decision-making and enable better patient retention.

The food distribution system presents an improvement in terms of the time needed for the provision of the service, at the same time, derived from the fact that fewer people use intrusive of information, decreases the probability of error product of over manipulation of the information. As a conclusion, the food distribution system for the San Rafael de Fusagasugá Hospital is able to be optimized using an intelligent software as a tool environmental intelligence system, it is concluded that the general objective has been met.

Table 6. Results Table

Selected Foods	Increased Fat and Protein Energy
Cooked rice	0.66
Boiled lentils	0.25
Boiled potatoes without skin	0.09

The process of the food distribution case study was widely and sufficiently known product of a field visit that lasted five days in days from 8 a.m. to 8 p.m. To design the system called the Intelligent Food Distribution System (SIDA abbreviated name) including to diseases as heart failures and AIDS, the food distribution system was divided into two sub systems the first one is the sub process of food allocation and the second is the sub process of distribution of foods. The sub-process of food allocation was designed with an Analytical Hierarchical Process (AHP) model for multicriteria decision making supported a parallel research that uses evolutionary theory to improve the coherence of the input matrices of the system. The second sub-process is essential to design how the automation system will be in order to determine the degree of optimization in the sub process of the physical distribution of the food.

In the comparison of the two models, a strong decrease in time was evidenced necessary for the system to operate if AIDS is implemented, these values were calculated from the hospital's maximum capacity to receive patients multiplied by the average time it takes for a doctor to assign a patient's diet added to the time it takes all the people who write the information from one format to another plus the time.

They take in transporting these documents. In the current system, there were many people who involved in the process of information management, and there were several minutes devoted to the transport of documents. With AIDS, only the doctor who assigns the diet intervenes in the process, eliminating transport, and information management times. The probability of error decreased considerably because the number of people that are involved in the process is minimal, the associated error will depend on the medical expertise and not from the monotony of the transcription of numerical data names and types of diet.

The AHP method was identified as the ideal mathematical tool for decision making, among the reasons for his choice is that he uses simple matrix algebra that is an ideal tool for multicriteria decision making that can accept a matrix of maxim input without needing large computing resources which allows to load an amount very large symptoms without the need to use expensive or sophisticated equipment.

A system test was done in Matlab, to select a diet from one, two and three symptoms, resulting in the most obvious diet assignment, the simulation is done to test the operation of the AHP tool, and is established as future research a whole stage of programming architecture design at user level and configuration with the different input and output peripherals of the system.

Finally, a simulation was performed in Excel that allowed to knowing from the data captured in the fieldwork as the current system behaves with respect to both variables of interest, time and probability of error, tables were also used in Excel to simulate the behavior of the proposed system numerically analyzing only the variables of interest. It is recommended as future work a simulation in some software like Simulate Promodel or Arena to study the whole system in a general way.

Improvements through macro-ergonomics in current organizations have been achieving changes with non-conventional applications or methods, as well as opening up new areas in which engineering and business administration can converge or the development of basic operations such as distribution of the food in a hospital, achieving with an industrial design product, the improvement in the process and avoiding errors that may affect the health or the improvement of the patients' health. It is concluded that the specific objectives are met in the design of an environmental intelligence system that optimizes a food distribution process for a public hospital using the San Rafael de Fusagasugá Hospital as a case study.

The future research should be aimed to improve the hybrid tool of artificial intelligence called virtual reality and Multiagent systems to feed a database that will be intelligent memory system called SIDA to specific changes in the horizon time and generate a balance in the distribution of food considering that the vents of time limit to less than an hour the distribution of food for each patient in 11 different floors of medical specialties.

In addition, in telemedicine, knowing the biomedical status of the patient in their usual environment of daily life and offering both the primary care physician and the specialist the diagnostic information of interest, allows us to draw a scenario of a hospital without barriers and virtually located in the patient's environment. In these scenarios, the patient is surrounded by multiple autonomous sensors that form ad hoc networks, whether BAN (Body Area Network), PAN (Personal AreaNetwork) and/or HAN (Home Area Network), acquire the information of interest; from portable scales or tensiometers to mobile electrocardiogram monitors, automatic defibrillators and devices that can be implanted in the skin or tissues, through environmental sensors of temperature, humidity, position, among other relevant aspects of the patient, this innovation will allow to increase the potentiality of use of Environmental Intelligence to improve the health in the society of a Smart City.

The relevance of this study is improving the waiting times to patients in diverse hospital in Mexico and other Latin-American societies.

REFERENCES

Anderson, D. (2008). *Statistics for Business and Economics* (10th ed.). Academic Press.

Barrett, C. B. (2006). *Food Aid as a Part of a Coherent Strategy to Advance Food Security Objectives.* ESA Working Paper no. 06-09. FAO. Retrieved from ftp://ftp.fao.org/docrep/fao/009/ag037e/ag037e00.pdf Bolis, I., & Sznelwar, L. (2016). A case study of the implementation of an ergonomics improvement committee in a Brazilian hospital – Challenges and benefits. *Applied Ergonomics, 53,* 181-189.

Caracuel, A. (2007). Alimentación hospitalaria: Del blanco y negro al color. *Anales de RACVAO, 20*(1), 43–67.

Cebola, M., Mahendra, A., Pombo, M., Marques, B., Pinto, H., Carolino, E., & Mendes, L. (2016). Comparison of nutritional status among elderly in hospital and in community environments. *European Geriatric Medicine, 7*(1), S195.

Escolano, F. (2003). *Artificial Intelligence Models, Techniques and Application Areas*. University of Alicante.

FAO. (1998). *Platos típicos de países de América Latina 1998.* Santiago de Chile, Chile: FAO.

FAO/LATINFOODS. (2009). *Tabla de Composición de Alimentos de América Latina*. Author.

García, G., & Lange, K. (2008). Macroergonomic study of food sector company distribution centres. *Applied Ergonomics, 39*, 439–449.

González Molero, I., Olveira Fuster, G., Liébana, M. I., Oliva, L., Laínez López, M., & Muñoz Aguilar, A. (2008). Influencia de la temperatura en la ingesta de pacientes hospitalizados. *Nutrición Hospitalaria*, *23*(1), 54–59. Retrieved from http://scielo.isciii.es/scielo.php?script=sci_arttext&pid=S0212-16112008000100009&lng=es&tlng=pt

Heizer, J. (2004). *Principles of operations management*. Pearson Prentice Hall Editorial.

Hendrick & Kleiner. (2002). *Macroergonomics. Theory, Methods, and Applications*. Lawrence Earlbaum Associates, Inc.

Lamata, F. (1998). *User administration and health management.* Doctoral Thesis.

Lee, J. (2000). *Operations management: strategy and analysis*. Addison-Wesley.

Maldonado, F., & Ochoa, A. (2016). Un estudio exploratorio de la competitividad en función de espacios de trabajo. *IX Hybrid Intelligent Systems Workshop, MICAI 2016.*

Chapter 13
Trends in Macroergonomics Applications for Improved Work Systems

Karina Cecilia Arredondo
Universidad Autonóma de Baja California, Mexico

Arturo Realyvásquez
Instituto Tecnológico de Tijuana, Mexico

Guadalupe Hernández-Escobedo
Instituto Tecnológico de Tijuana, Mexico

ABSTRACT

Macroergonomics is the subdiscipline of ergonomics that is concerned with the analysis, design, and evaluation of work systems. It means, macroergonomics focuses on harmonizing the organizational structure of a company and not only one workstation or one task, as microergonomics does. Macroergonomics is a top-down, middle-out, and bottom-up approach. In the top-down approach, the overall general work system structure may be prescribed to match the organization's sociotechnical characteristics. On the other hand, the middle-out approach focuses on the analysis of subsystems and work processes, which can be assessed both up and down the organizational hierarchy from intermediate levels, and also, up and down some changes may be done to ensure the work system design is harmonized. Finally, the bottom-up approach comprises an extensive participation of employees in the identification of problems. Currently, macroergonomics is considered an emergent subdiscipline, and there is the need to promote current theories and methods and propose new ones.

DOI: 10.4018/978-1-5225-7192-6.ch013

INTRODUCTION

Traditionally, work systems have been the focus of attention with the aim of increasing productivity. Here, it is principally from the economic perspective and is focused to obtain profits considering a purely consumption approach. However, this has evolved considering human and social aspects in such a way that economic advantages are put at the same level that the promotion of positive changes on individual and collective work environments. It is seen when organizations have adapted the work system conditions to the limitations, abilities and needs of their employees. One result of this view is Ergonomics, which arises supporting the design of products and systems enhancing human well-being and performance (Hassall, Xiao, Sanderson, & Neal, 2015).

Consequently, Macroergonomics, as a nested discipline of Ergonomics, emerges basically with the aim of design a fully "harmonized" work system enhancing the key performance indicators (organizational effectiveness, productivity and satisfaction). Herein the sociotechnical systems theory was the fundamental principle, and the industrial and organizational psychology viewpoints were also included (Lauren A. Murphy, Robertson, & Carayon, 2014). Thus, the macroergonomics approach mostly studies the interaction between the elements (machine, environment, software, job and organization) within the work system.

In this line, Hendrick & Kleiner (2002) identified diverse applications of this approach. For instance, reduction of work-related injuries, use of Information and Communication Technology (ICT) and their changes in work life, management of large scale organizational changes, improvement of some manufacturing aspects, development of training systems, relocation of technologies, increment of aviation safety and cut of large scale accidents are examples of these applications. Five years later, Hendrick (2007) analyzed the research trends on the management and organization of work systems, which acknowledged technology, demographic shifts, value changes, ergonomic-based litigation, world competition and failure of traditional Microergonomics as principal objects of study. These applications were frequently done in contexts as manufacturing, service organizations, healthcare, maintenance, telecommunications, maritime vessels, and industrial accident prevention. On the other hand and in an isolated way were in retail, military, typesetting, aviation operations, schools, and government.

The structure of this chapter is as follows. It starts by considering the historical contexts of Macroergonomics applications from 1994 to 2013, and the challenges that they posed for Macroergonomics. Next, it examines those works that took place from 2014 to 2018 and attempted to address the raised challenges and the emerged trends in the field. The chapter concludes with a set of predictions and challenges for macroergonomics in the near future.

HISTORICAL CONTEXTS

A search in the Scopus database of the word Macroergonomics in the article title, abstract or keywords, and filtering the period from 1994 to 2013 was done. 150 documents resulted from the search and were classified and summarized in 70 conference papers, 64 articles, four book chapters and two editorials. After, the analysis was conveniently limited to only articles and book chapters so that 68 documents were used in this phase. Table 1 presents the recurrent researchers in the field.

Table 1. Researchers with most published articles from 1994-2013 in the Macroergonomics field

Author	Frequency	Author	Frequency
Kleiner, B.M.	7	Azadeh, A.	2
Karsh, B.T.	5	Clegg, C.W.	2
Carayon, P.	4	Drury, C.G.	2
Newman, W.P.	4	James, C.G.	2
Robertson, M.M.	4	Nagamachi, M.	2
Sainfort, F.	4	Smith, M.J.	2
Erensal, Y.C.	3	Taveira, A.D.	2
Hendrick. H.W.	3	Village, J.	2
Albayrak, E.	2	Zink, K.J.	2
Alper, S.J.	2		

Source: (Scopus, 2018)

Subsequently, the results were organized using the number of citations obtained during the mentioned period. Therefore, Table 2 summarizes the influential documents serving as a basis to other researchers. Kleiner, Carayon, Hendrick, Robertson, Smith, Alpert, Punnet, Zink and Taveira have been considered the representatives of the Macroergonomics field. It is consequently by the quality and quantity of their publications. They addressed diverse issues as sociotechnical systems, design of work systems, healthcare and occupational ergonomics, manufacturing applications, industrial organization, and public administration.

Particularly, Hendrick & Kleiner (2002) pointed out the increase in the use of communication technologies so that work at home could be recognized as a new way of work. These anticipated and suggested mandatory changes in organizations and specifically, in their structure. They concluded that in the current era there should be developed methodologies, tools and curricula in higher education in order to overcome the challenges implied in the Macroergonomics applications.

EMERGING CONTEXTS

A second search in the Scopus database of the word Macroergonomics in the article title, abstract or keywords, and filtering the period from 2014 to 2018 was done. 48 documents resulted from the search and were classified and summarized in 31 articles, 12 conference papers, two reviews, one book and one book chapter. Similarly, the documents were classified according to the year of publication resulting 8 in 2018, 4 in 2017, 9 in 2016, 11 in 2015 and 16 in 2014. Later, all documents were used to uncover diverse fields that were addressed. The contexts were organized in general reviews; macroergonomics tools and methods; organizational analyzes and corporate strategies; sociotechnical systems approach; applications in home and community-based work and office jobs; hospitals and healthcare; manufacturing, and construction and trucking industry.

Table 2. Researchers publishing from 1994-2013 with most cited articles to the date in the Macroergonomics field

Author	Cited	Document Title
Pascale Carayon (2006)	169	Human factors of complex sociotechnical systems
Kleiner, Rajani, & Sadiq (2006)	73	Macroergonomics: Analysis and design of work systems
Smith & Carayon (1995)	69	New technology, automation, and work organization: stress problems and improved technology implementation strategies
Punnett, Cherniack, Henning, Morse, & Faghri (2009)	68	A conceptual framework for integrating workplace health promotion and occupational ergonomics programs
Alper & Karsh (2009)	64	A systematic review of safety violations in industry
Hendrick (2008)	60	Applying ergonomics to systems: Some documented "lessons learned"
Robertson, Huang, O'Neill, & Schleifer (2008)	49	Flexible workspace design and ergonomics training: Impacts on the psychosocial work environment, musculoskeletal health, and work effectiveness among knowledge workers
Husemann, Von MacH, Borsotto, Zepf, & Scharnbacher (2009)	46	Comparisons of musculoskeletal complaints and data entry between a sitting and a sit-stand workstation paradigm
Hendrick (1995)	42	Future directions in macroergonomics
Smith, Schoenbeck, & Clayton (2009)	35	Staff perceptions of work quality of a neonatal intensive care unit before and after transition from an open bay to a private room design
Neumann, Ekman, & Winkel (2009)	34	Integrating ergonomics into production system development - The Volvo Powertrain case
Karsh & Brown (2010)	33	Macroergonomics and patient safety: The impact of levels on theory, measurement, analysis and intervention in patient safety research
Brian M. Kleiner (2008)	33	Macroergonomics: Work system analysis and design
Neumann & Village (2012)	30	Ergonomics action research II: A framework for integrating HF into work system design
Zink (2000)	29	Ergonomics in the past and the future: From a german perspective to an international one
Holden, Or, Alper, Joy Rivera, & Karsh (2008)	28	A change management framework for macroergonomics field research
Wood, Stride, Wall, & Clegg (2004)	28	Revisiting the use and effectiveness of modern management practices
Korunka et al. (2007)	26	Customer orientation among employees in public administration: A transnational, longitudinal study
Haro & Kleiner (2008)	25	Macroergonomics as an organizing process for systems safety
Taveira, James, Karsh, & Sainfort (2003)	25	Quality management and the work environment: An empirical investigation in a public sector organization

Source: (Scopus, 2018)

General Reviews

This section contains nine documents addressing general reviews of Macroergonomics. First, Berlin, Neumann, Theberge, & Örtengren (2014) compared how the Canadian Industrial Engineers and Ergonomists work to influence the Human Factors and Ergonomics (HFE) issues. It was also considering the implied constraints and strategies on them. They found that the best strategy to show the benefits by HFE improvements should be presented in terms of business profits. Second, Parallel, Karsh, Waterson,

& Holden (2014) addressed the Mesoergonomics issue related to how the micro and macroergonomics should be linked and integrated into each other. In order to do so, they defined Mesoergonomics issue as an open system approach in the HFE theory and research. It results on study the relations between variables in which the HFE constructs are the dependent variables. They also proposed a framework for implementing that issue and illustrated its value within the theory through two case studies.

Moreover, Patrick Waterson et al. (2014) analyzed the current and future directions of Macroergonomics and the sociotechnical systems. Principally, their discussion was centered on the effectiveness of the Macroergonomics methods by facilitating the improvements on the workplace design. Specifically, some applications of them took the form of health and safe interventions in community settings in which there were chronically ill patients. Within the same type of contexts, Murphy et al. (2014) examined the relation of safety climate and Macroergonomics and how it affected the safety outcomes. Principally, they identified two subsystems (personnel –methods- and technology –processes-) as major factors affecting the achievement of a safety climate. It was seen in the management commitment on safety and communication highlights. Additionally, they proposed a conceptual model as a framework to guide organizational interventions on this matter. In the same line, Nathanael, Zarboutis, and Marmaras (2015) contributed to the intervention process in the Macroergonomics field with a methodology based on the dialectics notion of contradictions. Their proposal had a particular value on practical means by anticipating the effects and side effects of proposed changes during the Macroergonomics applications.

Furthermore, Kleiner, Hettinger, DeJoy, Huang, and Love (2015) proposed theoretical and practical approaches to identify sociotechnical attributes of safe and unsafe work systems. They also presented a variety of socio technical systems perspectives on intersections between social-organizational and technology-work process factors and the impact on analysis, design and operation of the work systems. In detail, Hassall et al. (2015) defined HFE as fields applying scientific information about human cognition and behavior to support the design of products and systems enhancing human well-being and performance. Their contribution consisted on defining the general concepts of HFE that included human-system interaction, HFE professionals, safety, and productivity, among others. Going beyond, Thatcher et al. (2016) link Macroergonomics and Sustainability concepts arguing that HFE professionals, due to their knowledge and skills) can support problem solving as pollution, climate change, renewable energy and land transformation.

Finally, Thatcher, Waterson, Todd, & Moray (2018) evaluated the impact of the Ergonomics discipline in the last 25 years to face up the global problems facing humanity and considered how Ergonomics might help in the future problems of society. Following this course, Marina Bischoff & Vianna Dos Santos (2018) presented a literature review regarding the Human Factors and Ergonomics and relationships with other areas as communication interactions research line, Macroergonomics, internal organization communication, and the process of communication auditing. To finish this section, Figure 1 exhibits the used keywords in the documents found and the number of them.

Macroergonomics Tools and Methods

Seim, Broberg, & Andersen (2014) studied the learning process for HFE professionals (researchers and practitioners) in an interactive research project. They proposed a new framework for intervention in the processes of workplace design in companies. This frame was based on applying the social learning theory. On the other hand, Lim, Village, Salustri, & Neumann (2014) developed and evaluated a method applying a process mapping as a tool for participative integration of human factors into work systems

Figure 1. Keywords in general review
Source: The authors

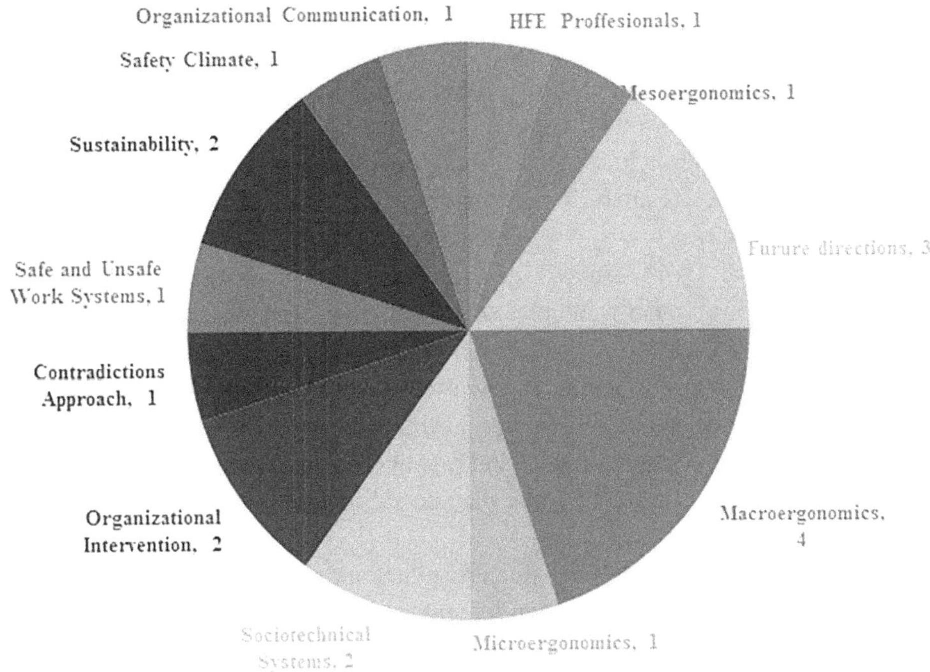

design within an electronic manufacturer. This process included a map creation, a map application and an evaluation of the mapping process. Similarly, the tools included a definition of the scope and level of detail in the mapping, contained errors, missing information, and map overlap.

Moreover, Barón Maldonado, Rivera, and Caro (2014) presented the development of a methodology to implement Lean Product Development (LPD) in apparel companies. This methodology incorporated concepts of Integrated Management Systems, Ecodesign and Macroergonomics. Four years after and in the same line, Silva, Nickel, & Dos Santos (2018) provided an evaluation of Macroergonomics methods for the application of organization analyzes in startups. They analyzed eight methods resulting that High Integration of Technology, Organization, and People (HITOP), Analysis and Macroergonomics Analyses of Structure (MAS) methods were only deemed applicable to startups. Also, Macroergonomics Analysis and Macroergonomics Design and Analysis (MEAD) may be applicable only in specific cases of startups. Figure 2 presents the main methods and tools used in Macroergonomics.

Figure 2. Keywords in macroergonomics methods and tools
Source: The authors

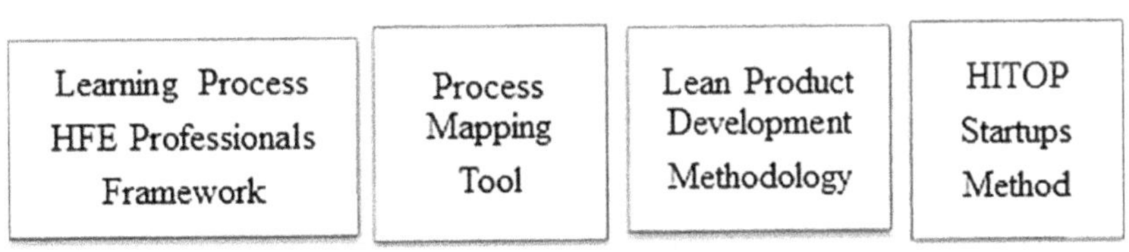

Organizational Analyzes and Corporate Strategies

Kluge (2014) studied as object the acquisition of knowledge and skills for task work and teamwork for controlling technical systems since the macroergonomics perspective. To do so, he proposed a theoretical model, which was focusing on how skill and knowledge acquisition for complex tasks was accomplished. In addition, he showed how this model should be used to derive training methods and instructional techniques. He also included the learning process, knowledge structure and skill requirements as fundamental variables of analysis and how they were associated with the effective operation and management of technology.

Furthermore, Valdez et al. (2014) studied the healthcare sector as a work system and this research was focused on how macroergonomics principles and models originally developed within an institutional context may be adapted for the contexts in which patients were embedded. In the study, they considered patients performing many non-paid activities in order to maintain and improve their health. These kinds of activities may be conceptualized as self-care and self-management work. They also focused in lessons related to the conceptual, methodological and practiced-related challenges on understanding and affecting patients' work and works systems that may be subsequently applied by other researchers on other contexts.

Concerning to team cohesion, Salas, Grossman, Hughes, and Coultas (2015) reviewed the literature finding relevant information about cohesion measurement, exploring developing measurement approaches, and providing theoretical and practical recommendations for optimizing cohesion measurements. They also reviewed empirical research uncovering specific information about cohesion conceptualization, measurement, and relationships with performance. This study ended with the current trends on team cohesion providing suggestions and possible solutions to guide future efforts in this field.

To increase the impact of ergonomics in a Colombian organization, Rodríguez Ruíz & Pérez Mergarejo (2016) proposed a Macroergonomics diagnosis, which performed an assessment on five Colombian organizations using the Ergonomics Maturity Model (EMM). The methods under this model were used to assess the capability of an organization by introducing, developing and implementing ergonomics. The results suggested that organizations could be classified in the lowest ignorance maturity level. Lack of trained personnel in the ergonomics field, reactive approach on solving problems and the limited ergonomics risk assessments were found as the main causes of this result.

In a specific context, Village, Salustri, and Neumann (2016) focused on how cognitive mapping links human factors to corporate strategies. Individual cognitive maps were drawn during one-hour interviews with seven senior directors of a large electronics firm. The objective was to harness individual tacit knowledge from senior directors about human factors relating to strategic goals and making explicit their shared managerial thinking with the aim of identifying improvement opportunities using those factors. They concluded on uncovering the strong connection between human factors and explicit strategic goals. Consequently, this connection can help organizations on identifying opportunities to improve human and business performance.

Additionally, Silva et al. (2018), as mentioned in the last section, used an analysis of macroergonomics methods on startups for study performance. It was discovered that methods employed only should be utilized on specific startups based their organizational objectives. As a result, they presented two types of applications of the methods, as explained above. Figure 3 shows the principal presented works in this section.

Figure 3. Main works for organizational analyzes and corporate strategies
Source: The authors

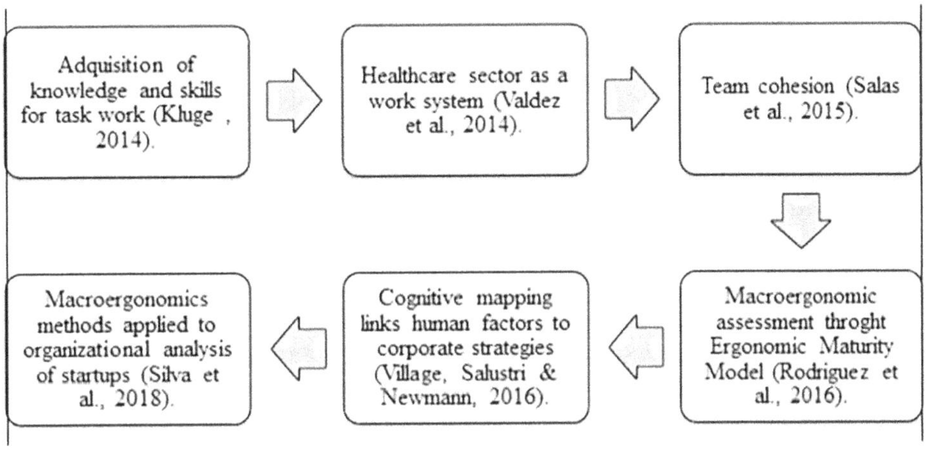

Sociotechnical Systems Approach

Waterson et al. (2015) provided an up-to-date review of Socio Technical Systems (STS) methods, tools, techniques and case studies focused on methods for workplace safety. They concluded with a roadmap for future work in the field. In the same line, Meshkati, Tabibzadeh, Farshid, Rahimi, & Alhanaee (2016) proposed a framework to ensure regional interoperability for safety, sustainability, and resilience of interdependent energy, water, and seafood sources in the Persian Gulf. They performed the integration of People-Technology-Ecosystem. This integration was built upon the Rasmussen's model, a macro system integrative framework that was based on the broader context of human factors considering as a meta-ergonomics paradigm. In other words, this was developed in order to use a great number of variables.

Applications in Home and Community Based Work and Office Jobs

Coelho, Tavares, Lima, and Lourenço (2017) analyzed the effect of different kinds of work on the psychosocial assessment of workers under the same management and organizational environment. The Copenhagen psychosocial questionnaire was used in a voluntary assessment in a utility company with the participation of 25 office workers and 32 field workers. The results suggested that variables as organizational design, management in system in place and cultural environment were more relevant in terms of decisive impact that specific factors.

On the other hand, Robertson, Huang, & Lee (2017) focused on a macroergonomics intervention in an office workplace. Two principal factors were considered which could produce beneficial effects for office and computer workers and their organizations. The effect of the intervention was stronger when the workers' ergonomics needs were taken into account by the management showing their responsibility on these issues.

Hospitals and Healthcare

Patel and Kannampallil (2014) examined the theoretical and foundational models of human factors and ergonomics (HFE) for achieving patient safety and quality and for evaluation of healthcare systems in the 2007-2013 period. They also investigated the potential of the approaches in biomedical informatics. Similarly, they discussed the opportunities for better integrating HFE in these contexts. In same kinds contexts, Barzideh, Choobineh, & Tabatabaee (2014) studied the job stress dimensions and examined their relationship to musculoskeletal disorders (MSDs) among nurses of Shiraz University of Medical Sciences hospitals. It was a cross-sectional study in which 385 randomly selected nurses participated. There, 89.9% of the participants experienced some form of MSDs during the last 12 months. The major lower back symptoms were the most prevalent problem reported (61.8%). In conclusion, the nurses were exposed to high levels of job stress. Similarly, Carayon et al. (2014) analyzed the Systems Engineering Initiative for Patient Safety (SEIPS) model. Initially, they described several research and practical applications for healthcare systems and process redesign in which the principles for redesign healthcare systems were included. In addition, Clack, Kuster, Giger, Giuliani, & Sax (2014) presented the case of infection risk associated with syringe manipulation that could easily be solved by introducing user-centered design solutions. Here, the organizational complexity made the implementation of such solutions hardly reachable.

Holden, Brown, Scanlon, Rivera, and Karsh (2015) presented the first study to quantitatively assess the short- and long-term impact of bar coded medication administration (BCMA) information technology on nurses' mental workload as well as on perceived medication safety. Findings raised questions about the achievement of sustainable performance improvement with health information technology and the balance between micro- and macroergonomics approaches for studying technology change. In a parallel way, the industry can learn of this case in the healthcare sector. It is because designers should consider how technology changes the cognitive work including mental work. Similarly and in the same year, Holden, Eriksson, Andreasson, Williamsson, & Dellve (2015) studied healthcare workers' perceptions of lean, a sensitive context, through mixed methods in three Swedish hospitals. They formally discussed strategies for implementing lean in healthcare and the importance of attending the levels, context and worker consequences on those perceptions. Furthermore, they presented some lines of future work on this topic.

McLaughlin, Ward, and Keene (2016) described a new surgical checklist for veterinary cardiologist based on the literature and application of human factors design. Bolis and Sznelwar (2016) also analyzed a case study of the implementation of an ergonomics improvement committee in a Brazilian hospital. This was considering the challenges and benefits. Moreover, Yang and Asan (2016) performed a literature review on how the designs of patient-facing Health Information Technologies (HITs) have been suggested and evaluated, and how they may potentially affect the doctor-patient communication and patient-centered care.

In same line, Holden, Valdez, Schubert, Thompson, and Hundt (2017) analyzed the macroergonomics factors in the patient work system with chronic illness and their informal caregivers. They identified 17 factors across physical, social and organizational domains and household and community levels. Health related activities of patients are embedded in and shaped by levels of the social, physical and organizational context. Finally, Steege, Pasupathy, & Drake (2018) analyzed the fatigue of hospital nurses. Similarly, they explored the relationships between nursing work system components and fatigue and recovery of nurses.

Manufacturing

Heras-Saizarbitoria, Cilleruelo, & Allur (2014) analyzed, from the macroergonomics perspective in the traditional industrial and service subsectors, the influence of the main global management meta-standard. This study was carried out in two Spanish organizations belonging to the subsector of residential care home for elderly persons and was focused on the quality of working life of the direct attention employees. In a parallel way, Sherehiy & Karwowski (2014) analyzed the relationship between work organization and workforce agility in small manufacturing enterprises. A total of 41 managers, 82 office workers, and 52 production workers from six small manufacturing companies participated in the study. The results should be used for optimizing work demands and conditions in the workplace environments in which it is required constant change and adaptation.

Furthermore, de Guimarães, Anzanello, Ribeiro, & Saurin (2015) presented a participatory intervention in a furniture manufacturing company in Southern Brazil. The study indicated that it was possible to balance both ergonomics and production demands. As a result, workload was reduced by 42% and productivity increased by 46%. On the other hand, Realyvásquez, Maldonado-Macías, García-Alcaraz, & Blanco-Fernández (2015) examined the effects of organizational macroergonomics compatibility elements over manufacturing systems performance from four companies in Mexico. Results indicated that macroergonomics organizational elements as organizational culture and organizational communication had different effect over clients and indirect effect over the manufacturing processes and growth of companies. Similarly, Cifuentes & Fulmer (2015) studied the use of treadmill workstations and psychosocial barriers to using them. They found that barriers were related to communication, need for motivation, peer pressure, and adaptation affecting performance. However, effectiveness and efficiency studies demonstrating improved energy expenditure in real workplaces have yet to build evidence to support use of such workstations.

Moreover, Realyvásquez, Maldonado-Macías, García-Alcaraz, Cortés-Robles, & Blanco-Fernández (2016) proposed a macroergonomics compatibility questionnaire that was statistically validated and 158 respondents at four manufacturing companies participated. They analyzed the effects of environmental elements on the psychological characteristics and performance of employees in manufacturing systems using the structural equation modeling. In this sense, Pacholski & Pawlewski (2017) used simulation technology for macroergonomics industrial systems improvement. They analyzed the buses' production process capability after the assembly hall expansion. This project also included the reengineering of the macroergonomics industrial system model for the new work organization, visualization and verification.

On a similar note, Miguez, Filho, Faustino, & Gonçalves (2018) developed a project in a 1,500-employee metallurgical factory, where a multidisciplinary team of certified ergonomist, engineers, managers and direct employees used concepts of Participatory Ergonomics and Lean Manufacturing. As a result, unnecessary activities were eliminated reducing the costs and lead-time and maintained the well being of workers. Finally, Realyvásquez, Maldonado-Macías, García-Alcaraz, & Arana (2018) proposed a Macroergonomics Compatibility Index (MCI) for manufacturing systems in Mexico. Here, the variables used to measure that index were Person, Organization, Technology and Tools, Task, and Environment as Macroergonomics factors.

Construction and Trucking Industry

Tavares et al. (2015) promoted the debate about work at heights in the construction industry. This type of work had been responsible for occupational accidents. In fact, it should be highlighted as a real problem because there is the lack of application of the occupational health and the safety ergonomic approach. In the same field, Zhao, McCoy, Kleiner, Smith-Jackson, & Liu (2016) analyzed fatal electrical injuries in the construction industry. This work contributed to the knowledge body of risk management, electrical safety and research methodologies. It was because the study used a triangulation approach for examining construction-fatality reports and considering effective risk management through the control over the whole sociotechnical system.

Murphy, Robertson, Huang, Jeffries, & Dainoff (2018) and Murphy, Huang, Robertson, Jeffries, & Dainoff (2018) applied the sociotechnical systems approach to enhance safety climate in the trucking industry. To do that, they developed a methodology to describe the results of the study. This methodology extended safety climate beyond an overall score by using the framework of Macroergonomics by examining the entire system in a more comprehensive manner. Also, it was applied to identify gaps in the specific work system and that information can be used to design interventions to change the safety climate and culture of an organization for reduce negative safety outcomes. Similarly, it included the critical incident technique, contextual inquiries, functional role diagrams, and affinity mapping. Key informants in the trucking industry helped on identifying 19 themes that affected safety. The themes ranged from balancing work and family/personal time, the company's policy vs. practice, respecting the job of the driver, and active listening and meaningful feedback.

FUTURE RESEARCH DIRECTIONS

According to the initial analysis carried out for the period from 1994 to 2014, there was evidence that studies had a greater impact in the actual research. This was based on the number of citations found in the search done in the Scopus database. These citations were related to the analysis of the work systems using the Sociotechnical Systems approach as principal view on the studies in which consequently there were diverse lessons learned. Principally, programs promoting the health and occupational safety, and ergonomic interventions in production systems and public administration were examples of its applications. On the other hand, the second study focused in the period from 2014 to 2018 uncovered different and interesting applications of the Macroergonomics approach.

For instance, the number of studies were proportionally increased in relation to the number of years used in this search. The search also put on manifest the varied interest of researchers on study the approach in other uncommon contexts. They were interested in knowing what is happening in hospitals, health care, manufacturing among other contexts. Moreover, they were done ergonomic interventions in organizations using various techniques and methods in order to prove their feasible on those contexts. Finally and using the information presented, it is concluded that applications of Macroergonomics would be found in diverse contexts using new techniques and methods in order to deal with the actual situations in which systems are immersed.

For instance, it is expected that researchers from diverse countries and continents would use the Macroergonomics approach to investigate topics related to the implementation of manufacturing philosophies, performance indexes, organizational policies and strategies, among others. Similarly, it is predicted the incorporation of theories, techniques, and methods employed in other Ergonomics approaches. Furthermore, the researchers will open those in other contexts in order to fill the gaps of knowledge discovered. In this line, it could be seen with the inclusion of rural contexts on research. In fact, contexts in primary sector is an opportunity of research using the Macroergomics approach, as emerging discipline.

CONCLUSION

It is considered that the future of Ergonomics is seen in the health sector and hospitals and manufacturing; however, it is important to highlight the necessity to focus on other aspects of them. For instance, biomechanics, anthropometry and physical chronic and occupational diseases are relevant topics of Ergonomics. These indirectly expanded the opportunity to investigate the mental health associated with the labor relations, teamwork, burnout, among other factors studied into the psychosocial factors provoking stress. As a result of the work, stress is considered a relevant factor, which should be studied to consequently reduce it. It is known that multiple causes obey to multiple variables attributable to organizational factors as structure and culture. This is because psychosocial factors represent a key element to maintain a good performance in each of the hierarchical levels of a company (and therefore at the organizational level).

In contrast, Macroergonomics is currently opening opportunities to research on other labor sectors. For example, a great number of research and applications are being carried out at the macro-ergonomic level in the agricultural sector, the education sector, and aviation, additionally to manufacturing and hospital contexts. Moreover, there is a tendency to carry out research on the mental burden and burnout of workers in the manufacturing industry. This opens the way for replicate the research in different labor sectors, as mentioned before. Furthermore, research is being carried out in the area of work psychology from a macro-ergonomic approach in workers with disabilities. This is in order to determine the behavior of those towards the demands, structure and culture of organizations and co-workers. In all these investigations, although the development and validation of a tool or a macroergonomic evaluation technique is not guaranteed, they could develop models or methodologies that could be applied in future and similar studies but in different contexts ensuring success in any macro-ergonomic research and applications by adopting the sociotechnical systems approach.

ACKNOWLEDGMENT

We would like to acknowledge to the Universidad Autónoma de Baja California, Tecnológico Nacional de México, Instituto Tecnológico de Tijuana and PRODEP for partially funding this chapter. The interpretations and views in this chapter, however, are solely those of the authors.

REFERENCES

Alper, S. J., & Karsh, B. T. (2009). A systematic review of safety violations in industry. *Accident; Analysis and Prevention*, *41*(4), 739–754. doi:10.1016/j.aap.2009.03.013 PMID:19540963

Barón Maldonado, D. I., Rivera, L., & Caro, M. (2014). *Improvement of new product development process in apparel SMEs using lean and life cycle* (Vol. 9-12-NaN-2014). AIDI - Italian Association of Industrial Operations Professors. Retrieved from https://www.scopus.com/record/display.uri?eid=2-s2.0-84982859268&origin=resultslist&sort=plf-f&src=s&st1=Macroergonomics&nlo=&nlr=&nls=&sid=cb1a157df15e8dea160496d8803d91dc&sot=b&sdt=cl&cluster=scopubyr%2C%222018%22%2Ct%2C%222017%22%2Ct%2C%222016%22%2Ct%2C%222015%22%2Ct%2C%222014%22%2Ct&sl=30&s=TITLE-ABS-KEY%28Macroergonomics%29&relpos=47&citeCnt=0&searchTerm=

Barzideh, M., Choobineh, A. R., & Tabatabaee, H. R. (2014). Job stress dimensions and their relationship to musculoskeletal disorders in Iranian nurses. *Work (Reading, Mass.)*, *47*(4), 423–429. doi:10.3233/WOR-121585 PMID:23324727

Berlin, C., Neumann, W. P., Theberge, N., & Örtengren, R. (2014). Avenues of entry: How industrial engineers and ergonomists access and influence human factors and ergonomics issues. *European Journal of Industrial Engineering*, *8*(3), 325. doi:10.1504/EJIE.2014.060999

Bolis, I., & Sznelwar, L. I. (2016). A case study of the implementation of an ergonomics improvement committee in a Brazilian hospital - Challenges and benefits. *Applied Ergonomics*, *53*, 181–189. doi:10.1016/j.apergo.2015.09.012 PMID:26464035

Carayon, P. (2006). Human factors of complex sociotechnical systems. *Applied Ergonomics, 37*(4), 525–535. doi:10.1016/j.apergo.2006.04.011

Carayon, P., Wetterneck, T. B., Rivera-Rodriguez, A. J., Hundt, A. S., Hoonakker, P., Holden, R., & Gurses, A. P. (2014). Human factors systems approach to healthcare quality and patient safety. *Applied Ergonomics*, *45*(1), 14–25. doi:10.1016/j.apergo.2013.04.023 PMID:23845724

Cifuentes, M., & Fulmer, S. (2015). Research Needs for and Barriers to Use of Treadmill Workstations. *Ergonomics in Design*, *23*(3), 25–30. doi:10.1177/1064804615588849

Clack, L., Kuster, S. P., Giger, H., Giuliani, F., & Sax, H. (2014). Low-hanging fruit for human factors design in infection prevention - Still too high to reach? *American Journal of Infection Control*, *42*(6), 679–681. doi:10.1016/j.ajic.2014.03.002 PMID:24837120

Coelho, D. A., Tavares, C. S. D., Lima, T. M., & Lourenço, M. L. (2017). Psychosocial and ergonomic survey of office and field jobs in a utility company. *International Journal of Occupational Safety and Ergonomics*, 1–12. doi:10.1080/10803548.2017.1331620 PMID:28589755

de Guimarães, L. B. M., Anzanello, M. J., Ribeiro, J. L. D., & Saurin, T. A. (2015). Participatory ergonomics intervention for improving human and production outcomes of a Brazilian furniture company. *International Journal of Industrial Ergonomics*, *49*, 97–107. doi:10.1016/j.ergon.2015.02.002

Haro, E., & Kleiner, B. M. (2008). Macroergonomics as an organizing process for systems safety. *Applied Ergonomics*, *39*(4), 450–458. doi:10.1016/j.apergo.2008.02.018 PMID:18407244

Hassall, M., Xiao, T., Sanderson, P., & Neal, A. (2015). *Human Factors and Ergonomics. International Encyclopedia of the Social &amp* (2nd ed.). Behavioral Sciences. doi:10.1016/B978-0-08-097086-8.22025-4

Hendrick, H. W. (1995). Future directions in macroergonomics. *Ergonomics*, *38*(8), 1617–1624. doi:10.1080/00140139508925213

Hendrick, H. W. (2007). Macroergonomics: The Analysis and Design of Work Systems. *Review of Human Factors and Ergonomics*, *3*(1), 44–78. doi:10.1518/155723408X299834

Hendrick, H. W. (2008). Applying ergonomics to systems: Some documented "lessons learned&quot. *Applied Ergonomics*, *39*(4), 418–426. doi:10.1016/j.apergo.2008.02.006 PMID:18374303

Hendrick, H. W., & Kleiner, B. M. (2002). *Macroergonomics. Theory, Methods, and Applications*. Mahwah, NJ: Lawrence Erlbaum Associates, Publishers. doi:10.1201/b12477

Heras-Saizarbitoria, I., Cilleruelo, E., & Allur, E. (2014). ISO 9001 and the quality of working life: An empirical study in a peripheral service industry to the standard's home market. *Human Factors and Ergonomics in Manufacturing*, *24*(4), 403–414. doi:10.1002/hfm.20392

Holden, R. J., Brown, R. L., Scanlon, M. C., Rivera, A. J., & Karsh, B.-T. (2015). Micro- and macroergonomic changes in mental workload and medication safety following the implementation of new health IT. *International Journal of Industrial Ergonomics*, *49*, 131–143. doi:10.1016/j.ergon.2014.04.003

Holden, R. J., Eriksson, A., Andreasson, J., Williamsson, A., & Dellve, L. (2015). Healthcare workers' perceptions of lean: A context-sensitive, mixed methods study in three Swedish hospitals. *Applied Ergonomics*, *47*, 181–192. doi:10.1016/j.apergo.2014.09.008 PMID:25479987

Holden, R. J., Or, C. K. L., Alper, S. J., Joy Rivera, A., & Karsh, B. T. (2008). A change management framework for macroergonomic field research. *Applied Ergonomics*, *39*(4), 459–474. doi:10.1016/j.apergo.2008.02.016 PMID:18417095

Holden, R. J., Valdez, R. S., Schubert, C. C., Thompson, M. J., & Hundt, A. S. (2017). Macroergonomic factors in the patient work system: Examining the context of patients with chronic illness. *Ergonomics*, *60*(1), 26–43. doi:10.1080/00140139.2016.1168529 PMID:27164171

Husemann, B., Von Mac, H. C. Y., Borsotto, D., Zepf, K. I., & Scharnbacher, J. (2009). Comparisons of musculoskeletal complaints and data entry between a sitting and a sit-stand workstation paradigm. *Human Factors: The Journal of the Human Factors and Ergonomics Society*, *51*(3), 310–320. doi:10.1177/0018720809338173 PMID:19750794

Karsh, B. T., & Brown, R. (2010). Macroergonomics and patient safety: The impact of levels on theory, measurement, analysis and intervention in patient safety research. *Applied Ergonomics*, *41*(5), 674–681. doi:10.1016/j.apergo.2009.12.007 PMID:20153456

Karsh, B.-T., Waterson, P., & Holden, R. J. (2014). Crossing levels in systems ergonomics: A framework to support "mesoergonomic" inquiry. *Applied Ergonomics*, *45*(1), 45–54. doi:10.1016/j.apergo.2013.04.021 PMID:23706573

Kleiner, B. M. (2008). Macroegonomics: Work system analysis and design. *Human Factors: The Journal of the Human Factors and Ergonomics Society*, *50*(3), 461–467. doi:10.1518/001872008X288501 PMID:18689054

Kleiner, B. M., Hettinger, L. J., DeJoy, D. M., Huang, Y.-H., & Love, P. E. D. (2015). Sociotechnical attributes of safe and unsafe work systems. *Ergonomics*, *58*(4), 635–649. doi:10.1080/00140139.2015.1009175 PMID:25909756

Kleiner, Y., Rajani, B., & Sadiq, R. (2006). Failure risk management of buried infrastructure using fuzzy-based techniques. *Journal of Water Supply: Research & Technology - Aqua*, *55*(2), 81–94. doi:10.2166/aqua.2006.075

Kluge, A. (2014). *The Acquisition of Knowledge and Skills for Taskwork and Teamwork to Control Complex Technical Systems. The Acquisition of Knowledge and Skills for Taskwork and Teamwork to Control Complex Technical Systems: A Cognitive and Macroergonomics Perspective* (Vol. 9789400750). doi:10.1007/978-94-007-5049-4

Korunka, C., Scharitzer, D., Carayon, P., Hoonakker, P., Sonnek, A., & Sainfort, F. (2007). Customer orientation among employees in public administration: A transnational, longitudinal study. *Applied Ergonomics*, *38*(3), 307–315. doi:10.1016/j.apergo.2006.04.019 PMID:16759625

Lim, A. J., Village, J., Salustri, F. A., & Neumann, W. P. (2014). Process mapping as a tool for participative integration of human factors into work system design. *European Journal of Industrial Engineering*, *8*(2), 273. doi:10.1504/EJIE.2014.060477

Marina Bischoff, B., & Vianna Dos Santos, F. A. N. (2018). *Ergonomics and organizational communication: Methods for business communication auditing* (Vol. 609). Advances in Intelligent Systems and Computing. doi:10.1007/978-3-319-60477-0_5

McLaughlin, A. C., Ward, J., & Keene, B. W. (2016). Development of a Veterinary Surgical Checklist. *Ergonomics in Design*, *24*(4), 27–34. doi:10.1177/1064804615621411

Meshkati, N., Tabibzadeh, M., Farshid, A., Rahimi, M., & Alhanaee, G. (2016). People-Technology-Ecosystem Integration: A Framework to Ensure Regional Interoperability for Safety, Sustainability, and Resilience of Interdependent Energy, Water, and Seafood Sources in the (Persian) Gulf. *Human Factors*, *58*(1), 43–57. doi:10.1177/0018720815623143 PMID:26857436

Miguez, S. A., Filho, J. F. A. G., Faustino, J. E., & Gonçalves, A. A. (2018). *A successful ergonomic solution based on lean manufacturing and participatory ergonomics* (Vol. 602). Advances in Intelligent Systems and Computing; doi:10.1007/978-3-319-60825-9_27

Murphy, L. A., Huang, Y.-H., Robertson, M. M., Jeffries, S., & Dainoff, M. J. (2018). A sociotechnical systems approach to enhance safety climate in the trucking industry: Results of an in-depth investigation. *Applied Ergonomics*, *66*, 70–81. doi:10.1016/j.apergo.2017.08.002 PMID:28958432

Murphy, L. A., Robertson, M. M., & Carayon, P. (2014). The next generation of macroergonomics: Integrating safety climate. *Accident; Analysis and Prevention*, *68*, 16–24. doi:10.1016/j.aap.2013.11.011 PMID:24368052

Murphy, L. A., Robertson, M. M., Huang, Y.-H., Jeffries, S., & Dainoff, M. J. (2018). A sociotechnical systems approach to enhance safety climate in the trucking industry: Development of a methodology. *Applied Ergonomics*, *66*, 82–88. doi:10.1016/j.apergo.2017.08.001 PMID:28958433

Nathanael, D., Zarboutis, N., & Marmaras, N. (2015). Contradiction analysis: Towards a dialectical approach in ergonomics field interventions. *Producao*, *25*(1), 223–231. doi:10.1590/S0103-65132014005000022

Neumann, W. P., Ekman, M., & Winkel, J. (2009). Integrating ergonomics into production system development - The Volvo Powertrain case. *Applied Ergonomics*, *40*(3), 527–537. doi:10.1016/j.apergo.2008.09.010 PMID:19019347

Neumann, W. P., & Village, J. (2012). Ergonomics action research II: A framework for integrating HF into work system design. *Ergonomics*, *55*(10), 1140–1156. doi:10.1080/00140139.2012.706714 PMID:22913397

Pacholski, L., & Pawlewski, P. (2017). *The usage of simulation technology for macroergonomic industrial systems improvement* (Vol. 487). Advances in Intelligent Systems and Computing; doi:10.1007/978-3-319-41688-5_1

Patel, V. L., & Kannampallil, T. G. (2014). Human factors and health information technology: Current challenges and future directions. *Yearbook of Medical Informatics*, *9*(1), 58–66. doi:10.15265/IY-2014-0005 PMID:25123724

Punnett, L., Cherniack, M., Henning, R., Morse, T., & Faghri, P. (2009). *A conceptual framework for integrating workplace health promotion and occupational ergonomics programs*. Academic Press.

Realyvásquez, A., Maldonado-Macías, A. A., García-Alcaraz, J., Cortés-Robles, G., & Blanco-Fernández, J. (2016). Structural model for the effects of environmental elements on the psychological characteristics and performance of the employees of manufacturing systems. *International Journal of Environmental Research and Public Health*, *13*(1), 104. doi:10.3390/ijerph13010104 PMID:26742054

Realyvásquez, A., Maldonado-Macías, A. A., García-Alcaraz, J. L., & Arana, A. (2018). *Macroergonomic compatibility index for manufacturing systems. A case study* (Vol. 606). Advances in Intelligent Systems and Computing. doi:10.1007/978-3-319-68684-4

Realyvásquez, A., Maldonado-Macías, A. A., García-Alcaraz, J. L., & Blanco-Fernández, J. (2015). Effects of Organizational Macroergonomic Compatibility Elements over Manufacturing Systems' Performance. *Procedia Manufacturing*, *3*, 5715–5722. doi:10.1016/j.promfg.2015.07.807

Robertson, M. M., Huang, Y. H., & Lee, J. (2017). Improvements in musculoskeletal health and computing behaviors: Effects of a macroergonomics office workplace and training intervention. *Applied Ergonomics*, *62*, 182–196. doi:10.1016/j.apergo.2017.02.017 PMID:28411728

Robertson, M. M., Huang, Y. H., O'Neill, M. J., & Schleifer, L. M. (2008). Flexible workspace design and ergonomics training: Impacts on the psychosocial work environment, musculoskeletal health, and work effectiveness among knowledge workers. *Applied Ergonomics*, *39*(4), 482–494. doi:10.1016/j.apergo.2008.02.022 PMID:18462704

Rodríguez Ruíz, Y., & Pérez Mergarejo, E. (2016). Macro-ergonomic diagnosis of Colombian organizations using ergonomics maturity model | Diagnóstico macro ergonómico de organizações colombianas usando o Modelo de Madureza de Ergonomia | Diagnóstico macroergonómico de organizaciones colombianas con el Mod. *Revista Ciencias de la Salud*, *14*, 11–25.

Salas, E., Grossman, R., Hughes, A. M., & Coultas, C. W. (2015). Measuring team cohesion: Observations from the science. *Human Factors*, *57*(3), 365–374. doi:10.1177/0018720815578267 PMID:25875429

Seim, R., Broberg, O., & Andersen, V. (2014). Ergonomics in design processes: The journey from ergonomist toward workspace designer. *Human Factors and Ergonomics in Manufacturing*, *24*(6), 656–670. doi:10.1002/hfm.20508

Sherehiy, B., & Karwowski, W. (2014). The relationship between work organization and workforce agility in small manufacturing enterprises. *International Journal of Industrial Ergonomics*, *44*(3), 466–473. doi:10.1016/j.ergon.2014.01.002

Silva, L. D., Nickel, E. M., & Dos Santos, F. A. N. V. (2018). *Evaluation of macroergonomic methods for the application of organization analyzes in startups* (Vol. 588). Advances in Intelligent Systems and Computing; doi:10.1007/978-3-319-60582-1_33

Smith, M. J., & Carayon, P. (1995). *New technology, automation, and work organization: stress problems and improved technology implementation strategies*. Retrieved from https://www.scopus.com/record/display.uri?eid=2-s2.0-0029582621&origin=resultslist&sort=cp-f&src=s&st1=Macroergonomics&nlo=&nlr=&nls=&sid=6ebdcdcaef482f3889ab4f6a72d2e8f7&sot=b&sdt=cl&cluster=scopubyr%2C%222013%22%2Ct%2C%222012%22%2Ct%2C%222011%22%2Ct%2C%222010%22%2Ct%2C%222009%22%2Ct%2C%222008%22%2Ct%2C%222007%22%2Ct%2C%222006%22%2Ct%2C%222005%22%2Ct%2C%222004%22%2Ct%2C%222003%22%2Ct%2C%222002%22%2Ct%2C%222001%22%2Ct%2C%222000%22%2Ct%2C%221999%22%2Ct%2C%221998%22%2Ct%2C%221997%22%2Ct%2C%221996%22%2Ct%2C%221995%22%2Ct%2C%221994%22%2Ct%2Bscosubtype%2C%22ar%22%2Ct%2C%22ch%22%2Ct&sessionSearchId=6ebdcdcaef482f3889ab4f6a72d2e8f7&relpos=2&citeCnt=69

Smith, T. J., Schoenbeck, K., & Clayton, S. (2009). *Staff perceptions of work quality of a neonatal intensive care unit before and after transition from an open bay to a private room design*. Academic Press. doi:10.3233/WOR-2009-0868

Steege, L. M., Pasupathy, K. S., & Drake, D. A. (2018). A work systems analysis approach to understanding fatigue in hospital nurses. *Ergonomics*, *61*(1), 148–161. doi:10.1080/00140139.2017.1280186 PMID:28064733

Tavares, A. S., Albuquerque, L. W. N. D., da Silva, J. C., Souza Júnior, C. B., Gálvez, C., & Soares, M. (2015). Work at Height: Neglect or Improvisation in Civil Construction in Brazil and Uruguay? *Procedia Manufacturing*, *3*, 6109–6115. doi:10.1016/j.promfg.2015.07.763

Taveira, A. D., James, C. A., Karsh, B. T., & Sainfort, F. (2003). Quality management and the work environment: An empirical investigation in a public sector organization. *Applied Ergonomics*, *34*(4), 281–291. doi:10.1016/S0003-6870(03)00054-1 PMID:12880738

Thatcher, A., Waterson, P., Hancock, P., Davis, M. C., Zink, K. J., & Hilliard, A. (2016). This changes everything: Macroergonomics and the future of sustainability. In *Proceedings of the Human Factors and Ergonomics Society* (pp. 870–874). Academic Press. 10.1177/1541931213601199

Thatcher, A., Waterson, P., Todd, A., & Moray, N. (2018). State of Science: Ergonomics and global issues. *Ergonomics*, *61*(2), 197–213. doi:10.1080/00140139.2017.1398845 PMID:29076757

Valdez, R. S., Holden, R. J., Hundt, A. S., Marquard, J. L., Montague, E., Nathan-Roberts, D., & Or, C. K. (2014). The Work and Work Systems of Patients: A New Frontier for Macroergonomics in Health Care. *Proceedings of the Human Factors and Ergonomics Society Annual Meeting*, *58*(1), 708–712. doi:10.1177/1541931214581165

Village, J., Salustri, F. A., & Neumann, W. P. (2016). Cognitive mapping links human factors to corporate strategies. *European Journal of Industrial Engineering*, *10*(1), 1–20. doi:10.1504/EJIE.2016.075126

Waterson, P., Robertson, M. M., Carayon, P., Hoonakker, P., Holden, R., Hettinger, L., … Waterson, P. (2014). Macroergonomics and Sociotechnical Methods Current and Future Directions. *Proceedings of the Human Factors and Ergonomics Society Annual Meeting*, *58*(1), 1536–1540. 10.1177/1541931214581320

Waterson, P., Robertson, M. M., Cooke, N. J., Militello, L., Roth, E., & Stanton, N. A. (2015). Defining the methodological challenges and opportunities for an effective science of sociotechnical systems and safety. *Ergonomics*, *58*(4), 565–599. doi:10.1080/00140139.2015.1015622 PMID:25832121

Wood, S. J., Stride, C. B., Wall, T. D., & Clegg, C. W. (2004). Revisiting the use and effectiveness of modern management practices. *Human Factors and Ergonomics in Manufacturing*, *14*(4), 415–432. doi:10.1002/hfm.20006

Yang, Y., & Asan, O. (2016). Designing patient-facing health information technologies for the outpatient settings: A literature review. *Journal of Innovation in Health Informatics*, *23*(1), 441–449. doi:10.14236/jhi.v23i1.185 PMID:27348487

Zhao, D., McCoy, A. P., Kleiner, B. M., Smith-Jackson, T. L., & Liu, G. (2016). Sociotechnical systems of fatal electrical injuries in the construction industry. *Journal of Construction Engineering and Management*, *142*(1), 04015056. doi:10.1061/(ASCE)CO.1943-7862.0001036

Zink, K. J. (2000). Ergonomics in the past and the future: From a german perspective to an international one. *Ergonomics*, *43*(7), 920–930. doi:10.1080/001401300409116 PMID:10929827

KEY TERMS AND DEFINITIONS

Ergonomics: Science that studies the best adaptation of the work system to the human.

Human Factors: The human factors (or ergonomics) is the study of how individuals react to certain environments, products, or services in physical and psychological terms.

Macroergonomics: Macroergonomics aims to design a fully "harmonized" work system enhancing the key performance indicators for organizational effectiveness, productivity and satisfaction; where the sociotechnical systems theory is the fundamental principle, including industrial and organizational psychology approaches.

Microergonomics: The Microergonomics approach study the interaction between machine, environment, software, and job with the human.

Organizational Performance: Measure of efficiency and effectiveness, with which administrators take advantage of resources to satisfy customers and achieve the goals of the organization.

Productivity: Production capacity per work unit.

Sociotechnical Systems Approach: It is the approach that recognizes the interaction between individuals and technology in workplaces so that it helps on the complex organizational work design.

Work Systems: It is a natural unit of analysis for thinking about systems in organizations. In organizational settings, work is the application of human, informational, physical, and other resources to produce products/services.

Chapter 14
Implementation of an Intelligent Model Based on Machine Learning in the Application of Macro-Ergonomic Methods in a Human Resources Process Based on ISO 12207

Edgar Cossio Franco
Universidad Enrique Díaz de León, Mexico

Jorge Alberto Delgado Cazarez
Universidad de Guadalajara, Mexico

Carlos Alberto Ochoa Ortiz Zezzatti
Universidad Autónoma de Ciudad Juárez, Mexico

ABSTRACT

The objective of this chapter is to implement an intelligent model based on machine learning in the application of macro-ergonomic methods in human resources processes based on the ISO 12207 standard. To achieve the objective, a method of constructing a Java language algorithm is applied to select the best prospect for a given position. Machine learning is done through decision trees and algorithm j48. Among the findings, it is shown that the model is useful in identifying the best profiles for a given position, optimizing the time in the selection process and human resources as well as the reduction of work stress.

DOI: 10.4018/978-1-5225-7192-6.ch014

INTRODUCTION

Nowadays, the success of many companies is priority. Among the important aspects and factors that impact on the success of companies is the active human. This means that employees should be comfortable in their workplace, this includes have an appropriate work area center for it can be adequately performed; lighting, ventilation, a comfortable chair or similar artifact to sit, the correct distance between the view and the monitor so you do not get tired and have enough tools. In relation with these specific tools, it is important that the worker has access and control of the elements that allow him to carry out their work. If industrial engineers, ergonomists, designers and managers take the elements described previously, they will be focusing about macroergonomics. The goal of this research is centered in the proposal of a specific model that allows to optimize the process of a company dedicated to the hiring of staff, and according to Lear (2011), a good way to reduce the stress in a job is with the automation of the tasks and process to save time. The process by which the companies in this field go through is eventually stressful and should be avoided. As established in (Palferman, 2011), the consequence of not avoiding this workload can be psychological damage. The scenario under which a situation of stress could happen is the following: applications or vacancies that are offered on a normal day may exceed the rate at which they can be evaluated, especially if the number of team members is reduced with respect to the number of candidates for each vacant. In this case, a bottleneck in which the profiles of the company are requested as soon as possible, the company does not have the names for when it is required. In this case, there is work pressure. For the improvement process in the human recourses area, the present work focuses in the ISO 12207 standard in the block of organizational processes. The proposal is applying a model described in two blocks. The first step is capturing the applicants and analyze them, using a tool developed in the Java language that implements an algorithm to making decision under uncertainty. Once the results are obtained, the second step is sending the data to a tool for automatic learning which implement the j48 selection algorithm in WEKA.

The Appendix shows the terminology presented in this work.

Macroergonomics

The goal of macroergonomics is create a pleasant environment in the workplace where intervene social aspects, but also technological (Hendrick, 1991). As established in Montero (2000), the socio-technical systems are characterized by their composition in 4 aspects as shown in Figure 1.

Each aspect shown in Figure 1 represents an axis that must be covered by Macroergonomics in order to guarantee productivity and quality in employees' life (Realyvásquez & Maldonado, 2018). In addition to the socio-technical aspects, Macroergonomics originates from Ergonomics, that is a field where takes care of three aspects: the healthy environment in the workplace, safety and efficiency. Figure 2 shows the areas to derive from Ergonomics.

Machine Learning

Learning is a process through which an agent receives information from their environment, that is processed and applied (Možina, 2018). In the present research, the knowledge that HR people have, although extensive, is insufficient when it is about to evaluating candidate by candidate for a job position; since there is not enough time. This is the reason why machine learning is used to optimize time and resources.

Figure 1. Socio-technical systems' aspects
Source: Montero (2000)

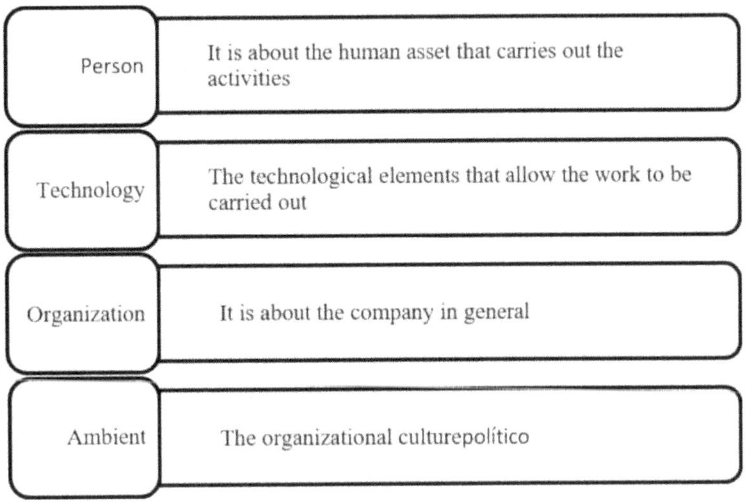

Figure 2. Areas of Ergonomics
Source: Silva, Nickel and dos Santos, 2018

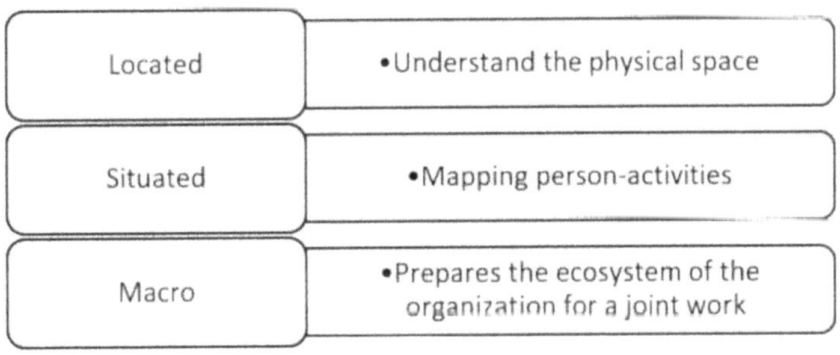

By means of machine learning it is possible provide knowledge to a computer by algorithms media and based in that knowledge to be able to process a bigger number of candidates in less time. Machine learning offers the algorithms automation to classify data (Arcila-Calderón et al., 2017) and based on this, it is possible to predict scenarios. The most basic and important concept is machine learning; it is a given representation to the process in general that computers work using algorithms. Machine learning has been applied to security systems, power, marketing, sales, medical diagnostics (Langley, & Simon, 1995), and in general, where it is possible to identify future scenarios through prediction (Goodfellow, Mcdaniel & Papernot, 2018). The prediction is achieved thanks to supervised and unsupervised algorithms. In supervised algorithms, it is applied to known information and in unsupervised algorithms the result is unknown, only the input data is known. In the study of Kononenko (2018) it is demonstrated that applying machine learning in complex processes and uncertainty it is possible to improve the accuracy. In other words, computer works better than humans do.

A Specific Standard to Process

In the international organization for standardization (ISO) there is a specific standard in its division of the International Electro-technical Commission (IEC) that is responsible for the life cycle of the software development process: the ISO 12207. This standard establishes a macro-process by which sub-processes are identified (main processes, organizational, support and adaptation processes) strategically direct to the areas that give life to the company (Baldassarre, Piattini, Pino & Visaggio, 2009). Table 1 shows the software life cycle processes according to ISO 12207, as well as the sub-processes that compose it (Piattini et al., 2007). The present research is focused on the human resources (HR) process, identified in Table 1, which provides the organization with adequate human resources and maintains the competence, consistent with the needs of the company. In this process, it is possible to identify the best profiles for a given position. Each company uses its own mechanisms to identify the best profiles. In next section it will be explained the problem that represents for any company the lack of a strategic hiring.

Problem Description

One problem that the Recruitment Company's Selection Department has is related to decides who is the candidate to choose for a specific job position. Said problem is modeled on the well-known problem of the secretary (Szajowski & Tamaki, 2016), which consists of what has been previously established. The more candidates are profiled to a diffuse job position, the problem becomes exponential in the strict sense to know what variables or characteristics the company requires for that position. It should not generate any problem when dealing with no more than 5 candidates who are only asked for a Java Senior certification. In this sense, the complexity is minimal. The problem arises when about 100 candidates are requested per day to different job positions for 20 companies and the profile or HR analysis team are only 10 people. At this point, the problem is in an advanced complexity. In previous scenarios, if a company does not have a solid strategy through which it is possible to optimize human resources

Table 1. Software life cycle processes according to ISO 12207

Main Processes	Support Processes
Acquisition	Documentation
Supply	Configuration management
Development	Quality assurance
Maintenance	Check
Organizational processes	Validation
Management	Review attached
Infrastructure	Audit
Improvement	Management of problem solving
Human resources	Usability
Asset management	Product evaluation
Reuse management programming	Management of exchange requests
Domain eEngineering	Adaptation processes

and time, the problems will come automatically and will be reflected in a brief period of time. Several times, when the candidate is already hired, the problem is expressed in diverse ways, for example, the profile was not adequate, the candidate does not have the necessary experience, he has bad behavior and disposition to work like a team, among others. The optimization of human resources is aimed at the productivity of the recruitment company where the current scenario is that a group of people evaluates the candidates through an application of exams, review, interviews, and studies. What happens if the HR team has only 10 people and the profiles to evaluate are 100 in a day? It should be considered that these 100 profiles must be submitted to the process of exams, review, interviews, and studies. The answer is chaos; the HR team will not have enough time to sit down and review each and every one of the profiles. Then, in order to comply with the review, the HR team will hire candidates who do not know the profile. This happens frequently. The worst scenario is when the candidate is fired for not complying with the profile and it involves spending time like a process is triggered insomuch as the cycle must be closed and a new profile started. It has been shown that it is possible to attack this problem by applying strategies, as shown in Broder et al. (2009), who applied Lake Wobegon strategies and the well-known case of Google where started in a garage and that day only the best are hired, a strategy to hire someone who is above average. The proposal in this chapter is to optimize time and resources and avoid scenario discussed previously. Through the application of this proposal, it will be possible to identify, in less time, the best profiles based on artificial intelligence and ideal profiles according to the request of the applicant companies. The proposal is described below.

Organization of This Chapter

The present research is composed of the following sections: INTRODUCTION, where the problem and the basic concepts are directed. The following section is METHODOLOGY, which is a design of the material and algorithms used to solve the problem. The section of RESULTS is related to the statistical analysis and to the results of the application of the algorithms. The sections of CONCLUSIONS and FUTURE RESEARCH contains the conclusions and researches derived from the present investigation.

METHODOLOGY

The methodology proposed in the present Research is divided into two blocks; First: the information analysis will be done in Java and then with machine learning. Figure 3 shows the methodology.

The proposed methodology is formed by three ways (user, algorithm and machine learning) and in each of them a process is carried out interacting with the others. The process starts in the algorithm way, because the processing of the tasks by which the CSV file is received and executed, the chronometer is started, criteria are assigned and, in general, the user interacts with the machine learning. The user's way feeds the CSV file. CSV is a simple file format used to store tabular data, such as a spreadsheet or database. The machine learning way applies artificial intelligence (AI) algorithms. The following sections explain the implementation of the methodology in detail.

Figure 3. Proposed methodology
Source: The authors

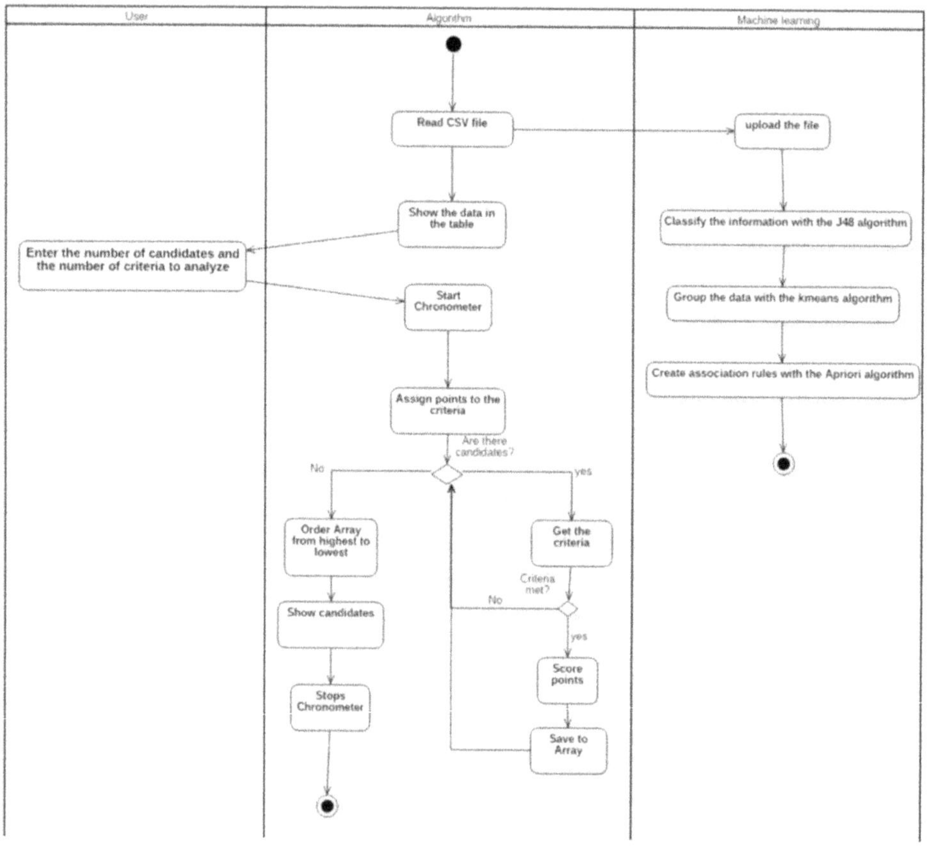

Java Application

Design of the Algorithm

For the Java application, the flowchart is shown as well as code segments in Figure 4. It is the design of the algorithm.

As specific software was built, the class diagram is shown in Figure 5.

Implementation

Following figures show code segments that make up the software in Java. This software was implemented with JAVA (JFrame) because it is fast, secure, reliable and stable. From laptops to data centers, from game consoles to supercomputers, from cell phones to the Internet, Java is everywhere. Figure 6 shows the main view of the proposed Java application. When starting with the software, a CSV (comma-separated values) file is read, it is a text file that stores the data in the form of columns, separated by a comma and the rows are distinguished by line breaks. This file contains all the information of all candidates for evaluation. As shown in Figure 7, seeing in code, the file is read, the name of the columns is obtained,

Figure 4. Design of the algorithm
Source: The authors

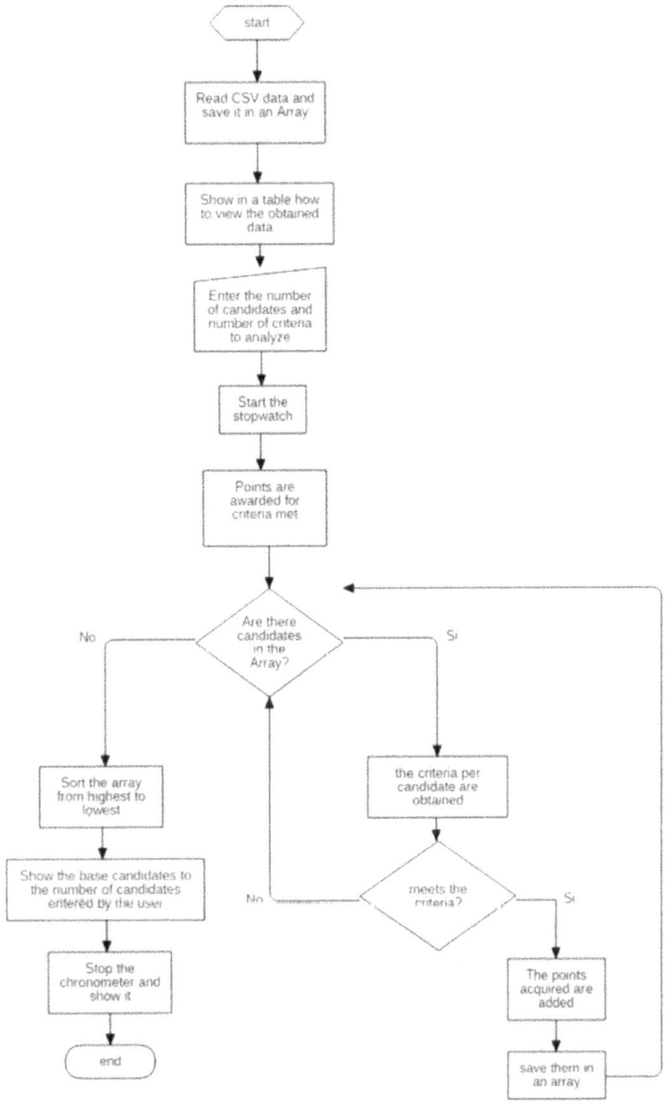

the information of the candidates and it is shown in the table of the software for the visualization of the user. Code for each line in Figure 7 are explained below.

Line 62: Candidates read by the CSV are obtained by means of the ReadCsv function
Line 63: Function where it fills the table with the data obtained from the file.
Line 64: Call the CreateTableResult () function; just create the columns to visualize the result.
Line 65: The Only Checkbox is marked.

The function shown in Figure 8 is very important because it is the obtaining of all the information about the candidates to be evaluated.

Figure 5. Class diagram
Source: The authors

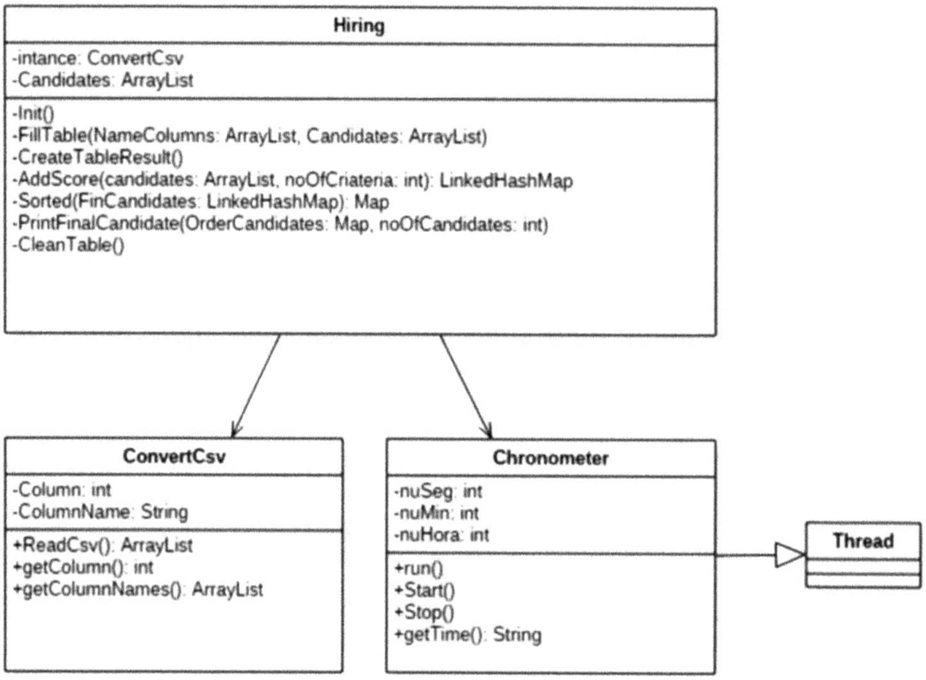

Figure 6. Main view of the proposed Java application
Source: The authors

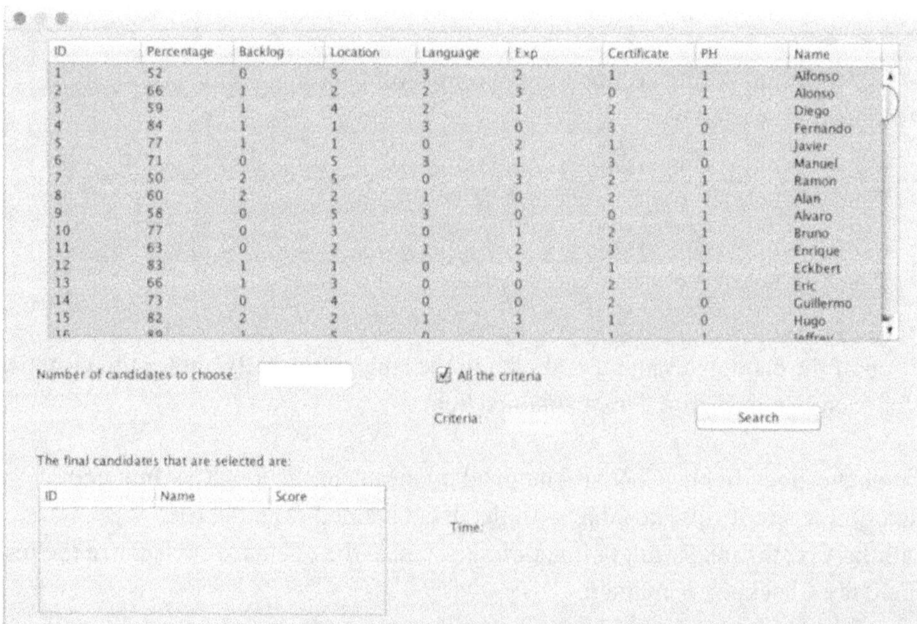

Figure 7. Function to read the candidates
Source: The authors

```
    private void Init() throws IOException{

        Candidates = data.ReadCsv();
        FillTable(data.getColumnNames(), Candidates);
        CreateTableResult();
        CheckCrite.setSelected(true);
    }
```

Each criterion is obtained, the qualification and the name of the person.

Line 35: The number of columns that the file has is obtained.
Line 36: The name of the columns is obtained.
Line 38 to 42: Take a tour to obtain all the criteria of the candidates and save them in an ArrayList. This data structure was used because it is much easier to keep information in list form, mining and handling of such information is more efficient.

The items are added to the list in the order they are inserted.

Figure 9 shows the separation of the names of the columns with the information of the candidates to be able to create the table and show it. It is done through a loop that interacts with each column and each evaluation criterion.

Line 36: Create the Model for the interaction of the table, where the candidates will be shown.
Line 38 to 40: The name of the columns crossing the array is added to the Model.
Line 42 to 44: The candidate data is added to the model to going through the array.
Line 47: The model is added to the view table.

Figure 10 shows the event where the user already requires searching the most suitable candidate for the desired needs. The parameters that the user needs to find their candidate are obtained, for example, the number of candidates that they need to search with the skills require, in the same way can be strict and search for the candidate who know all the criteria or only that know certain criteria, that will depend on the user.

Line 323 and 324: It makes the instance of the class of the Chronometer and starts it.
Line 326: Function to clean the results table.
Line 328: Get the number of candidates you want to search.
Line 334 to 338: Condition to check if "All criteria" was selected or enter a number of criteria to evaluate.
Line 340: Function to evaluate the criteria by the candidate.
Line 342: Function to sort by scores from highest to lowest of the candidates that approved the criteria.
Line 344: Function that shows the candidate (s) already ordered in the results table. It only shows the number of candidates who entered by the user that you want to search. (Line 328).
Line 349 and 350: Clean the fields.

Figure 8. Function to read the CSV file separated by comma
Source: The authors

```
public ArrayList ReadCsv() throws IOException{

    BufferedReader br = null;
    ArrayList list = new ArrayList();

    try {

       br =new BufferedReader(new FileReader(Read));

       String line = br.readLine();

       Column = line.split(",").length;
       ColumnName = line;

       while (null != line) {
          line = br.readLine();
          if(!line.equals(""))
            list.add(line);
       }

    } catch (Exception e) {

    } finally {
       if (null!=br) {
          br.close();
       }
    }

    return list;
}
```

Figure 9. Function where data are obtained from the fileSource: The authors

```
private void FillTable(ArrayList NameColumns, ArrayList Candidates){

    DefaultTableModel model = new DefaultTableModel();

    NameColumns.stream().forEach((Name) -> {
        model.addColumn(Name);
    });

    Candidates.stream().forEach((Data) -> {
        model.addRow(Data.toString().split(","));
    });

    TblCandidates.setBackground(Color.LIGHT_GRAY);
    TblCandidates.setModel(model);
}
```

Line 352 and 353: Stop the chronometer and show the time.

Figure 11 shows the candidate qualifies, reviews each criterion and depending on the qualification sought by the user, adds his score.

Line 70 to 76: Certain points are added to the criteria that will be evaluated by the candidate.
Line 82 to 125: The criteria of the candidate are reviewed and analyzed one by one if it is fulfilled, a certain score is given depending on what was assigned by criteria (Line 10 to 76).
Line 127 to 129: If the criteria were approved, it is saved in the array together with the score obtained.

Figure 12 shows the code for the ranking from highest to lowest and shows the candidate (s) suitable for the user. A data structure, "Map" is used because it allows the value to store with a key and it is much easier to find the desired value. In this case, it is easier to find the name of the candidate who obtained the best score in the evaluation.

Figure 13 shows the suitable candidates who were ordered, go through this function, where it shows in a table the name and the score obtained from the candidate that was the most suitable under the sought criteria. Similarly, if the user enters that he/she requires to search more than one candidate, this function will show the number of candidates with the highest score.

Line 155: The model of results table is obtained to be able to use it.

Figure 10. Event to looking candidates
Source: The authors

```
321     private void BtnSearchActionPerformed(java.awt.event.ActionEvent evt) {
322
323         Chronometer crono = new Chronometer();
324         crono.Start();
325
326         CleanTable();
327
328         int NumCandidates = Integer.parseInt(TxtNumCandi.getText());
329
330         try {
331
332             int NumCriateria = 0;
333
334             if(CheckCrite.isSelected()){
335                 NumCriateria = data.getColumn() - 2;
336             }else{
337                 NumCriateria = Integer.parseInt(TxtCriterios.getText());
338             }
339
340             LinkedHashMap finalCandidates = AddScore(Candidates, NumCriateria);
341
342             Map OrderCandidates =  Sorted(finalCandidates);
343
344             PrintFinalCandidate(OrderCandidates, NumCandidates);
345
346         } catch (Exception ex) {JOptionPane.showMessageDialog(null, ex.toString());}
347
348
349         TxtNumCandi.setText("");
350         TxtCriterios.setText("");
351
352         crono.Stop();
353         LblTime.setText(crono.getTime());
354
355     }
```

Figure 11. Function where add the scores to criteria
Source: The authors

```java
private LinkedHashMap AddScore(ArrayList candidates, int noOfCriateria){

    int Porcentaje = 25;
    int AcumTrabajo = 10;
    int Ubicacion = 10;
    int Lenguajes = 20;
    int Experiencia = 10;
    int Certificados = 20;
    int Discapacidad = 5;

    LinkedHashMap passedcandidates = new LinkedHashMap();
    int score = 0;
    int passedCriterias = 0;

    for (Object rows : candidates) {
        String [] row = rows.toString().split(",");

        //PERCENTAGE
        if (Integer.parseInt(row[1]) > 60){
            score += Porcentaje;
            passedCriterias += 1;
        }

        //BACKLOG
        if (Integer.parseInt(row[2]) == 0){
            score += AcumTrabajo;
            passedCriterias += 1;
        }

        //LOCATION
        if (Integer.parseInt(row[3]) < 4){
            score += Ubicacion;
            passedCriterias += 1;
        }

        //LANGUAGES
        if (Integer.parseInt(row[4]) > 1){
            score += Lenguajes;
            passedCriterias += 1;
        }

        //EXPERIENCE
        if (Integer.parseInt(row[5]) > 0){
            score += Experiencia;
            passedCriterias += 1;
        }

        //CERTIFICATION
        if (Integer.parseInt(row[6]) > 0){
            score += Certificados;
            passedCriterias += 1;
        }

        //PH
        if (Integer.parseInt(row[7]) < 1){
            score += Discapacidad;
            passedCriterias += 1;
        }

        if (passedCriterias >= noOfCriateria){
            passedcandidates.put(rows, score);
        }

        score = 0;
        passedCriterias = 0;

    }

    return passedcandidates;
}
```

Figure 12. Code for ordering candidates
Source: The authors

```
private Map Sorted(LinkedHashMap<String, Integer> FinCandidates){

    Map<String, Integer> Order = new LinkedHashMap<>();
    FinCandidates.entrySet().stream()
            .sorted(Map.Entry.<String, Integer>comparingByValue().reversed())
            .forEachOrdered(x -> Order.put(x.getKey(), x.getValue()));

    return Order;
}
```

Line 158 to 168: It obtains an Id, name and the score of the candidate adding to the model of the table to be able to visualize it.

Line 166: The cycle process is interrupted as long as the number of candidates that the user entered to view has already been saved.

Line 170: The model sends results to the table.

Screen *Shots*

Figure 14 shows all the candidates with their criteria read from the CSV file so be able to evaluate them.

Figure 15 shows the field where the number of candidates must be written to know the most optimal.

Figure 16 shows the selection criteria. If deactivate "All the criteria", the user must enter the number of minimum criteria to evaluate.

Figure 17 shows two candidates that were evaluated with all the criteria.

Necessarily, the user must comply with the seven criteria for this case.

Figure 18 shows the duration in which it casts the candidates. The format of the time is: Hours: Minutes: Seconds.

Figure 13. Candidates who approved the criteria
Source: The authors

```
152 private void PrintFinalCandidate(Map OrderCandidates, int noOfCandidates){
153
154     int cont = 1;
155     DefaultTableModel model = (DefaultTableModel) TblResult.getModel();
156
157     Iterator<Map.Entry<String, Integer>> it = OrderCandidates.entrySet().iterator();
158     while (it.hasNext()) {
159         Map.Entry<String, Integer> pair = it.next();
160
161         String [] Pair = pair.getKey().split(",");
162         String ID = Pair[0];
163         String Nombre = Pair[8];
164         model.addRow(new Object[]{ID, Nombre, pair.getValue()});
165
166         if(cont++ == noOfCandidates)
167             break;
168     }
169
170     TblResult.setModel(model);
171
172 }
```

Figure 14. Results from the CSV file
Source: The authors

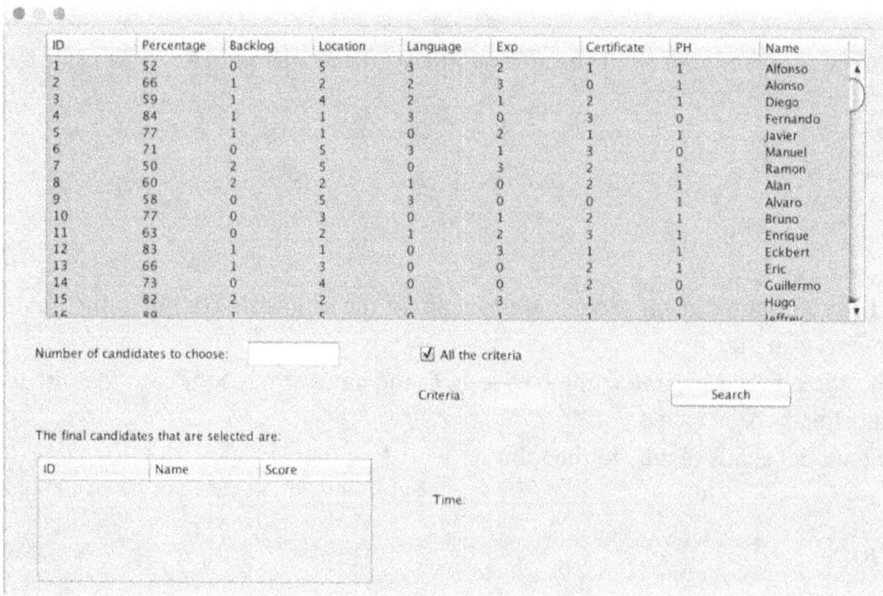

Figure 15. Selecting the number of candidates
Source: The authors

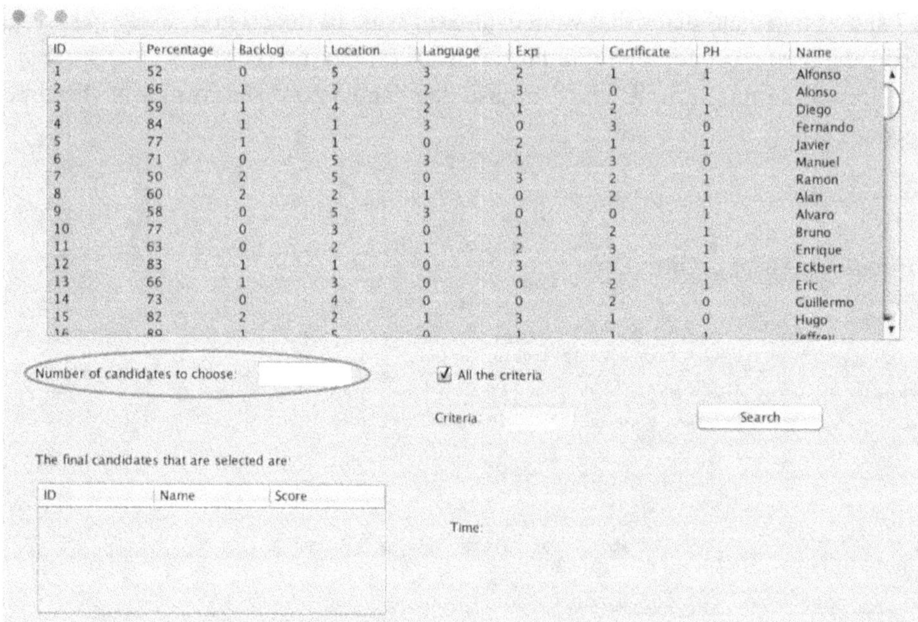

Figure 16. Choose criteria
Source: The authors

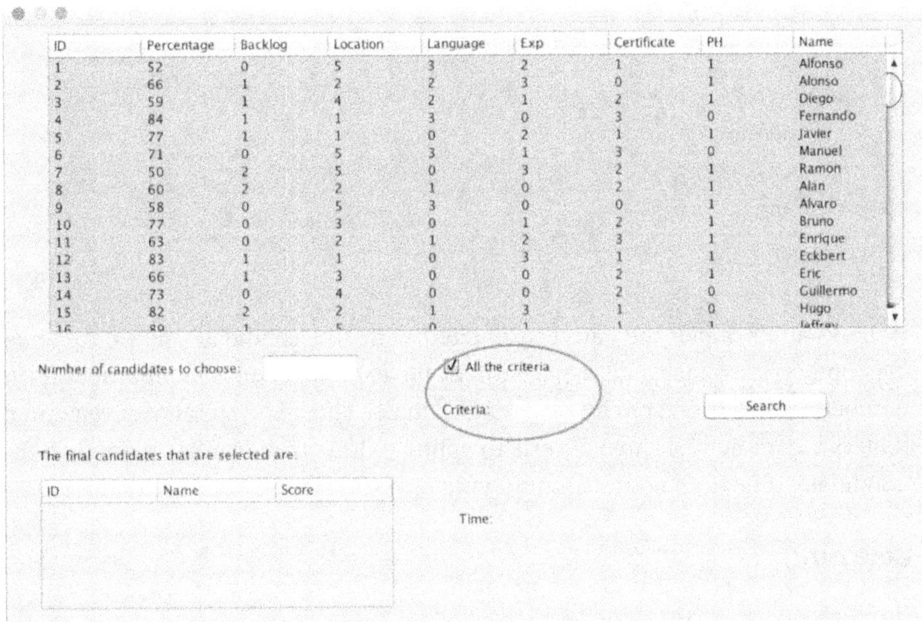

Figure 17. Results of the evaluation
Source: The authors

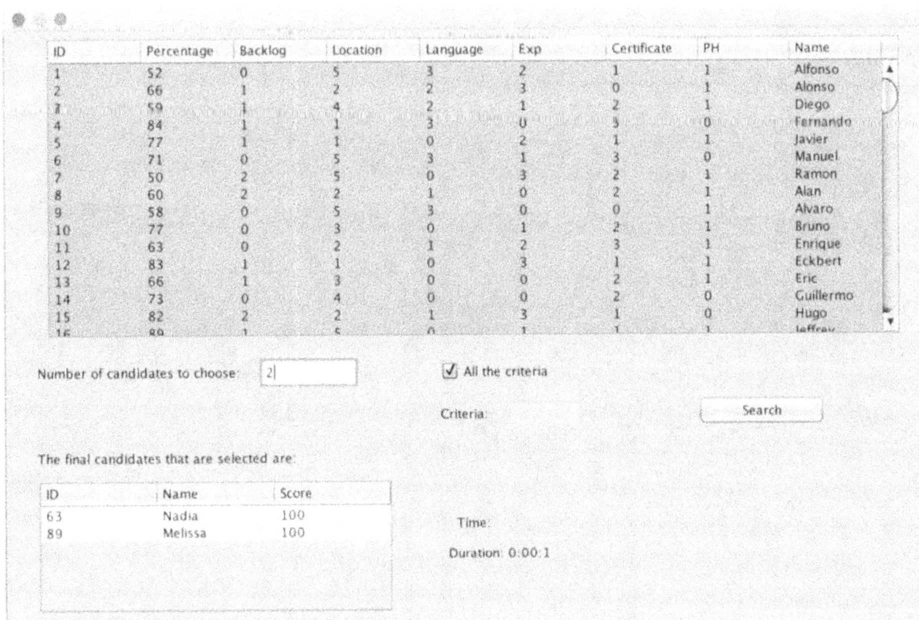

Figure 18. Timer
Source: The authors

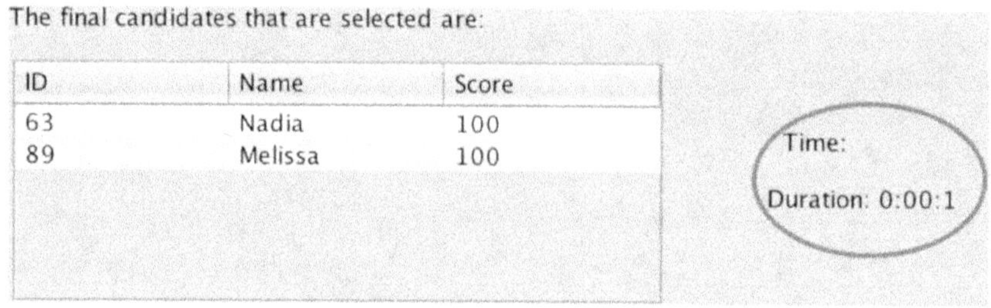

Figure 19 shows an example: the date was entered with five candidates and only three criteria to evaluate. In this option, it evaluates all the criteria of the candidate but only approves if the candidate exceeds the number of criteria entered by the user. Example: There are a total of seven criteria to evaluate; the user enters a minimum of three criteria to approve. If a candidate has more than three criteria, he is a final candidate, if he does not fulfill minimum 3, he is rejected.

Machine Learning

For the implementation of machine learning, the WEKA tool was used, which was developed by the University of Waikato. It is a software created in Java for data analysis. Figure 20 shows the process of the analysis.

Figure 19. Example
Source: The authors

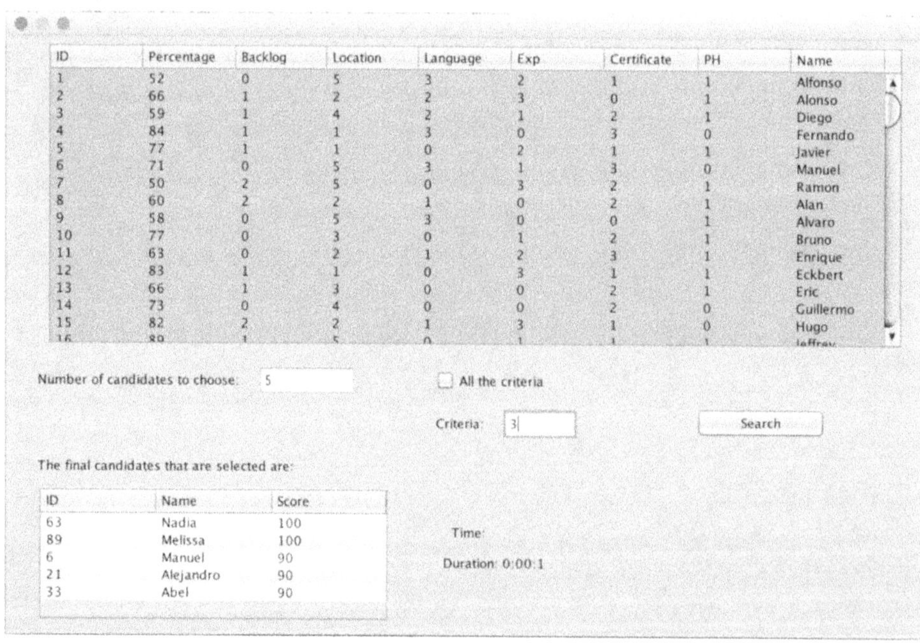

Figure 20. Analysis process

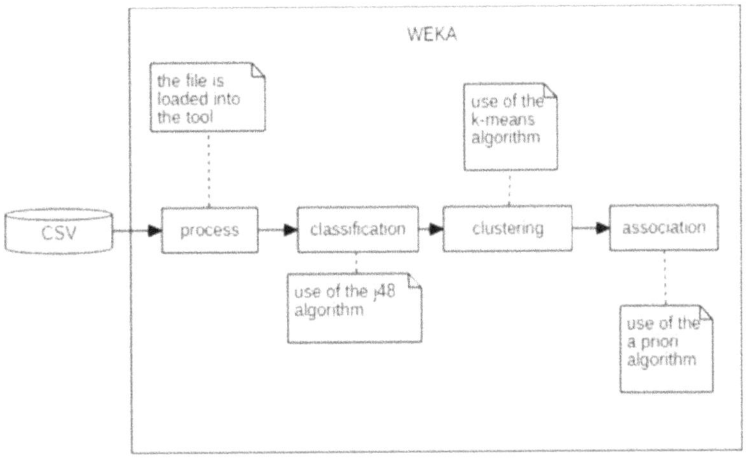

For the present research, the database obtained is submitted to the software to perform an analysis, as shown in the following diagram. Figure 3 shows the process by which the transformation is performed. It starts with the loading of the CSV file, to then move on to the first section of the tool that is the process. The next step is to classify the information using the j48 algorithm, then the information is grouped using the k-means algorithm and finally the association with the a priori algorithm is performed. J48 algorithm is a derivation of C4.5 algorithm, and its purpose is to construct decision and grouping threes in the field of data mining (Tello, Eslava & Tobias, 2013). The k-means algorithm is a grouping algorithm that is applied to the analysis for the prediction of scenarios (Reena & Selvi, 2016). The a priori algorithm is used to apply to learn in the operation rules of a data set (Randhir, Gupta & Selokar, 2013).

Process

The structure of the CSV file is composed of 100 records, and 7 criteria, as explained in the previous section. The metadata that makes up the structure is location; for the specific case of the present work, the case study is the city of Guadalajara, Mexico, which is divided into 7 zones: Downtown, Minerva, Huentitán, Oblatos, Tetlán, Technological and Industrial. The languages considered are Spanish, English, French and Chinese. The experience is of vital importance for the hiring of a prospect for a position. Respect the software, Java, c #, python, and ruby are the 4 technologies requested, the certificates support the experience of the supporter; it is evaluated how many it has, finally it is considered some possible physical impediment and if it is male or female. The model explained is shown in Figure 21.

Figure 21. Analysis scheme
Source: The authors

```
@attribute Location {centro,tecnológico,industrial,tetlán,minerva,huentitán,oblatos}
@attribute Language {spanish,french,chinese,english}
@attribute Exp {ruby,c#,python,java}
@attribute Certificate_binarized {0,1}
@attribute PH_binarized {0,1}
@attribute Sex {man,woman}
```

The distribution of the age variable is shown in Table 2. It can be seen that most people who apply live in Tetlán zone and those who are less interested in applying live in Oblatos zone. From the total of candidates, it is observed that 47 are men and 53 women, also observed that of the required experience, 36 know ruby, 12 c #, 26 python and 26 java; as is shown in Table 3. Of the languages, it is observed that the majority of people speak Spanish and only the minority speaks Chinese or another language, as is shown in Table 4.

RESULTS

The results of the present research are described in two moments; first, the performance of the application of the algorithm is shown according to the loads of candidates and evaluation criteria. The loads range from one hundred to one million candidates and from two to seven criteria. The result is shown in Table 5. The second moment is described by showing results of the analysis by applying machine learning.

Classification of Information

The j48 algorithm is used to construct a decision tree which has two approaches: construction and pruning (Muralidharan, & Sugumaran, 2013). The j48 algorithm is an implementation of the C4.5 algorithm that was proposed by Quinlan in 1993 and it is used to identify target values of a data set. The predicted values are the dependent variables (Ibrahim, Yazin, Udzir & Abdul, 2016). For classify the information,

Table 2. Location distribution

Number	Zone	Count
1	Centro	19
2	Tecnológico	16
3	Industrial	11
4	Tetlán	23
5	Minerva	9
6	Huentitán	14
7	Oblatos	8

Source: The authors

Table 4. Distribution of languages

Number	Language	Count
1	Spanish	31
2	French	22
3	Chinese	21
4	English	26

Source: The authors

Table 3. Distribution of experience

Number	Label	Count
1	ruby	36
2	c#	12
3	python	26
4	java	26

Source: The authors

Table 5. Performance of the algorithm with different data

Candidates	Criteria	Time
100	2	25 Milliseconds
100	5	26 Milliseconds
100	7	28 Milliseconds
1,000	2	31 Milliseconds
1,000	5	31 Milliseconds
1,000	7	32 Milliseconds
1,000,000	2	2 Seconds
1,000,000	5	3 Seconds
1,000,000	7	3 Seconds

Source: The authors

the j48 algorithm was applied to construct a tree for the analysis, the result is shown in Figure 22. It is observed that the algorithm performs information crossings with all the variables.

Clustering

The k-means calculation divides the data set into subsets or groups (M_1, M_2, M_3, … M_k). A group (M_i) is a group of vertices and centroids (μ_i) where i is an integer between 1 and k. The same vertex is not shared. By applying Equation (1), it is possible to minimize the distance of the centroids (Rezaei, & Gunpinar, 2018).

Figure 22. Result decision tree
Source: The authors

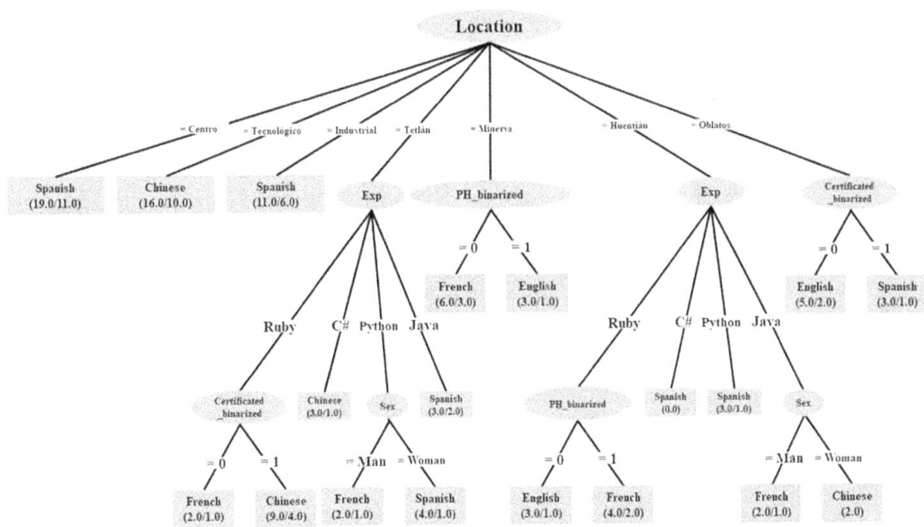

$$\min \sum_{i=1}^{k} \sum_{x \in M_1} || x - u_i ||^2 \quad (1)$$

The algorithm is used, which shows as a result that the information is grouped into two segments, in cluster 0 the highest percentage is concentrated with the 55 of the information where the men in the technologic zone know python and are certified, while in the 1 the 45 of the information is concentrated where women in the Tetlán zone known java and speak French. Figure 23 shows the result.

Association

The basis for the association rules was born in IBM in 1993 by R. Agrawal at Almaden research center. Originally, they were raised to identify a set of client operations in a database. The association rules have been applied to physical activities, business, finance, and medicine, among others. The association algorithms are distinguished in two moments; support and confidence. The support means the frequency with which an element appears in a list and is defined according to the Equation (2).

$$Support(A \Rightarrow B) = P(A \cap B) \quad (2)$$

The confidence consists of the generation of the association rules according to the Equation (3) (Li et al., 2017):

$$Confidence(A \Rightarrow B) = P(A \mid B) \quad (3)$$

For the association rules, the apriori algorithm was used, through which it was possible to build rules from the given information. The result is shown below and in Figure 24:

1. The men who are in the center are certified.
2. People who speak Chinese and have a disability are certified.

Figure 23. Clusters
Source: The authors

		Cluster#	
Attribute	Full Data	0	1
	(100.0)	(55.0)	(45.0)
Location	tetlán	tecnológico	tetlán
Language	spanish	spanish	french
Exp	ruby	python	java
Certificate_binarized	1	1	1
PH_binarized	1	1	0
Sex	woman	man	woman

3. Men who speak Chinese are certified.
4. The people of the industrial zone are certified.
5. The women of Huentitán are certified.
6. The experience in python is certified.
7. Men who know java are certified.
8. Men with some disability who know ruby are certified.
9. The Chinese language is certified.
10. Disabled people who know ruby are certified.

In Figure 24, the rules are shown with the degree of confidence

CONCLUSION

It is possible to reduce the stress levels of the personnel in charge of applying the surveys and analyze the results for a given position. This is achieved through the application of the model presented in the present work; using the algorithm we identify the best profiles. Through the application of artificial intelligence techniques, it is possible to predict behavior and group the profiles according to the results of the application. The importance of having information prior to the application of surveys allows visualizing scenarios for decision making.

FUTURE RESEARCH

As part of the future work of the present work, the development of a tool is contemplated through which it is possible to apply the surveys to the candidates for a job in and that the results are stored in real time through a mobile application.

Figure 24. Association rules and the level of trust
Source: The authors

```
 1. Location=centro Sex=man 12 ==> Certificate_binarized=1 12    <conf:(1)> lift:(1.28) lev:(0.03) [2] conv:(2.64)
 2. Language=chinese PH_binarized=1 12 ==> Certificate_binarized=1 11    <conf:(0.92)> lift:(1.18) lev:(0.02) [1] conv:(1.32)
 3. Language=chinese Sex=man 12 ==> Certificate_binarized=1 11    <conf:(0.92)> lift:(1.18) lev:(0.02) [1] conv:(1.32)
 4. Location=industrial 11 ==> Certificate_binarized=1 10    <conf:(0.91)> lift:(1.17) lev:(0.01) [1] conv:(1.21)
 5. Location=huentitan Sex=woman 11 ==> Certificate_binarized=1 10    <conf:(0.91)> lift:(1.17) lev:(0.01) [1] conv:(1.21)
 6. Exp=python PH_binarized=1 11 ==> Certificate_binarized=1 10    <conf:(0.91)> lift:(1.17) lev:(0.01) [1] conv:(1.21)
 7. Exp=java Sex=man 11 ==> Certificate_binarized=1 10    <conf:(0.91)> lift:(1.17) lev:(0.01) [1] conv:(1.21)
 8. Exp=ruby PH_binarized=1 Sex=man 11 ==> Certificate_binarized=1 10    <conf:(0.91)> lift:(1.17) lev:(0.01) [1] conv:(1.21)
 9. Language=chinese 21 ==> Certificate_binarized=1 19    <conf:(0.9)> lift:(1.16) lev:(0.03) [2] conv:(1.54)
10. Exp=ruby PH_binarized=1 20 ==> Certificate_binarized=1 18    <conf:(0.9)> lift:(1.15) lev:(0.02) [2] conv:(1.47)
```

REFERENCES

Arcila-Calderón, C., Ortega-Mohedano, F., Jiménez-Amores, J., & Trullenque, S. (2017). Análisis supervi-sado de sentimientos políticos en español: Clasificación en tiempo real de tweets basada en aprendizaje automático. *El Profesional de la Información*, *26*(5), 973–982. doi:10.3145/epi.2017.sep.18

Baldassarre, M. T., Piattini, M., Pino, F. J., & Visaggio, G. (2009). Comparing ISO/IEC 12207 and CMMI-DEV: Towards a mapping of ISO/IEC 15504-7. In *2009 ICSE Workshop on Software Quality* (pp. 59-64). ICSE. 10.1109/WOSQ.2009.5071558

Broder, A. Z., Kirsch, A., Kumar, R., Mitzenmacher, M., Upfal, E., & Vassilvitskii, S. (2009). The hiring problem and Lake Wobegon strategies. *SIAM Journal on Computing*, *39*(4), 1233–1255. doi:10.1137/07070629X

Goodfellow, I., Mcdaniel, P., & Papernot, N. (2018). Making Machine Learning Robust Against Adversarial Inputs. *Communications of the ACM*, *61*(7), 56–66. doi:10.1145/3134599

Hendrick, H. W. (1991). Ergonomics in organizational design and management. *Ergonomics*, *34*(6), 743–756. doi:10.1080/00140139108967348

Ibrahim, H., Yasin, W., Udzir, N. I., & Abdul Hamid, N. W. (2016). Intelligent Cooperative Web Caching Policies For Media Objects Based On J48 Decision Tree And Naïve Bayes Supervised Machine Learning Algorithms In Structured Peer-To-Peer Systems. *Journal Of Information & Communication Technology*, *15*(2), 85–116.

Kononenko, I. (2018). Early Machine Learning Research in Ljubljana. *Informatica*, *42*(1), 3-6.

Langley, P., & Simon, H. A. (1995). Applications of Machine Learning and Rule Induction. *Communications of the ACM*, *38*(11), 55–64. doi:10.1145/219717.219768

Lear, C. (2011). How can system administrators reduce stress and conflict in the workplace? *ACM Queue; Tomorrow's Computing Today*, *9*(1), 1–10.

Li, Q., Zhang, Y., Kang, H., Xin, Y., & Shi, C. (2017). Mining association rules between stroke risk factors based on the Apriori algorithm. *Technology and Health Care*, *25*, S197–S205. doi:10.3233/THC-171322 PMID:28582907

Montero, R. (2000). *Un paso hacia el futuro: el desarrollo de la Macroergonomia, en Factores Humanos*. Academic Press.

Možina, M. (2018). Arguments in Interactive Machine Learning. *Informatica*, *42*(1), 53-59.

Muralidharan, V., & Sugumaran, V. (2013). Selection Of Discrete Wavelets For Fault Diagnosis Of Monoblock Centrifugal Pump Using The J48 Algorithm. *Applied Artificial Intelligence*, *27*(1), 1–19. doi:10.1080/08839514.2012.721694

Palferman, D. (2011). Managing Conflict and Stress in the Workplace: Theory and Practice. *Legal Information Management*, *11*(2), 122–125. doi:10.1017/S1472669611000417

Piattini Velthuis, M., García Rubio, F., & Caballero Muñoz-Reja, I. (2007). *Calidad de sistemas informáticos*. Alfaomega.

Randhir, H. N., Gupta, R., & Selokar, G. R. (2013). Extract Knowledge and Association Rule from Free Log Data using an Apriori Algorithm. *International Journal Of Advanced Computer Research, 3*(12), 191–196.

Realyvásquez, A., & Maldonado-Macías, A. (2018). Measuring the Complex Construct of Macroergonomic Compatibility: A Manufacturing System Case Study. *Complexity, 2018*, 1–10. doi:10.1155/2018/7374307

Reena, R., & Selvi, R. T. (2016). Analyzing software defect prediction using k-means and expectation maximization clustering algorithm based on genetic feature selection. *Journal on Software Engineering, 11*(1), 28-36.

Rezaei, M., & Gunpinar, E. (2018). A k-means clustering based shape retrieval technique for 3d mesh models. *Selcuk University Journal of Engineering, Science & Technology, 6*(1), 114-128. doi:10.15317/Scitech.2018.119

Silva, L. D., Nickel, E. M., & dos Santos, F. A. N. V. (2018). Evaluation of Macroergonomic Methods for the Application of Organization Analyzes in Startups. In F. Rebelo & M. Soares (Eds.), *Advances in Ergonomics in Design. AHFE 2017. Advances in Intelligent Systems and Computing* (Vol. 588). Cham: Springer.

Szajowski, K., & Tamaki, M. (2016). Shelf life of candidates in the generalized secretary problem. *Operations Research Letters, 44*(4), 498–502. doi:10.1016/j.orl.2016.05.002

Tello, M. L., Eslava, H. J., & Tobías, L. B. (2013). Análisis y evaluación del nivel de riesgo en el otorgamiento de créditos financieros utilizando técnicas de minería de datos. *Visión Electrónica, 7*(1), 13–26.

KEY TERMS AND DEFINITIONS

AI: Artificial intelligence.

Array List: In Java it is a dynamic storage structure that allows to store data in memory, this has the similarity to a list.

C#: Microsoft object-oriented programming language.

Clustering: Grouping technique according to distance or similarity.

Csv: File separated by commas. Commonly used to manage data in an open data environment or to share non-native information between different software packages.

Data Structure: It is a particular way of organizing data in a computer so that they can be used efficiently.

Decision Tree: Intelligent model based on rules and used to predict.

ISO 12207: Standard for the life cycle specification system development of the International Standardization Organization.

J48 Algorithm: Algorithm that generates decision trees based on rules to classify.

Java: Set of computer programs. Also known as programming language or software.

Jframe: It is a graphic library to generate windows on which to add different objects with which you can interact with the user.

Loop: Also known as a cycle, it is a statement that executes a piece of code repeatedly, until the condition assigned to that loop is no longer fulfilled.

Machine Learning: Branch of artificial intelligence that allows computers to learn.

Macroergonomy: A field that takes care of three aspects: the healthy environment in the workplace, safety, and efficiency.

Map: Data structure that allows to store "key/value" pairs; in such a way that for a key we only have one value.

Object: Entity in the memory of the computer that has properties (attributes or data about itself stored by the object) and specific available operations (methods).

Object-Oriented Programming: Programming paradigm that uses objects and their interactions to design applications and computer programs.

Python: Interpreted programming language.

Ruby: Mixed programming language between interpreted and object-oriented.

Weka: Software for managing information that has to do with machine learning.

APPENDIX

Table 6. Extract of 19 of the 100 Records in the Database

ID	Percentage	Backlog	Location	Language	Exp	Certificate	PH	Sex	Age
1	52	0	Centro	Spanish	ruby	2	1	Man	20
2	66	1	Tecnológico	French	c#	3	1	Woman	30
3	59	1	Industrial	Spanish	python	3	1	Man	23
4	84	1	Tetlán	French	ruby	0	0	Man	24
5	77	1	Tetlán	Chinese	c#	2	1	Woman	28
6	71	0	Minerva	Chinese	java	2	0	Man	19
7	50	2	Huentitán	Chinese	java	1	1	Woman	24
8	60	2	Minerva	Spanish	java	0	1	Man	18
9	58	0	Tetlán	Spanish	java	0	1	Woman	34
10	77	0	Huentitán	Spanish	python	2	1	Woman	34
11	63	0	Tetlán	French	java	0	1	Woman	23
12	83	1	Centro	French	ruby	1	1	Man	22
13	66	1	Centro	English	ruby	3	1	Woman	21
14	73	0	Minerva	Spanish	python	1	0	Man	20
15	82	2	Tetlán	Spanish	python	0	0	Woman	28
16	89	1	Minerva	English	python	3	1	Woman	20
17	90	1	Industrial	Spanish	ruby	3	0	Woman	31
18	49	1	Centro	English	ruby	3	1	Man	27
19	83	1	Tetlán	Chinese	ruby	2	1	Man	26

Compilation of References

AA1000APS. (2008). *Account Ability Principles Standard.* Retrieved March 22, 2018, from http://www.accountability.org/standards/aa1000aps.html

Ackoff, R. (2002). The corporation as a community, not as a corpus. *Reflections: The SoL Journal, 4*(1), 14–21. doi:10.1162/152417302320467517

Adler, N. E. (2007). *Reaching for a Healthier Life: Facts on Socioeconomic Status and Health in the U.S.* J.D. and C.T. MacArthur Foundation Research Network on Socioeconomic Status and health.

Akao, Y. (1990). *Quality function deployment: Integrating customer requirements into product design.* Cambridge, MA: Journal of Productivity Press.

Akao, Y. (1997). QFD: Past, Present, and Future. *Proceedings of the International Symposium on QFD.*

Alaminos, A., & Castejón, J. L. (2006). *Elaboración, análisis e interpretación de encuestas, cuestionarios y escalas de opinión.* Universidad de Alicante. Retrieved January 15, 2018 from https://rua.ua.es/dspace/bitstream/10045/20331/1/Elaboraci%C3%B3n,%20an%C3%A1lisis%20e%20interpretaci%C3%B3n.pdf

Almagiá, E. B. (2014). Apoyo social, estrés y salud. *Psicología y salud, 14*(2), 237-243

Alper, S. J., & Karsh, B. T. (2009). A systematic review of safety violations in industry. *Accident; Analysis and Prevention, 41*(4), 739–754. doi:10.1016/j.aap.2009.03.013 PMID:19540963

American Institute of Stress (AIS). (2016). *Workplace Stress.* Retrieved December 20, 2017 from http://www.stress.org/workplace-stress/

Aminoff, E. M. (2009). The landscape of cognitive neuroscience: challenges, rewards, and new perspectives. In M. S. Gazaniga (Ed.), *The cognitive Neurosciences* (4th ed.; pp. 1255–1262). MIT.

Amir, E., Lev, V., & Sougiannis, T. (2003). Do Financial Analysts Get Intangibles? *European Accounting Review, 12*(4), 635–659. doi:10.1080/0963818032000141879

Anderson, D. (2008). *Statistics for Business and Economics* (10th ed.). Academic Press.

Appelbaum, S. H. (1997). Socio-technical systems theory: An intervention strategy for organizational development. *Management Decision, 35*(6), 452–463. doi:10.1108/00251749710173823

Aqlan, F., Lam, S. S., Testani, M., & Ramakrishnan, S. (2013). Ergonomic Risk Reduction to Enhance Lean Transformation. *Industrial and Systems Engineering Research Conference (ISERC).* Retrieved from https://www.highbeam.com/doc/1P3-3169576661.html

Aranda, C., Pando, M., & Salazar, J. G. (2015). Síndrome de burnout en trabajadores de diversas actividades económicas en México. *Revista iberoamericana de psicología: ciencia y tecnología, 8*(2), 23-28.

Aranda, C., Pando, M., Salazar, J. G., Torres, T. M., Aldrete, M. G., & Pérez, M. B. (2004). Factores psicosociales laborales y síndrome de burnout en médicos del primer nivel de atención. *Investigación en salud, 6*(1).

Arcila-Calderón, C., Ortega-Mohedano, F., Jiménez-Amores, J., & Trullenque, S. (2017). Análisis supervi-sado de sentimientos políticos en español: Clasificación en tiempo real de tweets basada en aprendizaje automático. *El Profesional de la Información, 26*(5), 973–982. doi:10.3145/epi.2017.sep.18

Argyris, C. (2001). *Sobre el aprendizaje organizacional.* Oxford University Press.

Ariely, D., Gneezy, U., Loewenstein, G., & Mazar, N. (2009). Large Stakes and Big Mistakes. *The Review of Economic Studies, 76*(2), 451–469. doi:10.1111/j.1467-937X.2009.00534.x

Arnold, H. J., & Feldman, D. C. (1981). Social desirability response bias in self-report choice situations. *Academy of Management Journal, 24*(2), 377–385. doi: 10.5465/255848

Arredondo-Soto, K. C., Carrillo-Gutiérrez, T., Solís-Quinteros, M., & Hernández-Escobedo, G. (2018). A Theoretical Framework About the Impact of Human Factors on Manufacturing Process Performance. In J. L. García-Alcaraz, G. Alor-Hernández, & A. A. Maldonado-Macías (Eds.), *New Perspectives on Applied Industrial Tools and Techniques* (pp. 327–352). Cham: Springer. doi:10.1007/978-3-319-56871-3_16

Ashkanasy, N. M., Wilderom, C. P. M., & Peterson, M. F. (2011). *The Handbook of Organizational Culture and Climate.* Sage.

Ashraf, G., & Abd, S. (2012). A Review on the Models of Organizational Effectiveness: A Look at Cameron's Model in Higher Education. *Educational Studies, 5*(2), 80–87.

Astrand, I. (1960). Aerobic work capacity in men and women with special reference to age. *Acta Physiologica Scandinavica, 49*(suppl 169), 45–60. PMID:13794892

Astrand, I. (1988). Physical demands in work life. *Scandinavian Journal of Work, Environment & Health, 14*(1), 10–13. PMID:3393858

Astrand, I., & Astrand, P.-O. (1978). *Aerobic work performance, a review.* New York: Academic Press Inc. doi:10.1016/B978-0-12-261350-0.50015-6

Astrand, P. O., Rodahl, K., Dahl, H., & Strømme, S. B. (1977). *Textbook of work physiology: physiological bases of exercise.* New York: McGraw Hill.

Astrand, P.-O. (1956). Human physical fitness with special reference to sex and age. *Physiological Reviews, 36*(3), 307–335. doi:10.1152/physrev.1956.36.3.307 PMID:13359126

Automotive Industry Action Group (AIAG). (2010). *Measurement Systems Analysis Reference Manual* (4th ed.). Chrysler Ford, General Motors Supplier Quality Requirements Task Force.

Autor, D. H. (2015). Why Are There Still So Many Jobs? The History and Future of Workplace Automation. *The Journal of Economic Perspectives, 29*(3), 3–30. doi:10.1257/jep.29.3.3

Aviation Accident Statistics. (2017). *National Transportation Safety Board.* Retrieved from www.ntsb.gov/aviation/aviation.htm

Ayabar, A., De la Riva, J., Sanchez, J., & Balderrama, C. (2015). Regression model to estimate standard time through energy consumption of workers in manual assembly lines under moderate workload. *Journal of Industrial Engineering*, 1-5.

Ayaz, H., Shewokis, P. A., Bunce, S., Izzetoglu, K., Willems, B., & Onaral, B. (2012). Optical brain monitoring for operator training and mental workload assessment. *NeuroImage*, *59*(1), 36–47. doi:10.1016/j.neuroimage.2011.06.023 PMID:21722738

Bakker, A., & Albrecht, S. (2018). Work engagement: Current trends. *Career Development International*, *23*(1), 4–11. doi:10.1108/CDI-11-2017-0207

Baldassarre, M. T., Piattini, M., Pino, F. J., & Visaggio, G. (2009). Comparing ISO/IEC 12207 and CMMI-DEV: Towards a mapping of ISO/IEC 15504-7. In *2009 ICSE Workshop on Software Quality* (pp. 59-64). ICSE. 10.1109/WOSQ.2009.5071558

Balderrama, C., Flores, J., & Maldonado, A. (2015). The effects of age to meet the standard production rate. *Work (Reading, Mass.)*, *51*(4), 827–837. doi:10.3233/WOR-141972 PMID:25425589

Balderrama, C., Ibarra, G., De La Riva, J., & López, L. (2010). Evaluation of three methodologies to estimate the VO2max in people of different ages. *Applied Ergonomics*, *42*(1), 162–168. doi:10.1016/j.apergo.2010.06.017 PMID:20650446

Bal, P. M., De Lange, A. H., Ybema, J. F., Jansen, P. G., & Van der Velde, M. E. (2011). Age and trust as moderators in the relation between procedural justice and turnover: A large-scale longitudinal study. *Applied Psychology*, *60*(1), 66–86. doi:10.1111/j.1464-0597.2010.00427.x

Banker, R. D., Potter, G., & Srinivasan, D. (2005). Association of Nonfinancial Performance Measures with the Financial Performance of a Lodging Chain. *The Cornell Hotel and Restaurant Administration Quarterly*, *46*(6), 394–412. doi:10.1177/0010880405275597

Banthia, J. K. (2004). *Census of India 2001 - Primary Census Abstracts*. Registrar General & Census Commissioner, Govt. of India.

Barkhuizen, N., Rothmann, S., & Vijver, F. J. (2014). Burnout and work engagement of academics in higher education institutions: Effects of dispositional optimism. *Stress and Health*, *30*(4), 322–332. doi:10.1002mi.2520 PMID:23949954

Barnden, A. W., & Darke, P. (2000). A Comparison of SSM with an Organisational Learning Model. *Proceedings of the International Conference on Systems Thinking in Management.*

Barón Maldonado, D. I., Rivera, L., & Caro, M. (2014). *Improvement of new product development process in apparel SMEs using lean and life cycle* (Vol. 9-12-NaN-2014). AIDI - Italian Association of Industrial Operations Professors. Retrieved from https://www.scopus.com/record/display.uri?eid=2-s2.0-84982859268&origin=resultslist&sort=plf-f&src=s&st1=Macroergonomics&nlo=&nlr=&nls=&sid=cb1a157df15e8dea160496d8803d91dc&sot= b&sdt=cl&cluster=scopubyr%2C%222018%22%2Ct%2C%222017%22%2Ct%2C%222016%22%2Ct%2C%222015%22%2Ct%2C%222014%22%2Ct&sl=30&s=TITLE-ABS-KEY%28Macroergonomics%29&relpos=47&citeCnt=0&searchTerm=

Barra, E. (2004). Apoyo social, estrés y salud. *Psicología y Salud, 238*(2), 237-243.

Barrett, C. B. (2006). *Food Aid as a Part of a Coherent Strategy to Advance Food Security Objectives.* ESA Working Paper no. 06-09. FAO. Retrieved from ftp://ftp.fao.org/docrep/fao/009/ag037e/ag037e00.pdf Bolis, I., & Sznelwar, L. (2016). A case study of the implementation of an ergonomics improvement committee in a Brazilian hospital – Challenges and benefits. *Applied Ergonomics, 53,* 181-189.

Bartlett, M. (1935). The Effect of Non-Normality on the t Distribution. *Mathematical Proceedings of the Cambridge Philosophical Society*, *31*(2), 223–231. doi:10.1017/S0305004100013311

Barzideh, M., Choobineh, A. R., & Tabatabaee, H. R. (2014). Job stress dimensions and their relationship to musculoskeletal disorders in Iranian nurses. *Work (Reading, Mass.)*, *47*(4), 423–429. doi:10.3233/WOR-121585 PMID:23324727

Batras, D., Duff, C., & Smith, B. J. (2014). Organizational change theory: Implications for health promotion practice. *Health Promotion International*, *31*(1), 231–241. PMID:25398838

Baumgartner, R. J., & Ebner, D. (2010). Corporate sustainability strategies: Sustainability profiles and maturity levels. *Sustainable Development*, *18*(2), 76–89. doi:10.1002d.447

Baxter, G., & Sommerville, I. (2011). Socio-technical systems: From design methods to systems engineering. *Interacting with Computers*, *23*(1), 4–17. doi:10.1016/j.intcom.2010.07.003

Beaudry, P., Green, D. A., & Sand, B. M. (2013). *The Great Reversal in the Demand for Skill and Cognitive Tasks* (Working Paper No. 18901). National Bureau of Economic Research. doi:10.3386/w18901

Bellamy, L. J., Geyer, T. A. W., & Wilkinson, J. (2008). Development of a functional model which integrates human factors, safety management systems and wider organizational issues. *Safety Science*, *46*(3), 461–492. doi:10.1016/j.ssci.2006.08.019

Berlin, C., Neumann, W. P., Theberge, N., & Örtengren, R. (2014). Avenues of entry: How industrial engineers and ergonomists access and influence human factors and ergonomics issues. *European Journal of Industrial Engineering*, *8*(3), 325. doi:10.1504/EJIE.2014.060999

Berns, M., Townend, A., Khayat, Z., Balagopal, B., Reeves, M., Hopkins, M., & Kruschwitz, N. (2009). *The business of sustainability: Findings and insights from the first annual business of sustainability survey and the global thought leaders' research project*. Massachusetts Institute of Technology. MIT Sloan Management Review. Retrieved March 22, 2018, from http://www.ideiasustentavel.com.br/pdf/mitsloan_sr2009-dl.pdf

Bertsch, V., Treitz, M., Geldermann, J., & Rentz, O. (2007). Sensitivity analyses in multi-attribute decision support for off-site nuclear emergency and recovery management. *International Journal of Energy Sector Management*, *1*(4), 342–365. doi:10.1108/17506220710836075

Beurskens, R., Haeger, M., Kliegl, R., Roecker, K., & Granacher, U. (2016). Postural control in dual-task situations: Does whole-body fatigue matter? *PLoS One*, *11*(1), e0147392. doi:10.1371/journal.pone.0147392 PMID:26796320

Bevilacqua, M., Ciarapica, F. E., Mazzuto, G., & Paciarotti, C. (2013). Visual Management implementation and evaluation through mental workload analysis. *IFAC Proceedings Volumes, 46*(7), 294–299. 10.3182/20130522-3-BR-4036.00065

Bigland-Ritchie, B., Jones, D. A., Hosking, G. P., & Edwards, R. H. (1978). Central and peripheral fatigue in sustained maximum voluntary contractions of human quadriceps muscle. *Clinical Science and Molecular Medicine*, *54*, 609–614. PMID:657729

Bink, B. (1962). The physical working capacity in relation to working time and age. *Ergonomics*, *5*(1-4), 25–28. doi:10.1080/00140136208930548

Bi, S., & Salvendy, G. (1994). Analytical modeling and experimental study of human workload in scheduling of advanced manufacturing systems. *The International Journal of Human Factors in Manufacturing*, *4*(2), 205–234. doi:10.1002/hfm.4530040207

Bittel, L., & Ramsey, J. (1988). Enciclopedia del Management: Organización y Administración de Empresas (Vol. 4). Madrid: Academic Press.

Blackman, D. (2006). How measuring learning may limit new knowledge creation. *Journal of Knowledge Management Practice*, *7*(3).

Boles, D. B., & Adair, L. P. (2001). The Multiple Resources Questionnaire (MRQ). *Proceedings of the Human Factors and Ergonomics Society Annual Meeting*, *45*(25), 1790–1794. doi:10.1177/154193120104502507

Boles, D. B., & Law, M. B. (1998). A simultaneous task comparison of differentiated and undifferentiated hemispheric resource theories. *Journal of Experimental Psychology. Human Perception and Performance*, *24*(1), 204–215. doi:10.1037/0096-1523.24.1.204 PMID:9483826

Bolis, I., & Sznelwar, L. I. (2016). A case study of the implementation of an ergonomics improvement committee in a Brazilian hospital - Challenges and benefits. *Applied Ergonomics*, *53*, 181–189. doi:10.1016/j.apergo.2015.09.012 PMID:26464035

Bommer, S. C., & Fendley, M. (2018). A theoretical framework for evaluating mental workload resources in human systems design for manufacturing operations. *International Journal of Industrial Ergonomics*, *63*, 7–17. doi:10.1016/j.ergon.2016.10.007

Borrás, F. X. (1995). Psiconeuroinmunología: Efectos del estrés psicológico sobre la función inmune en humanos sanos. *Ansiedad y Estrés*, *1*(1), 21–35.

Bouchard, D. R., & Trudeau, F. (2007). Reliability of the assessment of the oxygen/heart rate relationship during a workday. *Applied Ergonomics*, *38*(5), 491–497. doi:10.1016/j.apergo.2006.10.002 PMID:17368556

Bowling, N. A., Alarcon, G. M., Bragg, C. B., & Hartman, M. J. (2015). A meta-analytic examination of the potential correlates and consequences of workload. *Work and Stress*, *29*(2), 95–113. doi:10.1080/02678373.2015.1033037

Bridger, R. (2003). *Introduction to Ergonomics*. London: Taylor and Francis.

Broder, A. Z., Kirsch, A., Kumar, R., Mitzenmacher, M., Upfal, E., & Vassilvitskii, S. (2009). The hiring problem and Lake Wobegon strategies. *SIAM Journal on Computing*, *39*(4), 1233–1255. doi:10.1137/07070629X

Brodman, C. (1993). Social relationships and health-related behavior. *Journal of Behavioral Medicine*, *16*(4), 335–351. doi:10.1007/BF00844776 PMID:8411141

Brolin, A., Thorvald, P., & Case, K. (2017). Experimental study of cognitive aspects affecting human performance in manual assembly. *Production & Manufacturing Research*, *5*(1), 141–163. doi:10.1080/21693277.2017.1374893

Brunner, J. J. (2000) *Globalization and the future of education. Latin America and the Caribbean perspective on Education*. Santiago: UNESCO. Retrieved from Internet on January 2018: http://mt.educarchile.cl/archives/Futuro_EDU%25UNESCO-2000.pdf

BS OHSAS. (2007). *Occupational health and safety management systems*. Retrieved March 22, 2018, from http://www.isoqsltd.com/iso-certification/bs-ohsas-18001

Cacioppo, J. T., Bernston, G. G., & Decety, J. (2010). Social neuroscience and its relationship to social psychology. *Social Cognition*, *28*(6), 675–685. doi:10.1521oco.2010.28.6.675 PMID:24409007

Caetano, A., & Santos, S. C. (2017). The Gap Between Research and Professional Practice in Work and Organizational Psychology: Tensions, Beliefs, and Options. In E. R. Neiva, C. Vaz, & H. Mendonça (Eds.), *Organizational Psychology and Evidence-Based Management*. Springer. doi:10.1007/978-3-319-64304-5_1

Calton, J., & Kurland, N. (1995). *A theory of stakeholder enabling: Giving óbice to an emerging postmodern praxis of organizational discourse. In Postmodern management and organization theory* (pp. 154–179). Editorial Sage.

Cameron, K. (1980). Critical Questions in Assessing Organizational Effectiveness. *Organizational Dynamics*, *15*(2), 66–80. doi:10.1016/0090-2616(80)90041-8

Cameron, K. (1986). Effectiveness as paradox: Consensus and conflict in conceptions of organizational effectiveness. *Management Science*, *32*(5), 539–553. doi:10.1287/mnsc.32.5.539

Caracuel, A. (2007). Alimentación hospitalaria: Del blanco y negro al color. *Anales de RACVAO*, *20*(1), 43–67.

Carayon, P. (2006). Human factors of complex sociotechnical systems. *Applied Ergonomics*, *37*(4), 525–535. doi:10.1016/j.apergo.2006.04.011 PMID:16756937

Carayon, P., Hancockb, P., Levesonc, N., Noyd, I., Sznelware, L., & Hootegemf, G. V. (2015). Advancing a sociotechnical systems approach to workplace safety – developing the conceptual framework. *Ergonomics*, *58*(4), 548–564. doi:10.1080/00140139.2015.1015623 PMID:25831959

Carayon, P., Wetterneck, T. B., Rivera-Rodriguez, A. J., Shoofs Hundt, A., Hoonakker, P., Holden, R., & Gurses, A. P. (2014). Human factors systems approach to healthcare quality and patient safety. *Applied Ergonomics*, *45*(1), 14–25. doi:10.1016/j.apergo.2013.04.023 PMID:23845724

Carew, T. J., & Magsamen, S. H. (2010). Neuroscience and Education: An ideal partnership for producing evidence-based solutions to guide 21st century learning. *Neuron*, *65*(5), 685–688. doi:10.1016/j.neuron.2010.08.028 PMID:20826300

Carroll, A. B., & Shabana, K. M. (2010). The Business Case for Corporate Social Responsibility: A Review of Concepts, Research and Practice. *International Journal of Management Reviews*, *12*(1), 85–105. doi:10.1111/j.1468-2370.2009.00275.x

Cebola, M., Mahendra, A., Pombo, M., Marques, B., Pinto, H., Carolino, E., & Mendes, L. (2016). Comparison of nutritional status among elderly in hospital and in community environments. *European Geriatric Medicine*, *7*(1), S195.

Chan, L. K., & Wu, M. L. (2002). Quality function deployment: A literature review. *European Journal of Operational Research*, *143*(3), 463–497. doi:10.1016/S0377-2217(02)00178-9

Chan, L. K., & Wu, M. L. (2005). A systematic approach to quality function deployment with a full illustrative example. *International Journal of Management Sciences*, *33*, 119–139.

Checkland, P. (2000). Soft Systems Methodology: A thirty year retrospective. *Systems Research and Behavioral Science*, *17*(S1), 11–58. doi:10.1002/1099-1743(200011)17:1+<::AID-SRES374>3.0.CO;2-O

Cheng, S. (2014). The mediating role of organizational justice on the relationship between administrative performance appraisal practices and organizational commitment. *International Journal of Human Resource Management*, *25*(8), 1131–1148. doi:10.1080/09585192.2013.816864

Cheng, Y. C. (2012). *School Effectiveness And School-Based Management: A Mechanism For Development*. Taylor & Francis Group.

Chen, J., Song, X., & Lin, Z. (2016). Revealing the "Invisible Gorilla" in construction: Estimating construction safety through mental workload assessment. *Automation in Construction*, *63*, 173–183. doi:10.1016/j.autcon.2015.12.018

Chen, L. H., & Weng, M. C. (2006). An evaluation approach to engineering design in QFD processes using fuzzy goal programming models. *European Journal of Operational Research*, *172*(1), 230–248. doi:10.1016/j.ejor.2004.10.004

Cherns, A. (1976). The Principles of Socio-technical Design. *Human Relations*, *29*(8), 783–792. doi:10.1177/001872677602900806

Cherns, A. (1987). Principles of Socio-technical Design Revisited. *Human Relations*, *40*(3), 153–162. doi:10.1177/001872678704000303

Chiavenato, I. (2009). *Comportamiento organizacional* (2nd ed.). Mc-Graw Hill.

Chib, V. S., De Martino, B., Shimojo, S., & O'Doherty, J. P. (2012). Neural Mechanisms Underlying Paradoxical Performance for Monetary Incentives Are Driven by Loss Aversion. *Neuron*, *74*(3), 582–594. doi:10.1016/j.neuron.2012.02.038 PMID:22578508

Chung, J., & Monroe, G. S. (2003). Exploring social desirability bias. *Journal of Business Ethics*, *44*(4), 291–302. doi:10.1023/A:1023648703356

Cifuentes, M., & Fulmer, S. (2015). Research Needs for and Barriers to Use of Treadmill Workstations. *Ergonomics in Design*, *23*(3), 25–30. doi:10.1177/1064804615588849

Clack, L., Kuster, S. P., Giger, H., Giuliani, F., & Sax, H. (2014). Low-hanging fruit for human factors design in infection prevention - Still too high to reach? *American Journal of Infection Control*, *42*(6), 679–681. doi:10.1016/j.ajic.2014.03.002 PMID:24837120

Clegg, C. W. (2000). Sociotechnical principles for system design. *Applied Ergonomics*, *31*(5), 463–477. doi:10.1016/S0003-6870(00)00009-0 PMID:11059460

Coelho, D. A., Tavares, C. S. D., Lima, T. M., & Lourenço, M. L. (2017). Psychosocial and ergonomic survey of office and field jobs in a utility company. *International Journal of Occupational Safety and Ergonomics*, 1–12. doi:10.1080/10803548.2017.1331620 PMID:28589755

Cohen, B. E., Edmondson, D., & Kronish, I. M. (2015). State of the art review: Depression, stress, anxiety, and cardiovascular disease. *American Journal of Hypertension*, *28*(11), 1295–1302. doi:10.1093/ajh/hpv047 PMID:25911639

Cohen, L. (1995). *Quality function deployment: How to make QFD work for You*. Reading, MA: Addison- Wesley.

Colligan, T. W., & Higgins, E. M. (2005). Workplace stress: Etiology and consequences. *Journal of Workplace Behavioral Health*, *21*(2), 89–97. doi:10.1300/J490v21n02_07

Collins, J., & O'Sullivan, L. (2010). Psychosocial risk exposures and musculoskeletal disorders across working age males and females. *Ergonomics in Manufacturing & Service Industries*, *20*(4), 272–286. doi:10.1002/hfm.20220

Connell, N. (2001). Evaluating soft OR: Some reflections on an apparently 'unsuccessful' implementation using a Soft Systems Methodology (SSM) based approach. *The Journal of the Operational Research Society*, *52*(2), 150–160. doi:10.1057/palgrave.jors.2601054

Consejo Nacional de la Industria Maquiladora y Manufacturera de Exportación Index. (2016). *¿Cómo nace la Industria Maquiladora de exportación?* Retrieved from: http://www.index.org.mx/IMMEX/antecedentes.php

Cook, D. A., & Beckman, T. J. (2006). Current Concepts in Validity and Reliability for Psychometric Instruments: Theory and Application. *The American Journal of Medicine*, *119*(2), 166.e7–166.e16. doi:10.1016/j.amjmed.2005.10.036 PMID:16443422

Cooper, G. E., & Harper, R. P. Jr. (1969). *The use of pilot rating in the evaluation of aircraft handling qualities*. Neuilly-Sur-Seine, France: Advisory Group for Aerospace Research and Development.

Cooper, S. P., Burau, K. E., Frankowski, R., Shipp, E., Deljunco, D., Whitworth, R., ... Hanis, C. (2006). A cohort study of injuries in migrant farm worker families in South Texas. *Annals of Epidemiology*, *16*(4), 313–320. doi:10.1016/j.annepidem.2005.04.004 PMID:15994097

Copuroglu, C., Heybeli, N., Ozcan, M., Yilmaz, B., Ciftdemir, M., and Copuroglu, E. (2012). Major extremity injuries associated with farmyards accidents. *The Scientific World Journal*. doi:10.1100/2012/314038

Cristiano, J. J., Liker, J. K., & White, Ch. C. (2000). Customer-driven product development through quality function deployment in the U.S. and Japan. *Journal of Product Innovation Management*, *17*(4), 286–308. doi:10.1016/S0737-6782(00)00047-3

Cronbach, L. J., & Shavelson, R. J. (2004). My current thoughts on coefficient alpha and successor procedures. *Educational and Psychological Measurement*, *64*(3), 391–418. doi:10.1177/0013164404266386

Cullen, A. (1995). CE Credit: Burnout: Why Do We Blame the Nurse? *The American Journal of Nursing*, *95*(11), 23–28. doi:10.1097/00000446-199511000-00017 PMID:7485275

Cunliffe, A. (2008). Orientations to social constructionism: Relationally responsive social constructionism and its implications for knowledge and learning. *Management Learning*, *39*(2), 123–139. doi:10.1177/1350507607087578

Dalmau, I., & Ferrer, R. (2004). Revisión del concepto de carga mental: Evaluación, consecuencias y proceso de normalización. *Anuario de Psicología*, *35*(4), 521–546.

Dare, J. (2016). Will the Truth Set Us Free? An Exploration of CSR Motive and Commitment. *Business and Society Review*, *121*(1), 85–122. doi:10.1111/basr.12082

Davel, E., & Tremblay, D. G. (2003). Organizational culture and social performance: insights from the experience of family organizations. *Iberoamerican Academy of Management, 3rd International Conference.*

Davis, M. C., Challenger, R., Jayewardene, D. N. W., & Clegg, C. W. (2014). Advancing socio-technical systems thinking: A call for bravery. *Applied Ergonomics*, *45*(2), 171–180. doi:10.1016/j.apergo.2013.02.009 PMID:23664481

Dawson, S. L., Dash, S. R., & Jacka, F. N. (2016). The importance of diet and gut health to the treatment and prevention of mental disorders. *International Review of Neurobiology*, *131*, 325–346. doi:10.1016/bs.irn.2016.08.009 PMID:27793225

de Guimarães, L. B. M., Anzanello, M. J., Ribeiro, J. L. D., & Saurin, T. A. (2015). Participatory ergonomics intervention for improving human and production outcomes of a Brazilian furniture company. *International Journal of Industrial Ergonomics*, *49*, 97–107. doi:10.1016/j.ergon.2015.02.002

de Lange, D. E., Busch, T., & Delgado-Ceballos, J. (2012). Sustaining sustainability in organizations. *Journal of Business Ethics*, *110*(2), 157–172. doi:10.100710551-012-1425-0

de Vries, J. M., Hagemans, M. L. C., Bussmann, J. B. J., van der Ploeg, A. T., & van Doorn, P. A. (2010). Fatigue in neuromuscular disorders: Focus on Guillain-Barr syndrome and Pompe disease. *Cellular and Molecular Life Sciences*, *67*(5), 701–713. doi:10.100700018-009-0184-2 PMID:20196238

Dekker, S. W. A. (2006). Resilience engineering: Chronicling the emergence of confused consensus. In E. Hollnagel, D. D. Woods, & N. Leveson (Eds.), *Resilience Engineering: Concepts and Precepts* (pp. 77–92). Aldershot, UK: Ashgate Publishing Limited.

DeMuri, G. P., & Purschwitz, M. A. (2000). Farm injuries in children: A review. *Wisconsin Medical Journal*, *99*(9), 51–55. PMID:11220197

Denison, D. R., Janovics, J., Young, J., & Cho, H. J. (2005). *Diagnosing Organizational Cultures: Validating a Model and Method.* International Institute for Management Development, Working paper 2005-11.

Denison, D. R., & Mishra, A. K. (1995). Toward a Theory of Organizational Culture and Effectiveness. *Organization Science*, *6*(2), 204–223. doi:10.1287/orsc.6.2.204

Dent, E. B., & Umpleby, S. A. (1998). Underlying assumptions of several traditions in systems theory and cybernetics. *Cybernetics and Systems*, 513–518.

DesJardins, J. (2016). Is it Time to Jump off the Sustainability Bandwagon? *Business Ethics Quarterly*, *26*(1), 117–135. doi:10.1017/beq.2016.12

Di Maio, P. (2014). Towards a Metamodel to Support the Joint Optimization of Socio Technical Systems. *Systems Journal,* (2), 273-296.

Ding-Yu, L., & Sheue-Ling, H. (1998). The development of mental workload measurement in flexible manufacturing systems. *Human Factors and Ergonomics in Manufacturing & Service Industries*, *8*(1), 41–62. doi:10.1002/(SICI)1520-6564(199824)8:1<41::AID-HFM3>3.0.CO;2-E

Dixon, N. (2001). *El conocimiento común: Cómo prosperan las compañías que comparten lo que saben.* Oxford University Press.

Dobers, P., & Wolff, R. (2000). Competing with 'soft' issues - from managing the environment to sustainable business strategies. *Business Strategy and the Environment*, *9*(1), 143–150. doi:10.1002/(SICI)1099-0836(200005/06)9:3<143::AID-BSE239>3.0.CO;2-C

Dulac, N., & Leveson, N. (2004). An approach to design for safety in complex systems. *Int. Conference on System Engineering (INCOSE '04).* Retrieved March 22, 2018, from http://sunnyday.mit.edu/papers/incose-04.pdf

Dul, J., Broder, R., Buckle, P., Carayon, P., Talzon, P., Marras, W., ... Van Der Doelen, B. (2012). A strategy for human factors/ergonomics: Developing the discipline and the profession. *Ergonomics*, *55*(4), 377–395. doi:10.1080/00140139.2012.661087

Dyllick, T., & Hockerts, K. (2002). Beyond the business case for corporate sustainability. *Business Strategy and the Environment*, *11*(2), 130–141. doi:10.1002/bse.323

Easterby, M., Antonacopoulou, E., Simm, D., & Lyles, M. (2004). Constructing Contributions to Organizational Learning: Argyris and the Next Generation. *Management Learning*, *35*(4).

Edwards, K., & Jensen, P. L. (2014). Design of systems for productivity and well being. *Applied Ergonomics*, *45*(1), 26–32. doi:10.1016/j.apergo.2013.03.022 PMID:23631941

Eggleston, R. G., & Quinn, T. J. (1984). A Preliminary Evaluation of a Projective Workload Assessment Procedure. *Proceedings of the Human Factors Society Annual Meeting*, *28*(8), 695–699. 10.1177/154193128402800811

Ellis, G., & Roscoe, A. (1982). *The Airline Pilot's View of Flight Deck Workload: A Preliminary Study Using a Questionnaire*. Rarnborough, UK: Royal Aircraft Establishment.

Enderle, G. (2016). How Can Business Ethics Strengthen the Social Cohesion of a Society? *Journal of Business Ethics*. doi:10.100710551-016-3196-5

Erdogan, B. (2003). *The extent of information visualization in Turkish construction industry, A QFD approach*. The Middle East Technical University.

Eriksson, P., & Kovalainen, A. (2008). *Qualitative Methods in Business Research*. London: Sage Publications. doi:10.4135/9780857028044

Escolano, F. (2003). *Artificial Intelligence Models, Techniques and Application Areas*. University of Alicante.

Etzioni, A. (1960). Two approaches to Organizational Analysis: A Critique and a Suggestion. *Administrative Science Quarterly*, *5*(2), 257–278. doi:10.2307/2390780

Etzioni, A. (1987). Normative-Affective factors: Toward a new decision-making model. *Journal of Economic Psychology*, *9*(2), 125–150. doi:10.1016/0167-4870(88)90048-7

Eydi, H. (2015). Organizational Effectiveness Models: Review and Apply in Non-Profit Sporting Organizations. *American Journal of Economics, Finance and Management*, *1*(5), 460–467.

Fairhurst, G., & Grant, D. (2010). The social construction of leadership: A sailing guide. *Management Communication Quarterly*, *24*(2), 171–210. doi:10.1177/0893318909359697

FAO. (1998). *Platos típicos de países de América Latina 1998*. Santiago de Chile, Chile: FAO.

FAO/LATINFOODS. (2009). *Tabla de Composición de Alimentos de América Latina*. Author.

Farrer, F., Minaya, G., Niño, J., & Ruiz, M. (1997). *Manual de ergonomía*. Madrid: Mapfre.

Fernández, S., & Rosales, M. (2014). Administración educativa: la planificación estratégica y las prácticas gerenciales integrando la tecnología, su impacto en la educación. Congreso Iberoamericano de Ciencia, Tecnología, Innovación y educación. Artículo 1582. Buenos Aires Argentina.

Firstbeat Technologies. (2017). *VO2 estimation method based on heart rate measurement*. Retrieved from: https://assets.firstbeat.com/firstbeat/uploads/2015/10/white_paper_vo2_estimation.pdf

Fisher, C., & Schutta, J. T. (2003). *Developing New Service- Incorporating the Voice of the Customer into Strategic Service Development*. Milwaukee, WI: ASQ Quality Press.

Flueler, T. (2006). *Decision making for complex socio-technical systems: Robustness from Lessons Learned in Long-Term Radioactive Waste Governance. In Environment & Policy* (Vol. 42). Dordrecht, The Netherlands: Springer. doi:10.1007/1-4020-3529-2

Franklin, F. E. B. (2007). *Auditoría administrative, Strategic management of change* (2nd ed.). Editorial Pearson.

Freeman, R. E. (1994). The politics of stakeholder theory: Some future directions. *Business Ethics Quarterly*, *4*(4), 409–421. doi:10.2307/3857340

Freeman, R. E., Wicks, A., & Parmar, B. (2004). Stakeholder Theory and 'The Corporate Objective Revisited'. *Organization Science*, *15*(3), 364–369. doi:10.1287/orsc.1040.0066

Gallo, I. S., Cohen, A. L., Gollwitzer, P. M., & Oettingen, G. (2013). Neurophysiological correlates of the self-regulation of goal pursuit. In P. A. Hall (Ed.), *Social Neuroscience and Public Health* (pp. 19–34). New York: Springer. doi:10.1007/978-1-4614-6852-3_2

Gamero, C. (2010). Evaluación del coste por pérdida de jornadas laborales asociado al estrés laboral: Propuesta para España. *Estudios de Economía Aplicada*, *28*(3), 1–20.

Gandevia, S. C. (2001). Spinal and supraspinal factors in human muscle fatigue. *Physiological Reviews*, *81*(4), 1725–1789. doi:10.1152/physrev.2001.81.4.1725 PMID:11581501

García, G., & Lange, K. (2008). Macroergonomic study of food sector company distribution centres. *Applied Ergonomics*, *39*, 439–449.

Geary, R. C. (1947). *Testing for Normality. Biometrika*. Retrieved from https://pdfs.semanticscholar.org/2114/43e0d4bb0bcb6b200c263e250f5591a18458.pdf

Ghaffarian, V. (2011). The New Stream of Socio-Technical Approach and Main Stream Information Systems Research. *Computer Science*, *3*, 1499–1511.

Gibson, M., & Potvin, J. R. (2016). *An equation to calculate the recommended cumulative rest allowance across multiple subtasks*. Association of Canadian Ergonomists Conference, Niagara Falls, NY.

Gil Monte, P. (2008). El Síndrome de quemarse por el trabajo (burnout) cómo fenómeno transcultural. *Informació Psicològica*, *4*(11), 91–92. Retrieved from https://dialnet.unirioja.es/servlet/articulo?codigo=2551783

Gil-Monte, P., & Peiro, J. (1999). Perspectivas teóricas y modelos interpretativos para el estudio del síndrome de quemarse por el trabajo. *Anuales de Psicología, 15*(2), 261-268. Retrieved from http://www.um.es/analesps/v15/v15_2pdf/12v98_05Llag2.PDF

Gite, L. P., & Kot, L. S. (2003). *Accidents in Indian Agriculture. Technical Bulletin No. CIAE/2003/103.* Coordinating Cell, All India Coordinated Research Project on Human Engineering and Safety in Agriculture, Central Institute of Agricultural Engineering, Nabibagh, Bhopal.

Goh, Y. M., Brown, H., & Spickett, J. (2010). Applying systems thinking concepts in the analysis of major incidents and safety culture. *Safety Science, 45*(3), 302–309. doi:10.1016/j.ssci.2009.11.006

GOI. (2002). *India Vision 2020. Planning commission.* New Delhi: Govt. of India.

GOI. (2006). *Population projections for India and States 2001-2006, Report of the technical group on population projections constituted by the National Commission on Population.* New Delhi: Govt. of India.

Gómez-González, B., & Escobar, A. (2006). Estrés y sistema inmune. *Rev Mex Neuroci, 7*(1), 30–38.

González Molero, I., Olveira Fuster, G., Liébana, M. I., Oliva, L., Laínez López, M., & Muñoz Aguilar, A. (2008). Influencia de la temperatura en la ingesta de pacientes hospitalizados. *Nutrición Hospitalaria, 23*(1), 54–59. Retrieved from http://scielo.isciii.es/scielo.php?script=sci_arttext&pid=S0212-16112008000100009&lng=es&tlng=pt

González Munõz, E. L., & Gutiérrez, R. (2006). La carga de trabajo mental como factor de riesgo de estrés en trabajadores de la industria electrónica. *Revista Latinoamericana de Psicología, 38*(2), 259–270.

González, M. N. L. (2015). *Employer satisfaction assessment of Industrial Engineering graduates of classes of 2009 - 2014.* Professional Experience Report, ITCJ.

Gonzalez, M., & Landero, R. (2008). Confirmación de un modelo explicativo del estrés y de los síntomas psicosomáticos mediante ecuaciones estructurales. *Pan Am J Public Health, 23*(1), 7–18. doi:10.1590/S1020-49892008000100002

Goodfellow, I., Mcdaniel, P., & Papernot, N. (2018). Making Machine Learning Robust Against Adversarial Inputs. *Communications of the ACM, 61*(7), 56–66. doi:10.1145/3134599

Gooyers, C. E., & Stevenson, J. M. (2012). The impact of an increase in work rate on task demands for a simulated industrial hand tool assembly task. *International Journal of Industrial Ergonomics, 42*(1), 80–89. doi:10.1016/j.ergon.2011.11.005

Gracia, E. (2015). Stress, coping, and work engagement within the specific job context: Comment on Kaiseler, et al. (2014). *Psychological Reports, 116*(2), 377–380. doi:10.2466/01.20.PR0.116k22w1 PMID:25799121

Grandjean, E. (1988). *Fitting the task to the man.* New York: Taylor and Francis.

Grau, R., Salanova, M., & Peiró, J. M. (2012). Efectos moduladores de la autoeficacia en el estrés laboral. *Apuntes de Psicología, 30*(1-3), 311–321.

Greenwood, M., & van Buren, H. J. III. (2010). Trust and stakeholder theory: Trustworthiness in the organisation–stakeholder relationship. *Journal of Business Ethics, 95*(3), 425–438. doi:10.100710551-010-0414-4

Greimas, A. J. (1990). *Narrative semiotics and cognitive discourses.* Burns & Oates.

Greimas, A. J. (1990). *The social sciences: A semiotic view.* Minneapolis, MN: University of Minnesota Press.

GRI. (2014). *Global Reporting Initiatives: Sustainability Reporting Guidelines.* Retrieved March 22, 2018, from https://www.globalreporting.org/reporting/g4/Pages/default.aspx

Griffin, A., & Hauser, J. R. (1992). Patterns of communications among marketing, engineering and manufacturing. *Journal of Management and Science*, *38*(3), 360–373.

Griffin, A., & Hauser, R. J. (1991). The Voice of the Customer. *Marketing Science*, *12*(1), 1–27. doi:10.1287/mksc.12.1.1

Groover, M. P. (2007). *Work systems and the methods, measurement, and management of work*. Upper Saddle River, NJ: Pearson Prentice Hall.

Grote, G. (2014). Adding a strategic edge to human factors/ergonomics: Principles for the management of uncertainty as cornerstones for system design. *Applied Ergonomics*, *45*(1), 33–39. doi:10.1016/j.apergo.2013.03.020 PMID:23622735

Gruet, M., Temesi, J., Rupp, T., Levy, P., Millet, G. Y., & Verges, S. (2013). Stimulation of the motor cortex and corticospinal tract to assess human muscle fatigue. *Neuroscience*, *231*, 384–399. doi:10.1016/j.neuroscience.2012.10.058 PMID:23131709

Guerra, J., Martinez, I., Munduate, L., & Medina, F. (2004). A contingence perspective on the study of the consequences of conflict types: The role of organizational culture. *17th Conference of the International Association for Conflict Management*.

Guinta, L. R., & Praizler, N. C. (1993). *The QFD Book, The team approaches to solving problems and satisfying customers through quality function deployment*. New York: American Management Association.

Guiselle, M. G-V. (2016). *Desarrollo organizacional y los procesos de cambio en las instituciones educativas, un reto de la gestión de la educación*. Revista Educación Enero junio 2016, Volumen 40, Num. 1. Recuperado de: 26 enero 2018. https://revistas.ucr.ac.cr/index.php/educacion/article/view/22534

Guzmán, M. E., Del Castillo, A., & García, M. (2010). Factores psicosociales asociados al paciente con obesidad. In *Universidad Autónoma del Estado de Hidalgo. Obesidad: Un enfoque multidisciplinario*. Madrid: Colección Real, Museo Nacional del Prado.

Hakulinen, C., Pulkki-Råback, L., Jokela, M., Ferrie, J. E., Aalto, A. M., Virtanen, M., ... Elovainio, M. (2016). Structural and functional aspects of social support as predictors of mental and physical health trajectories: Whitehall II cohort study. *Journal of Epidemiology and Community Health*, *70*(7), 710–715. doi:10.1136/jech-2015-206165 PMID:26767407

Hale, A., & Borys, D. (2014). Working to rule, or working safely? Part 1: A state of the art review. *Safety Science*, *55*(1), 207–221.

Haro, E., & Kleiner, B. M. (2008). Macroergonomics as an organizing process for systems safety. *Applied Ergonomics*, *39*(4), 450–458. doi:10.1016/j.apergo.2008.02.018 PMID:18407244

Harrison, J. S., Freeman, R. E., & Abreu, M. C. S. (2015). Stakeholder theory as an ethical approach to effective management: Applying the theory to multiple contexts. *Review of Business Management*, *17*(55), 858–869. doi:10.7819/rbgn.v17i55.2647

Harrison, J. S., & Wicks, A. C. (2013). Stakeholder Theory, Value, and Firm Performance. *Business Ethics Quarterly*, *23*(1), 97–124. doi:10.5840/beq20132314

Hart, S. G., & Staveland, L. E. (1988). Development of NASA-TLX (Task Load Index): Results of Empirical and Theoretical Research. In P. A. Hancock & N. Meshkati (Eds.), *Advances in Psychology* (Vol. 52, pp. 139–183). North-Holland. doi:10.1016/S0166-4115(08)62386-9

Hassall, M., Xiao, T., Sanderson, P., & Neal, A. (2015). *Human Factors and Ergonomics. International Encyclopedia of the Social &amp* (2nd ed.). Behavioral Sciences. doi:10.1016/B978-0-08-097086-8.22025-4

Hauser, J., & Clausing, D. (1989). The house of quality. *Harvard Business Review*, *66*(3), 63–73. PMID:10303477

Hébert, L. (2014). *The Actantial Model.* Retrieved March 22, 2018, from http://www.signosemio.com/greimas/actantial-model.asp

Heckscher, C. (1994). *Defining the post bureaucratic type. In The post-bureaucratic organization: new perspectives on organizational change*. Editorial Sage.

Heizer, J. (2004). *Principles of operations management.* Pearson Prentice Hall Editorial.

Helkamp, J., & Lundstrom, W. (2002). Tractor Related Deaths Among West Virginia Farmers. *Annals of Epidemiology*, *12*(7), 510. doi:10.1016/S1047-2797(02)00344-7

Hellriegel, D., Jackson, S., & Solcum, J. (2005). *Management. A Competency-Based Approach* (10th ed.). McGraw-Hill de México.

Hendrick & Kleiner. (2002). *Macroergonomics. Theory, Methods, and Applications.* Lawrence Earlbaum Associates, Inc.

Hendrick, H. W. (1991). Ergonomics in organizational design and management. *Ergonomics*, *34*(6), 743–756. doi:10.1080/00140139108967348

Hendrick, H. W. (1995). Future directions in macroergonomics. *Ergonomics*, *38*(8), 1617–1624. doi:10.1080/00140139508925213

Hendrick, H. W. (2002a). An overview of macroergonomics. In H. W. Hendrick & B. M. Kleiner (Eds.), *Macroergonomics: Theory, methods, and applications* (pp. 1–23). Mahwah, NJ: Lawrence Erlbaum Associates. doi:10.1201/b12477-2

Hendrick, H. W. (2002b). Macroergonomics methods: Assessing work system structure. In H. W. Hendrick & B. M. Kleiner (Eds.), *Macroergonomics: Theory, methods, and applications* (pp. 45–66). Mahwah, NJ: Lawrence Erlbaum Associates. doi:10.1201/b12477-4

Hendrick, H. W. (2007). Macroergonomics: The Analysis and Design of Work Systems. *Review of Human Factors and Ergonomics*, *3*(1), 44–78. doi:10.1518/155723408X299834

Hendrick, H. W. (2008). Applying ergonomics to systems: Some documented "lessons learned". *Applied Ergonomics*, *39*(4), 418–426. doi:10.1016/j.apergo.2008.02.006 PMID:18374303

Hendrick, H. W., & Kleiner, B. M. (1999). *Macroergonomics: An introduction to work system design.* Santa Monica, CA: Human Factors and Ergonomics Society.

Henri, J. F. (2004). Performance Measurement and Organizational Effectiveness: Bringing the gap. *Managerial Finance*, *30*(6), 93–123. doi:10.1108/03074350410769137

Heras-Saizarbitoria, I., Cilleruelo, E., & Allur, E. (2014). ISO 9001 and the quality of working life: An empirical study in a peripheral service industry to the standard's home market. *Human Factors and Ergonomics in Manufacturing*, *24*(4), 403–414. doi:10.1002/hfm.20392

Hoerr, J., Fogel, J., & Van Voorhees, B. (2017). Ecological correlations of dietary food intake and mental health disorders. *Journal of Epidemiology and Global Health*, *7*(1), 81–89. doi:10.1016/j.jegh.2016.12.001 PMID:28003094

Holden, R. J., Brown, R. L., Scanlon, M. C., Rivera, A. J., & Karsh, B.-T. (2015). Micro- and macroergonomic changes in mental workload and medication safety following the implementation of new health IT. *International Journal of Industrial Ergonomics*, *49*, 131–143. doi:10.1016/j.ergon.2014.04.003

Holden, R. J., Eriksson, A., Andreasson, J., Williamsson, A., & Dellve, L. (2015). Healthcare workers' perceptions of lean: A context-sensitive, mixed methods study in three Swedish hospitals. *Applied Ergonomics*, *47*, 181–192. doi:10.1016/j.apergo.2014.09.008 PMID:25479987

Holden, R. J., Or, C. K. L., Alper, S. J., Joy Rivera, A., & Karsh, B. T. (2008). A change management framework for macroergonomic field research. *Applied Ergonomics*, *39*(4), 459–474. doi:10.1016/j.apergo.2008.02.016 PMID:18417095

Holden, R. J., Valdez, R. S., Schubert, C. C., Thompson, M. J., & Hundt, A. S. (2017). Macroergonomic factors in the patient work system: Examining the context of patients with chronic illness. *Ergonomics*, *60*(1), 26–43. doi:10.1080/00140139.2016.1168529 PMID:27164171

Holt-Lunstad, J., & Uchino, B. (2015). Social support and health. *Health behavior: Theory, research and practice*, 183-204.

Hope, A., Kelleher, C., Holmes, L., & Hennessy, T. (1999). Health and safety practices among farmers and other workers: A needs assessment. *Occupational Medicine (Philadelphia, Pa.)*, *49*, 231–235. doi:10.1093/occmed/49.4.231 PMID:10474914

Hughes, H. P. N., Clegg, C. W., Bolton, L. E., & Machon, L. C. (2017). Systems scenarios: A tool for facilitating the socio-technical design of work systems. *Ergonomics*, *60*(10), 1319–1335. doi:10.1080/00140139.2017.1288272 PMID:28277171

Huh, J., Delorme, D. E., & Reid, L. N. (2006). Perceived third-person effects and consumer attitudes on prevetting and banning DTC advesiting. *The Journal of Consumer Affairs*, *40*(1), 90–116. doi:10.1111/j.1745-6606.2006.00047.x

Huiyun, X., Zengzhen, W., Lorann, S., Thomas, J. K., Xuzhen, H., & Xianghua, F. (2000). Agricultural work-related injuries among farmers in Hubei, People's Republic of China. *American Journal of Public Health*, *90*(8), 1269–1276. doi:10.2105/AJPH.90.8.1269 PMID:10937008

Husemann, B., Von Mac, H. C. Y., Borsotto, D., Zepf, K. I., & Scharnbacher, J. (2009). Comparisons of musculoskeletal complaints and data entry between a sitting and a sit-stand workstation paradigm. *Human Factors: The Journal of the Human Factors and Ergonomics Society*, *51*(3), 310–320. doi:10.1177/0018720809338173 PMID:19750794

Ibrahim, H., Yasin, W., Udzir, N. I., & Abdul Hamid, N. W. (2016). Intelligent Cooperative Web Caching Policies For Media Objects Based On J48 Decision Tree And Naïve Bayes Supervised Machine Learning Algorithms In Structured Peer-To-Peer Systems. *Journal Of Information & Communication Technology*, *15*(2), 85–116.

Ilmarinen, J. (1984). Physical load on the cardiovascular system in different work tasks. *Scandinavian Journal of Work, Environment & Health*, *10*(6), 403–408. doi:10.5271jweh.2303 PMID:6535243

Ilmarinen, J. (1988). Physiological criteria for retirement age. *Scandinavian Journal of Work, Environment & Health*, *14*(1), 88–89. PMID:3393893

Ilmarinen, J. (1992). Job design for the aged with regard to decline in their maximal aerobic capacity: Part I—guidelines for the practitioner. *International Journal of Industrial Ergonomics*, *10*(1), 53–63. doi:10.1016/0169-8141(92)90048-5

Ilmarinen, J. (2001). Aging Workers. *Occupational and Environmental Medicine*, *58*(8), 546–548. doi:10.1136/oem.58.8.546 PMID:11452053

Ilmarinen, J., & Rutenfranz, J. (1980). Occupationally induced stress, strain and peak loads as related to age. *Scandinavian Journal of Work, Environment & Health*, *6*(4), 274–282. doi:10.5271jweh.2607 PMID:7233116

Imbeau, D., Desjardins, L., Dessureault, P. C., Riel, P., & Fraser, R. (1995). Oxygen consumption during scaffold assembling and disassembling work: Comparison between field measurements and estimation from heart rate. *International Journal of Industrial Ergonomics*, *15*(4), 247–259. doi:10.1016/0169-8141(94)00040-A

IMSS advierte sobre los riesgos del estrés laboral. (2015). *Noticias Tijuana.* Retrieved from http://tijuananoticias.info/bc/imss-advierte-sobre-los-riesgos-del-estres-laboral/

Instituto Nacional de Estadística y Geografía (INEGI). (2016). *Banco de información económica, Manufacturas.* Retrieved from http://www.inegi.org.mx/sistemas/bie/

Instituto Tecnológico de Ciudad Juárez. (2015). *2013-2018 Institutional Innovation and Development Program.* TNM.

International Civil Aviation Organization. (1998). Human Factors Training Manual (1st ed.). Doc. ICAO 9683-AN/950. Montreal, Canada: Author.

International Civil Aviation Organization. (2002). Human Factors Guidelines for Safety Audits Manual (1st ed.). Doc. ICAO 9806-AN/763. Montreal, Canada: Author.

International Civil Aviation Organization. (2004). *Cross-Cultural Factors in Aviation Safety: Human Factors Digest Nº 16. Circ. ICAO 302-AN/175.* Montreal, Canada: Author.

International Civil Aviation Organization. (2005). *Global Air Traffic Management Operational Concept. Doc. ICAO 9854.* Montreal, Canada: Author.

International Civil Aviation Organization. (2009). *Manual on Global Performance of the Air Navigation System. Doc. ICAO 9883.* Montreal, Canada: Author.

International Civil Aviation Organization. (2013a). Safety Management Manual (SMM) (3rd ed.). Doc. ICAO 9859-AN 474. Montreal, Canada: Author.

International Civil Aviation Organization. (2013b). *State of Global Aviation Safety.* Montreal, Canada: Author.

International Civil Aviation Organization. (2014). *Safety Report.* Montreal, Canada: Author.

ISO 14001. (2004). *Environmental management systems.* International Organization for Standardization. Retrieved March 22, 2018, from http://www.isoqsltd.com/iso-certification/iso-14001

ISO 9001. (2008). *Quality management systems.* International Organization for Standardization. Retrieved March 22, 2018, from http://www.isoqsltd.com/iso-certification/iso-9001

ISO-8996. (2004). *Ergonomics of Thermal Environment. Determination of Metabolic Rate.* International Organization of Standardization.

Ittner, C. D., & Larcker, D. F. (2001). Determinants of performance measure choices in worker incentive plans. *Journal of Labor Economics, 20*(2), S58–S90.

Jackson, M. C. (2003). *Systems thinking: creative holism for managers.* John Wiley & Sons Ltd. Retrieved July 18, 2018, from http://webcourses.ir/dl/Systems%20Thinking.pdf

Jackson, M. C. (2003). *Systems Thinking: Creative Holism for Managers.* West Sussex, UK: John Wiley & Sons Ltd. Retrieved March 22, 2018, from http://citeseerx.ist.psu.edu/viewdoc/download?doi=10.1.1.476.2858&rep=rep1&type=pdf

Johnson, R. B., Onwuegbuzie, A. J., & Turner, L. A. (2007). Toward a definition of mixed-methods research. *Journal of Mixed Methods Research, 1*(1), 112–133. doi:10.1177/1558689806298224

Jorgensen, K. (1985). Permissible Loads Based on Energy Expenditure Measurements. *Ergonomics, 28*(1), 365–369. doi:10.1080/00140138508963145 PMID:3996374

Kanawaty, G. (1992). *Introduction to Work Study. Ginebra.* International Labor Organization.

Kang, M., Koh, S., Cha, B. S., Park, J. K., Woo, J. M., & Chang, S. J. (2004). Association between job stress on heart rate variability and metabolic syndrome in shipyard male workers. *Yonsei Medical Journal*, *45*(5), 838–846. doi:10.3349/ymj.2004.45.5.838 PMID:15515194

Kaplan, R. M., Sallis, J. F., & Patterson, T. L. (1993). *Health and human behavior*. New York: Mcgrath-Hill.

Karasek, R., & Theorell, T. (1990). *Healthy work. Stress, productivity, and the reconstruction of working life*. New York: Basic Books.

Karsh, B. T., & Brown, R. (2010). Macroergonomics and patient safety: The impact of levels on theory, measurement, analysis and intervention in patient safety research. *Applied Ergonomics*, *41*(5), 674–681. doi:10.1016/j.apergo.2009.12.007 PMID:20153456

Karsh, B.-T., Waterson, P., & Holden, R. J. (2014). Crossing levels in systems ergonomics: A framework to support 'mesoergonomic' inquiry. *Applied Ergonomics*, *5*(1), 45–54. doi:10.1016/j.apergo.2013.04.021 PMID:23706573

Karthikeyan, C., Veeraragavathatham, D., Karpagam, D., & Ayisha, F. S. (2009). Traditional tools in agricultural practices. *Indian Journal of Traditional Knowledge*, *8*(2), 212–217.

Keating, C., Fernandez, A. A., Jacobs, D. A., & Kauffmann, P. (2001). A methodology for analysis of complex socio-technical processes. *Business Process Management Journal*, *7*(1), 33–50. doi:10.1108/14637150110383926

Kemira Corporation. (2016a). *Responsibility Report 2015*. Retrieved March 22, 2018, from http //www.kemira.com/en/newsroom/publications/sustainability-report/pages/default.aspx

Kemira Corporation. (2016b). *Annual Report 2015*. 220 pages. Retrieved March 22, 2018, from http://www.kemira.com/en/newsroom/publications/annual-report/pages/default.aspx

Kemira Corporation. (2016c). *High performance engagement model applied in Kemira corporation: A survey in collaboration between Kemira and IBM*. Helsinki: Kemira.

Kharchenko, V., Shmelova, T., & Sikirda, Y. (2012a). *Decision-making by an operator in the air navigation system: A monograph*. Kirovograd: KFA of NAU.

Kharchenko, V., Shmelova, T., & Sikirda, Y. (2011a). Graphanalytical models of decision-making by air navigation system's human-operator. *Proceedings of the National Aviation University*, *1*, 5–17.

Kharchenko, V., Shmelova, T., & Sikirda, Y. (2011b). Methodology for analysis of decision making in air navigation system. *Proceedings of the National Aviation University*, *3*, 85–94.

Kharchenko, V., Shmelova, T., & Sikirda, Y. (2012b). Modeling of behavioral activity of air navigation system's human-operator in flight. *Proceedings of the National Aviation University*, *2*, 5–17.

Kharchenko, V., Shmelova, T., & Sikirda, Y. (2016). *Decision-making in socio-technical systems: A monograph*. Kyiv: NAU.

Kharchenko, V., Shmelova, T., & Sikirda, Y. (2018). The methodology of Research and Training in Air Navigation Socio-technical System. *Proceedings of the National Aviation University*, *1*(74), 8–23.

Kim, J. S., Gračanin, D., Yang, T., & Quek, F. (2015). Action-transferred navigation technique design approach supporting human spatial learning. *ACM Transactions on Computer-Human Interaction*, *22*(6), 30. doi:10.1145/2811258

Kivimäki, M., & Kawachi, I. (2015). Work stress as a risk factor for cardiovascular disease. *Current Cardiology Reports*, *17*(9), 74. doi:10.100711886-015-0630-8 PMID:26238744

Kleiner, B. M. (1999). Analysis and Design for Improved Safety and Quality Performance. *Occupational Safety and Ergonomics, 5*(2), 217-245.

Kleiner, B. M. (2008). Macroergonomics: Work system analysis and design. *Human Factors, 50*(3), 461–467. doi:10.1518/001872008X288501 PMID:18689054

Kleiner, B. M., Hettinger, L. J., DeJoy, D. M., Huang, Y.-H., & Love, P. E. D. (2015). Sociotechnical attributes of safe and unsafe work systems. *Ergonomics, 58*(4), 635–649. doi:10.1080/00140139.2015.1009175 PMID:25909756

Kleiner, Y., Rajani, B., & Sadiq, R. (2006). Failure risk management of buried infrastructure using fuzzy-based techniques. *Journal of Water Supply: Research & Technology - Aqua, 55*(2), 81–94. doi:10.2166/aqua.2006.075

Kluge, A. (2014). *The Acquisition of Knowledge and Skills for Taskwork and Teamwork to Control Complex Technical Systems. The Acquisition of Knowledge and Skills for Taskwork and Teamwork to Control Complex Technical Systems: A Cognitive and Macroergonomics Perspective* (Vol. 9789400750). doi:10.1007/978-94-007-5049-4

Knapp, L. W. (1966). Occupational and Rural Accidents. *Archives of Environmental Health, 13*(4), 501–506. doi:10.1080/00039896.1966.10664604 PMID:5922020

Knapp, L. W. Jr. (1965). Agricultural Injuries Prevention. *Journal of Occupational Medicine., 7*(11), 553–745. doi:10.1097/00043764-196511000-00001 PMID:5831719

Kogure, M., & Akao, Y. (1983). Quality function deployment and CWQC in Japan. *Journal of Quality and Progress, 16*(10), 25–29.

Kononenko, I. (2018). Early Machine Learning Research in Ljubljana. *Informatica, 42*(1), 3-6.

Korunka, C., Scharitzer, D., Carayon, P., Hoonakker, P., Sonnek, A., & Sainfort, F. (2007). Customer orientation among employees in public administration: A transnational, longitudinal study. *Applied Ergonomics, 38*(3), 307–315. doi:10.1016/j.apergo.2006.04.019 PMID:16759625

Krause, N., Brand, R. J., Kaplan, G. A., Kauhanen, J., Malla, S., Tuomainen, T., & Salonen, J. T. (2007). Occupational physical activity, energy expenditure and 11-year progression of carotid atherosclerosis. *Scandinavian Journal of Work, Environment & Health, 33*(6), 405–424. doi:10.5271jweh.1171 PMID:18327509

Kroenke, K., Wood, D. R., Mangelsdorff, A. D., Meier, N. J., & Powell, J. B. (1988). Chronic fatigue in primary care. Prevalence, patient characteristics, and outcome. *Journal of the American Medical Association, 260*(7), 929–934. doi:10.1001/jama.1988.03410070057028 PMID:3398197

Ksiazek, I., Stefaniak, T. J., Stadnyk, M., & Ksiazek, J. (2011). Burnout syndrome in surgical oncology and general surgery nurses: A cross-sectional study. *European Journal of Oncology Nursing, 15*(4), 347–350. doi:10.1016/j.ejon.2010.09.002 PMID:20951089

Kuchar, J. K., & Yang, L. C. (2000, December). A review of conflicts detection and resolution modeling methods. *IEEE Transactions on Intelligent Transportation Systems, 1*(4), 179–189. doi:10.1109/6979.898217

Kumar, A., Singh, J. K., Mohan, D., & Varghese, M. (2008). Farm hand tools injuries: A case study from northern India. *Safety Science, 46*(1), 54–65. doi:10.1016/j.ssci.2007.03.003

Kumar, A., Varghese, M., & Mohan, D. (2000). Equipment-related injuries in agriculture: An international perspective. *Injury Control and Safety Promotion, 7*(3), 1–12. doi:10.1076/1566-0974(200009)7:3;1-N;FT175

Kushner, R. J. (2002). *Action Research Validation of an Inventory of Effectiveness Measures.* Nonprofit Organizational Effectiveness and Performance.

Kvale, S., & Brinkmann, S. (2009). *Interviews: Learning the Craft of Qualitative Research Interviewing* (2nd ed.). London: Sage Publications.

Kweku, F. N., & Mishra, M. (2018). Influence of Human Resource Development (HRD) Practices on Organizational Effectiveness: The Role of Employee Competencies. *International Journal of Management Studies*, *2*(6), 110–124.

Lagraauw, H. M., Kuiper, J., & Bot, I. (2015). Acute and chronic psychological stress as risk factors for cardiovascular disease: Insights gained from epidemiological, clinical and experimental studies. *Brain, Behavior, and Immunity*, *50*, 18–30. doi:10.1016/j.bbi.2015.08.007 PMID:26256574

Lamata, F. (1998). *User administration and health management.* Doctoral Thesis.

Landsbergis, P. A., Cahill, J., & Schnall, P. (1999). The impact of lean production and related new systems of work organization on worker health. *Journal of Occupational Health Psychology*, *4*(2), 108–130. doi:10.1037/1076-8998.4.2.108 PMID:10212864

Langley, P., & Simon, H. A. (1995). Applications of Machine Learning and Rule Induction. *Communications of the ACM*, *38*(11), 55–64. doi:10.1145/219717.219768

Larun, L., Brurberg, K. G., Odgaard-Jensen, J., & Price, J. R. (2017). Exercise therapy for chronic fatigue syndrome. *BJPsych Advances*, *23*(3), 144–144. doi:10.1192/apt.23.3.144 PMID:28444695

Latorre, M. J. M., & Javier, B.-E. F. (2013). *Strategic Management as Key to Improve the Quality of Education*, 1st World Congress of Administrative & Political Sciences (ADPOL-2012). *Procedia: Social and Behavioral Sciences*, *81*, 270–274. doi:10.1016/j.sbspro.2013.06.426

Lazarus, R. S., & Folkman, S. (1984). *Stress, Appraisal and Coping*. New York: Springer.

Lazarus, R., & Folkman, S. (1984). *Stress, Appraissal and Coping*. New York: Springer. Retrieved from https://books.google.com.mx/books?hl=es&lr=&id=i-ySQQuUpr8C&oi=fnd&pg=PR5&dq=Lazarus+R.S.+%26+Folkman,+S.+(1984).+Stress,+Appraisal+and+Coping.+cita&ots=DeJQjrciQf&sig=VDJ2kDOeTPHWbRlELA9nClz36oU#v=onepage&q&f=false

Lear, C. (2011). How can system administrators reduce stress and conflict in the workplace? *ACM Queue; Tomorrow's Computing Today*, *9*(1), 1–10.

Lee, J. (2000). *Operations management: strategy and analysis*. Addison-Wesley.

Lee, M. D. (2008). A review of the theories of corporate social responsibility: Its evolutionary path and the road ahead. *International Journal of Management Reviews*, *10*(1), 53–73. doi:10.1111/j.1468-2370.2007.00226.x

Lefebvre, B. (2008, January-June). Intentional-reflective model of the agent. *Reflexive Processes and Control, 1,* 69-78.

Lefebvre, B., & Adams-Webber, J. (2002, January-June). Functions of fast reflexion in bipolar choice. *Reflexion Processes and Control, 1,* 29-40.

Lefebvre, V. A. (2001). Algebra of conscience (2nd enlarged ed.). Dordrecht, The Netherlands: Kluwer Publishers.

Leijten, F. R. M., Van den Heuvel, S. G., Ybema, J. F., Robroek, S. J. W., & Burdorf, A. (2013). Do work factors modify the association between chronic health problems and sickness absence among older employees? *Scandinavian Journal of Work, Environment & Health*, *39*(5), 477–485. doi:10.5271jweh.3353 PMID:23440271

Leka, S., Griffiths, A., & Cox, T. (2004). Estrategias sistemáticas de solución de problemas para empleadores, personal directivo y representantes sindicales. Institute of Work, Health & Organisations, (3), 5.

Lev, B., & Radhakrishnan, S. (2005). The Valuation of Organization Capital. In C. Corrado, J. Haltiwanger, & D. Sichel (Eds.), *Measuring Capital in a New Economy* (pp. 73–99). National Bureau of Economic Research and University of Chicago Press. doi:10.7208/chicago/9780226116174.003.0004

Leveson, N. (2002). *Model-based Analysis of Socio-Technical Risk.* Technical Report. Engineering Systems Division, Massachusetts Institute of Technology. Retrieved March 22, 2018, from http://sunnyday.mit.edu

Leveson, N. G. (2005). *Safety in integrated systems health engineering and management*. NASA Ames Integrated System Health Engineering and Management Forum (ISHEM). Retrieved from http://sunnyday.mit.edu

Leychenko, S., Malishevskiy, A., & Mikhalic, N. (2006). *Human factors in aviation: a monograph in two books. Book 1st.* Kirovograd: YMEKS.

Lim, A. J., Village, J., Salustri, F. A., & Neumann, W. P. (2014). Process mapping as a tool for participative integration of human factors into work system design. *European Journal of Industrial Engineering*, *8*(2), 273. doi:10.1504/EJIE.2014.060477

Lindgreen, A., & Swaen, V. (2009). Corporate social responsibility. *International Journal of Management Reviews*, *12*(1), 1–7. doi:10.1111/j.1468-2370.2009.00277.x

Li, Q., Zhang, Y., Kang, H., Xin, Y., & Shi, C. (2017). Mining association rules between stroke risk factors based on the Apriori algorithm. *Technology and Health Care*, *25*, S197–S205. doi:10.3233/THC-171322 PMID:28582907

Lombardi, M., & Fargnoli, M. (2018). Prioritization of hazards by means of a QFD- based procedure. *International Journal of Safety and Security Engineering*, *8*(2), 342–353. doi:10.2495/SAFE-V8-N2-342-353

Lord, R. G., & Dinh, J. E. (2012). Aggregation Processes and Levels of Analysis as Organizing Structures for Leadership Theory. In D. V. Day & J. Antonakis (Eds.), *The Nature ofLeadership* (pp. 29–65). Thousand Oaks, CA: Sage Publications.

Louhevaara, V. (1999). Is the physical workload equal for ageing and young blue-collar workers? *International Journal of Industrial Ergonomics*, *24*(5), 559–564. doi:10.1016/S0169-8141(98)00061-4

Maak, T. (2007). Responsible leadership, stakeholder engagement, and the emergence of social capital. *Journal of Business Ethics*, *74*(4), 329–343. doi:10.100710551-007-9510-5

Maier, M. W. (1998). Architecting principles for systems-of-systems. *Systems Engineering*, *1*(4), 267–284. doi:10.1002/(SICI)1520-6858(1998)1:4<267::AID-SYS3>3.0.CO;2-D

Makarov, R., Nidziy, N., & Shishkin, G. (2000). *Psychological foundations of didactics in-flight education: A monograph.* Moscow: MAPCHAK.

Maldonado, F., & Ochoa, A. (2016). Un estudio exploratorio de la competitividad en función de espacios de trabajo. *IX Hybrid Intelligent Systems Workshop, MICAI 2016.*

Manero, R., & Manero, J. (1991). *Dos alternativas para el estudio y promoción de la capacidad física de los trabajadores*. MAPFRE Seguridad.

Marina Bischoff, B., & Vianna Dos Santos, F. A. N. (2018). *Ergonomics and organizational communication: Methods for business communication auditing* (Vol. 609). Advances in Intelligent Systems and Computing. doi:10.1007/978-3-319-60477-0_5

Maritan, D., & Panizzolo, R. (2009). Identifying business priorities through quality function deployment. *Marketing Intelligence & Planning*, *27*(5), 714–728. doi:10.1108/02634500910977917

Marsanasco, A. M., & García, P. S. (2013). The management of knowledge and the learning process in SMES clusters: A study case. *Economía, Sociedad y Territorio*, *13*(41), 1–36.

Martínez, R., Vera, J. G. S., & Vera, M. A. M. (2013). Leadership styles and organizational effectiveness in small construction businesses in Puebla, Mexico. *Global Journal of Business Research*, *7*(5), 47–56.

Martínez, V., Ramos, J., & Moliner, C. (2015). *Psicología de las organizaciones*. Madrid: Síntesis.

Maslach, C. (2003). Burnout the cost of caring. Malor Books.

Maslow, A. (1954). *Motivation and Personality*. Harper.

Maslow, A. (2016). *El hombre autorrealizado: hacia una psicología del ser*. Editorial Kairós.

McDonald, N. (2006). Organizational Resilience and Industrial Risk. In E. Hollnagel, D. D. Woods, & N. Leveson (Eds.), *Resilience Engineering: Concepts and Precepts* (pp. 155–180). Aldershot, UK: Ashgate Publishing Limited.

McLaughlin, A. C., Ward, J., & Keene, B. W. (2016). Development of a Veterinary Surgical Checklist. *Ergonomics in Design*, *24*(4), 27–34. doi:10.1177/1064804615621411

Mescon, M., Albert, M., & Khedouri, F. (2008). *Management: A monograph*. Moscow: Vilyams.

Meshkati, N., Tabibzadeh, M., Farshid, A., Rahimi, M., & Alhanaee, G. (2016). People-Technology-Ecosystem Integration: A Framework to Ensure Regional Interoperability for Safety, Sustainability, and Resilience of Interdependent Energy, Water, and Seafood Sources in the (Persian) Gulf. *Human Factors*, *58*(1), 43–57. doi:10.1177/0018720815623143 PMID:26857436

México, Instituto Nacional de Salud Pública. (2016). *Encuesta Nacional de Salud y Nutrición de Medio Camino (ENSANUT MC) Informe final de resultados*. Retrieved from https://www.gob.mx/cms/uploads/attachment/file/209093/ENSANUT.pdf

México, primer lugar en estrés laboral: OMS. (2015). *Diario el Independiente*. Retrieved from https://www.diarioelindependiente.mx/2015/05/mexico-primer-lugar-en-estres-laboral-oms

Meyer, F. (1991). *Motion and Time Study: For Lean Manufacturing*. New York: Prentice Hall.

Michie, S. (2002). Causes and management of stress at work. *Occupational and Environmental Medicine*, *59*(1), 67–72. doi:10.1136/oem.59.1.67 PMID:11836475

Miguez, S. A., Filho, J. F. A. G., Faustino, J. E., & Gonçalves, A. A. (2018). *A successful ergonomic solution based on lean manufacturing and participatory ergonomics* (Vol. 602). Advances in Intelligent Systems and Computing; doi:10.1007/978-3-319-60825-9_27

Mikkelsen, K., Stojanovska, L., Polenakovic, M., Bosevski, M., & Apostolopoulos, V. (2017). Exercise and mental health. *Maturitas*, *106*, 48–56. doi:10.1016/j.maturitas.2017.09.003 PMID:29150166

Mishra, D., & Satapathy, S. (2018). Drudgery Reduction of Farm Women of Odisha by Improved Serrated Sickle. *International Journal of Mechanical Engineering and Technology*, *9*(2), 53–61.

Mital, A., & Pennathur, A. (2004). Advanced technologies and humans in manufacturing workplaces: An interdependent relationship. *International Journal of Industrial Ergonomics*, *33*(4), 295–313. doi:10.1016/j.ergon.2003.10.002

Mital, A., & Shell, R. L. (1984). A compressive metabolic energy model for determining rest allowances for physical tasks. *Journal of Methods-Time Measurement*, *11*, 2–8.

Mittal, V. K., Bhatia, B. S., & Ahuja, S. S. (1996). *A Study of the Magnitude, Causes, and Profile of Victims of Injuries with Selected Farm Machines in Punjab. Final Report of ICAR adhoc Research Project*. Ludhiana, IN: Department of Farm Machinery and Power Engineering, Punjab Agricultural University.

Mohan, D., & Patel, R. (1992). Design of safer agricultural equipment: Application of ergonomics and epidemiology. *International Journal of Industrial Ergonomics*, *10*(4), 301–309. doi:10.1016/0169-8141(92)90097-J

Monfardini, E., Reynaud, A. J., Prado, J., & Meunier, M. (2017). Social modulation of cognition: Lessons from Rhesus macaques relevant to education. *Neuroscience and Biobehavioral Reviews*, *82*, 45–57. doi:10.1016/j.neubiorev.2016.12.002 PMID:27923731

Montalvo, F., Kozachuk, J., Rupp, M. A., Michaelis, J. R., McConnell, D. S., & Smither, J. A. (2016). *Examining methods to induce cognitive fatigue*. Academic Press.

Montero, R. (2000). *Un paso hacia el futuro: el desarrollo de la Macroergonomia, en Factores Humanos*. Academic Press.

Moon, T., Hur, W., Ko, S., Kim, J., & Yoon, S. (2014). Bridging corporate social responsibility and compassion at work: Relations to organizational justice and affective organizational commitment. *Career Development International*, *19*(1), 49–72. doi:10.1108/CDI-05-2013-0060

Moreno Jiménez, B., Rodríguez Carvajal, R., & Escobar Redonda, E. (2001). La evaluación del burnout profesional. Factorialización del MBI-GS. Un análisis preliminar. *Ansiedad y Estrés*, *7*(1), 69–78. Retrieved from https://www.researchgate.net/publication/230596331_La_Evaluacion_del_Burnout_profesional_Factorializacion_del_MBI-GS_Un_analisis_preliminar

Moriano-León, J. A. (2005). La comunicación organizacional. In Psicología de las organizaciones. Pearson, Prentince Hall.

Morse, J. M., Barrett, M., Mayan, M., Olson, K., & Spiers, J. (2002). Verification Strategies for Establishing Reliability and Validity in Qualitative Research. *International Journal of Qualitative Methods*, *1*(2), 13–22. doi:10.1177/160940690200100202

Možina, M. (2018). Arguments in Interactive Machine Learning. *Informatica, 42*(1), 53-59.

Mumford, E. (2006). The story of socio-technical design: Reflections on its successes, failures and potential. *Information Systems Journal*, *16*(4), 317–342. doi:10.1111/j.1365-2575.2006.00221.x

Muralidharan, V., & Sugumaran, V. (2013). Selection Of Discrete Wavelets For Fault Diagnosis Of Monoblock Centrifugal Pump Using The J48 Algorithm. *Applied Artificial Intelligence*, *27*(1), 1–19. doi:10.1080/08839514.2012.721694

Murdaugh J., (1998). *Organizational Effectiveness and Executive Succession: Conclusions About and Implications for Florida's Municipal Police Chiefs*. Senior Leadership Program Publications, SLP-5.

Murphy, D. J. (1992). *Safety and Health for Production Agriculture*. St. Joseph, MI: American Society of Agricultural Engineers.

Murphy, L. A., Huang, Y.-H., Robertson, M. M., Jeffries, S., & Dainoff, M. J. (2018). A sociotechnical systems approach to enhance safety climate in the trucking industry: Results of an in-depth investigation. *Applied Ergonomics*, *66*, 70–81. doi:10.1016/j.apergo.2017.08.002 PMID:28958432

Murphy, L. A., Robertson, M. M., & Carayon, P. (2014). The next generation of macroergonomics: Integrating safety climate. *Accident; Analysis and Prevention*, *68*, 16–24. doi:10.1016/j.aap.2013.11.011 PMID:24368052

Murphy, L. A., Robertson, M. M., Huang, Y.-H., Jeffries, S., & Dainoff, M. J. (2018). A sociotechnical systems approach to enhance safety climate in the trucking industry: Development of a methodology. *Applied Ergonomics*, *66*, 82–88. doi:10.1016/j.apergo.2017.08.001 PMID:28958433

Murrel, K. F. (1965). *Human performance in industry*. New York: Reinhold.

Nachreiner, F. (1995). Standards for ergonomics principles relating to the design of work systems and to mental workload. *Applied Ergonomics*, *26*(4), 259–263. doi:10.1016/0003-6870(95)00029-C PMID:15677027

Nag, P. K., & Nag, A. (2004). Drudgery, accidents and injuries in Indian agriculture. *Industrial Health*, *42*(4), 149–162. doi:10.2486/indhealth.42.149 PMID:15128164

Nakai, M., Chen, D., Nishimura, K., & Miyamoto, Y. (2014). Comparative study of four methods in missing value imputations under missing completely at random mechanism. *Open Journal of Statistics*, *4*(01), 27–37. doi:10.4236/ojs.2014.41004

Naranjo-Flores, A. A., & Ramírez-Cárdenas, E. (2014). Human factors and ergonomics for lean manufacturing applications. In J. L. García-Alcaraz, A. A. Maldonado-Macías, & G. Cortes-Robles (Eds.), *Lean Manufacturing in the Developing World: Methodology, Case Studies and Trends from Latin America* (pp. 281–299). Springer International Publishing Switzerland. doi:10.1007/978-3-319-04951-9_13

Nathanael, D., Zarboutis, N., & Marmaras, N. (2015). Contradiction analysis: Towards a dialectical approach in ergonomics field interventions. *Producao*, *25*(1), 223–231. doi:10.1590/S0103-65132014005000022

National Health and Nutrition Survey. (2012). *Resultados Nacionales*. Cuernavaca, Mexico: Instituto Nacional de Salud Pública. Retrieved from http://ensanut.insp.mx/informes/BajaCalifornia-OCT.pdf

Negruti, A., Hristova, P., Larsen, K., & Krumov, K. (2015). *Positive Organizational Psychology: Advances in Creating Improved Workplaces and Employee Well-Being*. Kassel University Press GmbH.

Neumann, W. P., Ekman, M., & Winkel, J. (2009). Integrating ergonomics into production system development - The Volvo Powertrain case. *Applied Ergonomics*, *40*(3), 527–537. doi:10.1016/j.apergo.2008.09.010 PMID:19019347

Neumann, W. P., & Village, J. (2012). Ergonomics action research II: A framework for integrating HF into work system design. *Ergonomics*, *55*(10), 1140–1156. doi:10.1080/00140139.2012.706714 PMID:22913397

Niebel, B. W., & Freivalds, A. (1998). *Methods, standards, and work design*. New York: WCB/McGraw-Hill.

Nilsson, K. S., Pinzke, S. P., & Lundqvist, P. (2010). Occupational Injuries to Senior Farmers in Sweden. *Journal of Agricultural Safety and Health*, *16*(1), 19–29. doi:10.13031/2013.29246 PMID:20222268

Niskanen, T. (2018). A Resilience Engineering -related approach applying a taxonomy analysis to a survey examining the prevention of risks. *Safety Science*, *101*, 108–120. doi:10.1016/j.ssci.2017.08.016

Noda, T., & Bower, J. L. (1996). Strategy making as iterated processes of resource allocation. *Strategic Management Journal*, *17*(S1), 159–192. doi:10.1002mj.4250171011

Norheim, K. B., Jonsson, G., & Omdal, R. (2011). Biological mechanisms of chronic fatigue. *Rheumatology*, *50*(6), 1009–1018. doi:10.1093/rheumatology/keq454 PMID:21285230

Norton, M., Mochon, D., & Ariely, D. (2010). *The IKEA Effect: When labor leads to love*. Universidad de Hardvard.

Ntuen, C. A. (1999). The application of fuzzy set theory to cognitive workload evaluation of electronic circuit board inspectors. *Human Factors and Ergonomics in Manufacturing & Service Industries*, *9*(3), 291–301. doi:10.1002/(SICI)1520-6564(199922)9:3<291::AID-HFM6>3.0.CO;2-S

O'Riordan, L., & Fairbrass, J. (2014). Managing CSR Stakeholder Engagement: A New Conceptual Framework. *Journal of Business Ethics*, *125*(1), 121–145. doi:10.100710551-013-1913-x

Oah, S., Na, R., & Moon, K. (2018). The Influence of Safety Climate, Safety Leadership, Workload, and Accident Experiences on Risk Perception: A Study of Korean Manufacturing Workers. *Safety and Health at Work*. doi:10.1016/j.shaw.2018.01.008

OECD. (2011). *OECD Guidelines for Multinational Enterprises*. OECD Publishing. Retrieved March 22, 2018, from http://www.oecd-ilibrary.org/governance/oecd-guidelines-for-multinational-enterprises_9789264115415-en

Oficial Mexicana, N. O. R. M. A. (2012). *Servicios básicos de salud. Promoción y educación para la salud en materia alimentaria. Criterios para brindar orientación* (NOM-043-SSA2-2012). Ciudad de México. Retrieved from http://dof.gob.mx/nota_detalle.php?codigo=5285372&fecha=22/01/2013

Oja, P., Ilmarinen, J., & Louhevaara, V. (1982). Heart rate as an estimator of oxygen consumption during manual postal delivery. *Scandinavian Journal of Work, Environment & Health*, *8*(1), 29–36. doi:10.5271jweh.2499 PMID:7134920

Omar, A. (2015). Justicia organizacional. In H. F. Littlewoord & S. A. Vega (Eds.), *Psicología Industrial-Organizacional. Una visión latinoamericana*. Instituto Tecnológico y de Estudios Superiores de Monterrey.

Organización Internacional del Trabajo (OIT). (2016). *Estrés en el trabajo, Día mundial de la seguridad y la salud en el trabajo 28 de abril*. Retrieved from file:///C:/Users/end%20user/Desktop/Papers%20readed/Estres%20en%20el%20trabajo%20OIT.pdf

Orlitzky, M., Schmidt, F. L., & Rynes, S. L. (2003). Corporate social and financial performance: A metaanalysis. *Organization Studies*, *24*(3), 403–441. doi:10.1177/0170840603024003910

Ostroff, C., & Bowen, D. E. (2000). Moving HR to a higher level: HR practices and organizational effectiveness. In K. J. Klein & S. W. J. Kozlowski (Eds.), *Multilevel theory, research, and methods in organizations: Foundations, extensions, and new directions* (pp. 211–266). San Francisco, CA: Jossey-Bass.

Öztürkoğlu, Y. Y., & Bulfin, R. L. (2012). Scheduling jobs to consider physiological factors. *Human Factors and Ergonomics in Manufacturing & Service Industries*, *22*(2), 113–120. doi:10.1002/hfm.20257

Pacholski, L., & Pawlewski, P. (2017). *The usage of simulation technology for macroergonomic industrial systems improvement* (Vol. 487). Advances in Intelligent Systems and Computing; doi:10.1007/978-3-319-41688-5_1

Padula, R. S., Chiavegato, L. D., Cabral, C. M. N., Almeid, T., Ortiz, T., & Carregaro, R. L. (2012). Is occupational stress associated with work engagement? *Work (Reading, Mass.)*, *41*(Supplement 1), 2963–2965. PMID:22317170

Pai, P., Cudney, E. A., & Murray, S. L. (2009). An Analysis of Integrating of Lean and Safety. *Proceedings of the 30th Annual National Conference of the American Society for Engineering Management: ASEM 2009*.

Palferman, D. (2011). Managing Conflict and Stress in the Workplace: Theory and Practice. *Legal Information Management*, *11*(2), 122–125. doi:10.1017/S1472669611000417

Parker, C. P., Baltes, B., Young, S., Altmann, R., LaCost, H., Huff, J., & Roberts, J. E. (2003). Relationships between psychological climate perceptions and work outcomes: A meta-analytic review. *Journal of Organizational Behavior*, *24*(4), 389–416. doi:10.1002/job.198

Patel, S. K., Varma, M. R., & Kumar, A. (2010). Agricultural injuries in Etawah district of Uttar Pradesh in India. *Safety Science*, *48*(2), 222–229. doi:10.1016/j.ssci.2009.08.003

Patel, V. L., & Kannampallil, T. G. (2014). Human factors and health information technology: Current challenges and future directions. *Yearbook of Medical Informatics*, *9*(1), 58–66. doi:10.15265/IY-2014-0005 PMID:25123724

Patterson, M., West, M., Shackleton, V. J., Dawson, J. F., Lawthom, R., Maitlis, S., ... Wallace, A. (2005). Validating the organizational climate measure: Links to managerial practices, productivity and innovation. *Journal of Organizational Behavior, 26*(4), 379–40. doi:10.1002/job.312

Pawitra, T. A., & Tan, K. C. (2003). Tourist satisfaction in Singapore-a perspective from Indonesian tourists. *Managing Service Quality, 13*(5), 339–411. doi:10.1108/09604520310495868

Pearson, E. S. (1931). The Analysis of variance in case of non-normal variation. *Biometrika, 23*(1-2), 114–133. doi:10.1093/biomet/23.1-2.114

Pedersen, E. B. (2006). Making Corporate Social Responsibility (CSR) Operable: How Companies Translate Stakeholder Dialogue into Practice. *Business and Society Review, 111*(2), 137–163. doi:10.1111/j.1467-8594.2006.00265.x

Peiró, J. M. (1983-1984). *Psicología de la organización.* UNED.

Pellegrin, K., & Currey, H. (2011). Demistifyin and improving organizational culture in health-care. In J. Wolf, M. Moir, H. Hanson, L. Friedman, & G. Savage (Eds.), Organization Development in Healthcare: Conversations on Research and Strategies. Emerald Group Publishing.

Pellicer, O., Salvador, A., & Benet, I. A. (2002). Efectos de un estresor académico sobre las respuestas psicológica e inmune en jóvenes. *Psicothema, 14*(2).

Peña-Ochoa, M., & Durán Palacio, N. M. (2016). Justicia organizacional, desempeño laboral y discapacidad. *Revista Colombiana de Ciencias Sociales, 7*(1), 201–222. doi:10.21501/22161201.1540

Phillips, J. B., & Boles, D. B. (2004). Multiple Resources Questionnaire and Workload Profile: Application of competing models to subjective workload measurement. *Proceedings of the Human Factors and Ergonomics Society Annual Meeting, 48*(16), 1963–1967. doi:10.1177/154193120404801636

Piattini Velthuis, M., García Rubio, F., & Caballero Muñoz-Reja, I. (2007). *Calidad de sistemas informáticos.* Alfaomega.

Pirson, M., & Malhotra, D. (2008). Unconventional insights for managing stakeholder trust. *MIT Sloan Management Review.* Retrieved July 18, 2018, from http://www.hbs.edu/research/pdf/08-057.pdf

Portuondo, Y., & Portuondo, J. (2011). La repetibilidad y reproducibilidad en el aseguramiento de la calidad de los procesos de medición. *Tecnología Química, 30*(2), 117-121. Retrieved from http://www.redalyc.org/articulo.oa?id=445543770014

Post, J., Preston, L., & Sachs, S. (2002). *Redefining the corporation: Stakeholder management and organizational wealth.* Stanford, CA: Stanford University Press.

Pourdehnad, J., Wexler, E. R., & Wilson, D. V. (2011). *Systems & Design Thinking: A Conceptual Framework for TeirIntergration.* Organizational Dynamics Working Papers.

Prado, J. F. U. (2014). *Clima y ambiente organizacional: trabajo, salud y factores psicosociales.* Editorial el manual moderno.

Prasanna Kumar, G. V., & Dewangan, K. N. (2009). Agricultural accidents in north eastern region of India. *Safety Science, 47*(2), 199–205. doi:10.1016/j.ssci.2008.03.007

Punnett, L., Cherniack, M., Henning, R., Morse, T., & Faghri, P. (2009). *A conceptual framework for integrating workplace health promotion and occupational ergonomics programs.* Academic Press.

Puteh, F., Kaliannan, M., & Alam, N. (2015). Learning for professional development via peers: A System Theory approach. *Procedia: Social and Behavioral Sciences, 172*, 88–95.

Quiñones, V., Otarla, S., Ramos, F., & Aguilera, N. (2008). *Organizational Development*. Retrieved from Internet on January 18, 2018: https://www.bing.com/search?q=Desarrollo+Organizacional.+Blogspot%2C+Grupo15.&PC=U316&FORM=CHROMN

Randhir, H. N., Gupta, R., & Selokar, G. R. (2013). Extract Knowledge and Association Rule from Free Log Data using an Apriori Algorithm. *International Journal Of Advanced Computer Research*, *3*(12), 191–196.

Rasmussen, J. (1986). *Information Processing and Human-Machine Interaction: An Approach to Cognitive Engineering. North Holland series in system science and engineering 12*. New York: Elsevier Science Publishing Co. Inc.

Rasmussen, J. (1997). Risk management in a dynamic society: A modelling problem. *Safety Science*, *27*(2-3), 183–213. doi:10.1016/S0925-7535(97)00052-0

Rasmussen, J., Pejtersen, A. M., & Goodstein, L. P. (1994). *An Introduction to Systems Engineering*. New York: John Wiley & Sons, Inc.

Rasmussen, J., & Svedung, I. (2000). *Proactive Risk Management in a Dynamic Society*. Karlstad: Räddningsverket.

Rautiainen, R. H., & Reynolds, S. J. (2002). Mortality and morbidity in agriculture in the United States. *Journal of Agricultural Safety and Health*, *8*(3), 259–276. doi:10.13031/2013.9054 PMID:12363178

Realyvásquez, A., & Maldonado-Macías, A. (2018). Measuring the Complex Construct of Macroergonomic Compatibility: A Manufacturing System Case Study. *Complexity*, *2018*, 1–10. doi:10.1155/2018/7374307

Realyvásquez, A., Maldonado-Macías, A. A., García-Alcaraz, J. L., & Arana, A. (2018). *Macroergonomic compatibility index for manufacturing systems. A case study* (Vol. 606). Advances in Intelligent Systems and Computing. doi:10.1007/978-3-319-68684-4

Realyvásquez, A., Maldonado-Macías, A. A., García-Alcaraz, J. L., & Blanco-Fernández, J. (2015). Effects of Organizational Macroergonomic Compatibility Elements over Manufacturing Systems' Performance. *Procedia Manufacturing*, *3*, 5715–5722. doi:10.1016/j.promfg.2015.07.807

Realyvásquez, A., Maldonado-Macías, A. A., García-Alcaraz, J., Cortés-Robles, G., & Blanco-Fernández, J. (2016). Structural Model for the Effects of Environmental Elements on the Psychological Characteristics and Performance of the Employees of Manufacturing Systems. *International Journal of Environmental Research and Public Health*, *13*(1), 104. doi:10.3390/ijerph13010104 PMID:26742054

Realyvásquez-Vargas, A., García-Alcaraz, J. L., & Blanco-Fernández, J. (2016). Desarrollo y validación de un cuestionario de compatibilidad macroergonómica. *Contaduría y Administración*, *61*(3), 478–498. doi:10.1016/j.cya.2016.04.002

Realyvásquez-Vargas, A., Maldonado-Macias, A. A., & García-Alcaraz, J. L. (2017). *Macroergonomics for Manufacturing Systems: An Evaluation Approach*. Springer.

Reena, R., & Selvi, R. T. (2016). Analyzing software defect prediction using k-means and expectation maximization clustering algorithm based on genetic feature selection. *Journal on Software Engineering, 11*(1), 28-36.

Reichers, A. E., & Schneider, B. (1990). Climate and culture: An evolution of constructs. In B. Schneider (Ed.), *Organizational climate and culture* (pp. 5–39). San Francisco, CA: Jossey-Bass.

Reid, G. B., & Nygren, T. E. (1988). The Subjective Workload Assessment Technique: A Scaling Procedure for Measuring Mental Workload. In P. A. Hancock & N. Meshkati (Eds.), *Advances in Psychology* (Vol. 52, pp. 185–218). North-Holland; doi:10.1016/S0166-4115(08)62387-0

Reynaldo, J., & Santos, A. (1999). Cronbach's alpha: A tool for assessing the reliability of scales. *Journal of Extension, 37*(2), 11–14.

Rezaei, M., & Gunpinar, E. (2018). A k-means clustering based shape retrieval technique for 3d mesh models. *Selcuk University Journal of Engineering, Science & Technology, 6*(1), 114-128. doi:10.15317/Scitech.2018.119

Richardson J., (2014). Double Loop Learning: A Powerful Force for Organizational Excellence. *Excerpt from PNSQC 2014 Proceedings.*

Robertson, M. M., Huang, Y. H., & Lee, J. (2017). Improvements in musculoskeletal health and computing behaviors: Effects of a macroergonomics office workplace and training intervention. *Applied Ergonomics, 62*, 182–196. doi:10.1016/j.apergo.2017.02.017 PMID:28411728

Robertson, M. M., Huang, Y. H., O'Neill, M. J., & Schleifer, L. M. (2008). Flexible workspace design and ergonomics training: Impacts on the psychosocial work environment, musculoskeletal health, and work effectiveness among knowledge workers. *Applied Ergonomics, 39*(4), 482–494. doi:10.1016/j.apergo.2008.02.022 PMID:18462704

Rodríguez Ruíz, Y., & Pérez Mergarejo, E. (2016). Macro-ergonomic diagnosis of Colombian organizations using ergonomics maturity model | Diagnóstico macro ergonómico de organizações colombianas usando o Modelo de Madureza de Ergonomia | Diagnóstico macroergonómico de organizaciones colombianas con el Mod. *Revista Ciencias de la Salud, 14*, 11–25.

Rolo González, G., Díaz Cabrera, D., & Hernández Fernaud, E. (2009). Desarrollo de una Escala Subjetiva de Carga Mental de Trabajo (ESCAM). *Revista de Psicología del Trabajo y de las Organizaciones, 25*(1), 29–37. doi:10.4321/S1576-59622009000100004

Roscoe, A. H., & Ellis, G. A. (1990). *A subjective rating scale for assessing pilot workload in flight: A decade of practical use*. Rarnborough, UK: Royal Aircraft Establishment.

Rubio, S., Díaz, E. M., & Martín, J. (2001). Aspectos metodológicos de la evaluación subjetiva de la carga mental de trabajo. *Archivos de Prevencion de Riesgos Laborales, 4*(4), 160–168.

Rubio, S., Díaz, E., Martín, J., & Puente, J. M. (2004). Evaluation of Subjective Mental Workload: A Comparison of SWAT, NASA-TLX, and Workload Profile Methods. *Applied Psychology, 53*(1), 61–86. doi:10.1111/j.1464-0597.2004.00161.x

Russell, A. L. (1979). *Redesigning the future* (13th ed.). Editorial Limusa.

Rutenfranz, J., Ilmarinen, J., Klimmer, F., & Kylian, H. (1990). Work load and demanded physical performance capacity under different industrial working conditions. Fitness for the Aged, Disabled and Industrial Worker. *International Series on Sport Sciences, 20*, 217–238.

Salanova, M. (2009). Prevención de riesgos laborales en tiempos de crisis. *Gestión Práctica de Riesgos Laborales, 58.*

Salanova, M. (2008). Organizaciones saludables y desarrollo de recursos humanos. *Estudios Financieros, 303*, 179–214.

Salanova, M., Martínez, I. M., & Llorens, S. (2005). Psicología Organizacional Positiva. In *Psicología de la Organización* (pp. 349–376). Madrid: Pearson Prentice Hall.

Salanova, M., Martínez, I. M., & Llorens, S. (2014). Una mirada más "positiva" a la salud ocupacional desde la psicología organizacional positiva en tiempos de crisis: Aportaciones desde el equipo de investigación WoNT. *Papeles del Psicólogo, 35*(1).

Salas-Arbeláez, L., Solarte, M. G., & Vargas, G. M. (2017). Efecto de la cultura organizacional en el rendimiento de las PYMES de Cali. *Suma de Negocios, 8*(18), 88–95. doi:10.1016/j.sumneg.2017.11.006

Salas, E., Grossman, R., Hughes, A. M., & Coultas, C. W. (2015). Measuring team cohesion: Observations from the science. *Human Factors*, *57*(3), 365–374. doi:10.1177/0018720815578267 PMID:25875429

Salgado, P., & Mejía, R. (2008). Estrés en ejecutivos de medianas y grandes empresas mexicanas: Un enfoque de desarrollo humano organizacional. *Estudios Gerenciales*, *24*(108), 15–36. doi:10.1016/S0123-5923(08)70042-1

Sandoval, J., & Ramos, J. (1995). *Tablas de Indicadores Antropométricos y Fisiológicos en Trabajadores Mexicanos: Instituto Mexicano del Seguro Social*. IMSS.

SAS. (2005). *SAS User's Guide. Version 9*. Cary: SAS Institute Inc.

Satapathy, S. (2014). ANN, QFD and ISM approach for framing electricity utility service in India for consumer satisfaction. *International Journal of Services and Operations Management*, *18*(4), 404–428. doi:10.1504/IJSOM.2014.063243

Savinainen, M., Nygård, C. H., & Ilmarinen, J. (2004). Workload and physical capacity among ageing municipal employees—a 16-year follow-up study. *International Journal of Industrial Ergonomics*, *34*(6), 519–533. doi:10.1016/j.ergon.2004.06.006

Schaufeli, W. B., Salanova, M., González-Romá, V., & Bakker, A. B. (2002). The measurement of engagementand burnout and: A confirmative analytic approach. *Journal of Happiness Studies*, *3*(1), 71–92. doi:10.1023/A:1015630930326

Sealetsa, O. J., & Thatcher, A. (2011). Ergonomics issues among sewing machine operators in the textile manufacturing industry in Botswana. *Work (Reading, Mass.)*, *38*(3), 279–289. doi:10.3233/WOR-2011-1131 PMID:21447888

Secretaria de salud (SS). (2013). *Servicios básicos de salud. Promoción y educación para la salud en materia alimentaria. Criterios para brindar orientación* (NOM-043-SSA2-2012). Diario Oficial de la Federación.

Sedgwick, P., & Greenwood, N. (2015). Understanding the Hawthorne effect. *British Medical Journal*, *351*, h4672. doi:10.1136/bmj.h4672 PMID:26341898

Seim, R., Broberg, O., & Andersen, V. (2014). Ergonomics in design processes: The journey from ergonomist toward workspace designer. *Human Factors and Ergonomics in Manufacturing*, *24*(6), 656–670. doi:10.1002/hfm.20508

Seligman, M., & Csikszentmihalyi, M. (2000). Positive Psychology: An introduction. *The American Psychologist*, *55*(1), 5–14. doi:10.1037/0003-066X.55.1.5 PMID:11392865

Selye, H. (1936). A síndrome produced by diverse nocouous agents. *Nature*, *138*(3479), 32. doi:10.1038/138032a0

Selye, H. (1950). *Stress: the physiology and pathology of exposure to stress*. Montreal: Acta Inc.

Senge, P. (1994). *The Fifth Discipline: The art and practice of the learning organization*. Random House.

Senge, P. (1998). *La quinta disciplina*. Granica.

Senge, P. M. (1994). *The Fifth Discipline: the Art & Practice of the Learning Organization* (2nd ed.). New York: Bantam Doubleday/Currency Publishing Group.

Senge, S., Smith, B., Nina Kruschwitz, N., Laur, J., & Schley, S. (2008). *The necessary revolution: How individuals and organizations are working together to create a sustainable world*. London: Nicholas Brealey Publishing.

Sepahvand, R., Aghdam, S. R., Alavi, S. A., & Rezaei, M. (2016). Investigating the Impact of Organizational Learning Capability on Organizational Intelligence in Knowledge Based Organizations. *Proceedings of International Conference on Science, Technology, Humanities and Business Management*.

Serrano, M. A., & Costa, R. (2018). *Stressin the stress or the complexity of the human factor:* psychobiological *consequences of distress*. In J. L. García-Alcaraz, G. Alor-Hernández, A. A. Maldonado-Macías, & C. Sánchez-Ramírez (Eds.), *New Perspectives on applied industrial tools and techniques* (pp. 431–447). Springer. doi:10.1007/978-3-319-56871-3_21

Serrat, O. (2010). *A Primer on organizational learning*. Asian Development Bank.

Shah, R., & Ward, P. (2003). Lean manufacturing: Context, practice bundles, and performance. *Journal of Operations Management*, *21*(2), 129–149. doi:10.1016/S0272-6963(02)00108-0

Shephard, R. J. (2000). Aging and productivity: Some physiological issues. *International Journal of Industrial Ergonomics*, *25*(5), 535–545. doi:10.1016/S0169-8141(99)00036-0

Sherehiy, B., & Karwowski, W. (2014). The relationship between work organization and workforce agility in small manufacturing enterprises. *International Journal of Industrial Ergonomics*, *44*(3), 466–473. doi:10.1016/j.ergon.2014.01.002

Shmelova, T., & Sikirda, Y. (2016). Calculation the scenarios of the flight situation development using GERT's and Markov's networks. In *Proceedings of the 5th International Scientific and Practical Conference "Management of High-Speed Moving Objects and Professional Training of Operators of Complex Systems"*. Kirovograd: KFA NAU.

Shmelova, T., & Sikirda, Y. (2012). Analysis of human-operator's decision-making in the air navigation system. *Radioelectronic and Computer Systems*, *7*(59), 319–324.

Shmelova, T., & Sikirda, Y. (2017). Model of optimization of transport flows taking into account combinatorics and regression analysis methods. In *Proceedings of the 13th International Scientific and Technical Conference "AVIA-2017"*. Kyiv: NAU.

Shmelova, T., Sikirda, Y., Assaul, A., & Stasiuk, O. (2015). Structural Analysis of Management Environment of Aviation Enterprise from the Point of Systematic Approach. *Proceedings of the National Aviation University*, *2*, 3–12.

Shmelova, T., Sikirda, Y., & Assaul, Y. (2015). Influence of Factors of Management Environment of Aviation Enterprise on the Level of Aviation Safety. *Technological Audit and Production Reserves*, *2/3*(22), 17–24. doi:10.15587/2312-8372.2015.41440

Shmelova, T., Sikirda, Y., & Jafarzade, T. (2013). Models of flight situations development while decision-making by air navigation system's human-operator. *Information Processing Systems*, *8*(115), 136–142.

Shmelova, T., Sikirda, Y., & Sunduchkov, K. (2013). A sociotechnical analysis of air navigation system. *Science and Technics of Air Forces of Armed Forces of Ukraine*, *4*(13), 34–39.

Shvartz, E., & Reibold, R. (1990). Aerobic fitness norms for males and females aged 6 to 75 years: A review. *Aviation, Space, and Environmental Medicine*, *6*, 3–11. PMID:2405832

Siemieniuch, C. E., & Sinclair, M. A. (2014). Extending systems ergonomics thinking to accommodate the sociotechnical issues of Systems of Systems. *Applied Ergonomics*, *45*(1), 85–98. doi:10.1016/j.apergo.2013.03.017 PMID:24011651

Sikirda, Yu., Shmelova, T., & Tkachenko, D. (2017). Evaluation of the organizational factors influence on flight safety in air traffic control. *Collections of Scientific Works of the Kharkiv National University of Air Force, No.*, *3*(52), 39–44.

Silva, L. D., Nickel, E. M., & Dos Santos, F. A. N. V. (2018). *Evaluation of macroergonomic methods for the application of organization analyzes in startups* (Vol. 588). Advances in Intelligent Systems and Computing; doi:10.1007/978-3-319-60582-1_33

Silva, L. D., Nickel, E. M., & dos Santos, F. A. N. V. (2018). Evaluation of Macroergonomic Methods for the Application of Organization Analyzes in Startups. In F. Rebelo & M. Soares (Eds.), *Advances in Ergonomics in Design. AHFE 2017. Advances in Intelligent Systems and Computing* (Vol. 588). Cham: Springer.

Silverman, M. N., Heim, C. M., Nater, U. M., Marques, A. H., & Sternberg, E. M. (2010). Neuroendocrine and immune contributors to fatigue. *PM & R*, *2*(5), 338–346. doi:10.1016/j.pmrj.2010.04.008 PMID:20656615

Simon, H. A. (2000). Barriers and bounds to Rationality. *Structural Change and Economic Dynamics*, *11*(1-2), 243–253. doi:10.1016/S0954-349X(99)00022-3

Slomp, J., & Ruël, G. C. (2000). *A socio-technical approach for the design of a production control system: towards controllable production units*. University of Groningen.

Smith, M. J., & Carayon, P. (1995). *New technology, automation, and work organization: stress problems and improved technology implementation strategies*. Retrieved from https://www.scopus.com/record/display.uri?eid=2-s2.0-0029582621&origin=resultslist&sort=cp-f&src=s&st1=Macroergonomics&nlo=&nlr=&nls=&sid=6ebdcdcaef482f3889ab4f6a72d2e8f7&sot=b&sdt=cl&cluster=scopubyr %2C%222013%22%2Ct%2C%222012%22%2Ct%2C%222011%22%2Ct%2C%222010%22%2Ct%2C%222009%22%2Ct%2C%222008%22%2Ct%2C%222007%22%2Ct%2C%222006%22%2Ct%2C%222005%22%2Ct%2C%222004%22%2Ct%2C%222003%22%2Ct%2C%222002%22%2Ct%2C%222001%22%2Ct%2C%222000%22%2Ct%2C%221999%22%2Ct%2C%221998%22%2Ct%2C%221997%22%2Ct%2C%221996%22%2Ct%2C%221995%22%2Ct%2C%221994%22%2Ct%2Bscosubtype%2C%22ar%22%2Ct%2C%22ch%22%2Ct&sessionSearchId=6ebdcdcaef482f3889ab4f6a72d2e8f7&relpos=2&citeCnt=69

Smith, T. J., Schoenbeck, K., & Clayton, S. (2009). *Staff perceptions of work quality of a neonatal intensive care unit before and after transition from an open bay to a private room design*. Academic Press. doi:10.3233/WOR-2009-0868

Smolander, J., Juuti, T., Kinnunen, M. L., Laine, K., Louhevaara, V., Männikkö, K., & Rusko, H. (2008). A new heart rate variability-based method for the estimation of oxygen consumption without individual laboratory calibration: Application example on postal workers. *Applied Ergonomics*, *39*(3), 325–331. doi:10.1016/j.apergo.2007.09.001 PMID:17950689

Sonnentag, S., Mojza, E. J., Demerouti, E., & Bakker, A. B. (2012). Reciprocal relations between recovery and work engagement: The moderating role of job stressors. *The Journal of Applied Psychology*, *97*(4), 842–853. doi:10.1037/a0028292 PMID:22545619

Sorensen, L., & Valqui, R.V. (2006). Evaluating six soft approaches. *Economic Analysis Working Papers, 7*(9), 1-20.

Sorensen, J. (2002). The strength of corporate culture and the reliability of firm performance. *Administrative Science Quarterly*, *47*(1), 70–91. doi:10.2307/3094891

Stanton, N., Hedge, A., Brookhuis, K., Salas, E., & Hendrick, H. W. (Eds.). (2004). *Handbook of Human Factors and Ergonomics Methods*. CRC Press. doi:10.1201/9780203489925

Stanton, N., Salmon, P. M., & Rafferty, L. A. (2013). *Human Factors Methods: A Practical Guide for Engineering and Design*. Ashgate Publishing, Ltd. doi:10.1201/9781315587394

Staub, F., & Bogousslavsky, J. (2001). Post-stroke depression or fatigue. *European Neurology*, *45*(1), 3–5. doi:10.1159/000052081 PMID:11205620

Steege, L. M., Pasupathy, K. S., & Drake, D. A. (2018). A work systems analysis approach to understanding fatigue in hospital nurses. *Ergonomics*, *61*(1), 148–161. doi:10.1080/00140139.2017.1280186 PMID:28064733

Stein, D. J., He, Y., Phillips, A., Sahakian, B. J., Williams, J., & Patel, V. (2017). Global mental health and neuroscience: Potential synergies. *Lancet*, *2*(2), 178–185. PMID:26359754

Sterman, J. D. (1994). Learning in and about complex systems. *System Dynamics Review*, *10*(2-3), 291–330. doi:10.1002dr.4260100214

Stringfellow, M. V. (2010). *Accident analysis and hazard analysis for human and organizational factors* (Doctoral dissertation). Massachusetts Institute of Technology. Retrieved March 22, 2018, from http://sunnyday.mit.edu/safer-world/MaggieStringfellowDissertation.pdf

Sunstein, C. R., & Thaler, R. H. (2003). Libertarian paternalism is not an oxymoron. *The University of Chicago Law Review. University of Chicago. Law School*, *70*(4), 1159–1202. doi:10.2307/1600573

Szajowski, K., & Tamaki, M. (2016). Shelf life of candidates in the generalized secretary problem. *Operations Research Letters*, *44*(4), 498–502. doi:10.1016/j.orl.2016.05.002

Tang, K.-H., Koubek, R. J., Lightner, N. J., & Salvendy, G. (1999). Development and validation of a theoretical model for cognitive skills acquisition. *International Journal of Industrial Ergonomics*, *24*(1), 25–38. doi:10.1016/S0169-8141(98)00085-7

Tan, K. C., & Pawitra, T. A. (2001). Integrating SERVQUAL and Kana's model into QFD for service excellence development. *Managing Service Quality*, *11*(6), 418–430. doi:10.1108/EUM0000000006520

Tavares, A. S., Albuquerque, L. W. N. D., da Silva, J. C., Souza Júnior, C. B., Gálvez, C., & Soares, M. (2015). Work at Height: Neglect or Improvisation in Civil Construction in Brazil and Uruguay? *Procedia Manufacturing*, *3*, 6109–6115. doi:10.1016/j.promfg.2015.07.763

Taveira, A. D., James, C. A., Karsh, B. T., & Sainfort, F. (2003). Quality management and the work environment: An empirical investigation in a public sector organization. *Applied Ergonomics*, *34*(4), 281–291. doi:10.1016/S0003-6870(03)00054-1 PMID:12880738

Taylor, F. W. (1984). *Management científico*. Madrid: Orbis.

Tecnológico Nacional de México. (1999). *Technological Institutes Procedures and Guidelines Manual*. Academic Press.

Tecnológico Nacional de México. (2010). *Authorized Plans and Programs for ITCJ, 2012, 2010 y 2016*. Academic Press.

Tecnológico Nacional de México. (2015). *Academic-administrative Guidelines Manual*. Academic Press.

Teddlie, C. (2005). Methodological issues related to causal studies of leadership: A mixed methods perspective from the USA. *Educational Management Administration & Leadership*, *33*(2), 211–217. doi:10.1177/1741143205051054

Tello, M. L., Eslava, H. J., & Tobías, L. B. (2013). Análisis y evaluación del nivel de riesgo en el otorgamiento de créditos financieros utilizando técnicas de minería de datos. *Visión Electrónica*, *7*(1), 13–26.

Thaler, R. H. (2016). Behavioral economics: Past, present, and future. *The American Economic Review*, *106*(7), 1577–1600. doi:10.1257/aer.106.7.1577

Thatcher, A., Waterson, P., Hancock, P., Davis, M. C., Zink, K. J., & Hilliard, A. (2016). This changes everything: Macroergonomics and the future of sustainability. In *Proceedings of the Human Factors and Ergonomics Society* (pp. 870–874). Academic Press. 10.1177/1541931213601199

Thatcher, A., Waterson, P., Todd, A., & Moray, N. (2017). State of Science: Ergonomics and global issues. *Ergonomics*, *61*(2), 197–213. doi:10.1080/00140139.2017.1398845 PMID:29076757

Thies, C. G. (2002). A pragmatic guide to qualitative historical analysis in the study of international relations. *International Studies Perspectives*, *3*(4), 351–372. doi:10.1111/1528-3577.t01-1-00099

Thomas, K., Goodall, S., Stone, M., Howatson, G., Gibson, A. S. C., & Ansley, L. (2015). Central and peripheral fatigue in male cyclists after 4-, 20-, and 40-km time trials. *Medicine and Science in Sports and Exercise*, *47*(3), 537–546. doi:10.1249/MSS.0000000000000448 PMID:25051388

Tiwari, P. S., Gite, L. P., Dubey, A. K., & Kot, L. S. (2002). Agricultural injuries in Central India: Nature, magnitude and economic impact. *Journal of Agricultural Safety and Health*, *8*(1), 95–111. doi:10.13031/2013.7221 PMID:12002378

Tolvanen, J. P. (1998). *Incremental Method Engineering with Modeling Tools*. University of Jyväskylä.

Tran, Y., Craig, A., Wijesuriya, N., & Nguyen, H. (2010). Improving Classification Rates for Use in Fatigue Countermeasure Devices using Brain Activity. In *Proceedings of 32nd Annual International Conference of the IEEE Engineering in Medicine and Biology Society* (pp. 4460-4463). Buenos Aires, Argentina: IEEE.

Trist, E., & Bamforth, K. (1951). Some Social and Psychological Consequences of the Longwall Method of Coal Getting. *Human Relations*, *4*(1), 3–38. doi:10.1177/001872675100400101

Tsang, P. S., & Velazquez, V. L. (1996). Diagnosticity and multidimensional subjective workload ratings. *Ergonomics*, *39*(3), 358–381. doi:10.1080/00140139608964470 PMID:8849491

Tünnerman, C. (2000). Pertinencia social y principios básicos para orientar el diseño de políticas de educación superior. *Educación Superior Sociedad*, 181-196. Recuperado de http://ess.iesalc.unesco.org.ve/index.php/ess/article/viewArticle/364

Tversky, A., & Kahneman, D. (1974). Judgment under Uncertainty: Heuristics and Biases. *Science*, *185*(4157), 1124–1131. doi:10.1126cience.185.4157.1124 PMID:17835457

Uhl-Bien, M., Maslyn, J., & Ospina, S. (2012). The nature of relational leadership: A multitheoretical lens on leadership relationships and processes. In D. V. Day & J. Antonakis (Eds.), *The Nature of Leadership* (pp. 289–330). Thousand Oaks, CA: Sage Publications.

Universidad Veracruzana. (2014). *Plan de Desarrollo del Centro de Idiomas Poza Rica 2009-2013*. Universidad Veracruzana. Retrieved from Internet on August 13, 2018: http://www.uv.mx/pozarica/ci/files/2012/12/C.-I.-POZA-RICA-CORREGIDO.pd

Uusitalo, A., Mets, T., Martinmäki, K., Mauno, S., Kinnunen, U., & Rusko, H. (2011). Heart rate variability related to effort at work. *Applied Ergonomics*, *42*(6), 830–838. doi:10.1016/j.apergo.2011.01.005 PMID:21356531

Vaca, C. A. M., Vaca, L. O. M., & Quintero, J. N. (2015). El clima organizacional y la satisfacción laboral: Un análisis cuantitativo riguroso de su relación. *AD-Minister*, (26): 5–15.

Valdez, R. S., Holden, R. J., Hundt, A. S., Marquard, J. L., Montague, E., Nathan-Roberts, D., & Or, C. K. (2014). The Work and Work Systems of Patients: A New Frontier for Macroergonomics in Health Care. *Proceedings of the Human Factors and Ergonomics Society Annual Meeting*, *58*(1), 708–712. doi:10.1177/1541931214581165

Valqui, R.V. (2005a). *Operational Research: A multidisciplinary discipline*. Informatics and Mathematical Modelling, Technical University of Denmark, DTU.

Valquil, R. V. (2005b). Soft OR approaches. *Engevista*, *7*(1), 4–20.

van Dijk, T. A. (1995). Discourse Semantics and Ideology. *Discourse & Society*, *6*(2), 243–289. doi:10.1177/0957926595006002006

van Luijk, H. J. L. (1997). Business ethics in Western and Northern Europe: A search for effective alliances. *Journal of Business Ethics*, *16*(14), 1579–1587. doi:10.1023/A:1005819216111

Veen, P., & Korver, T. (1998). Theories of organization. In Handbook of work and organizational psychology: Vol. 3: Organizational psychology. Psychology Press.

Vega, M. M. C., Gálvez, S. A. H., & Santamaría, J. (2017). Organizational Climate and Psychological Health: An Organizational Duality. *Dimensión empresarial, 15*(1), 63-76.

Vergara Garibaldi María Olivia. (2012), *Plan estratégico para el centro de actualización profesional e idiomas del Instituto Tecnológico de la Paz*. División de Estudios de Posgrado e Investigación, Thesis of Administration degree. México, 2012.

Vicente, K. J., & Rasmussen, J. (1992). Ecological Interface Design: Theoretical Foundations. *IEEE Transactions on Systems, Man, and Cybernetics*, *22*(4), 589–606. doi:10.1109/21.156574

Vidulich, M. A. (1989). The Use of Judgment Matrices in Subjective Workload Assessment: The Subjective Workload Dominance (SWORD) Technique. *Proceedings of the Human Factors Society Annual Meeting*, *33*(20), 1406–1410. 10.1177/154193128903302009

Vidulich, M. A., & Tsang, P. S. (2012). Mental Workload and Situation Awareness. In G. Salvendy (Ed.), *Handbook of Human Factors and Ergonomics* (pp. 243–273). John Wiley & Sons, Inc. doi:10.1002/9781118131350.ch8

Vidullch, M. A., Ward, G. F., & Schueren, J. (1991). Using the Subjective Workload Dominance (SWORD) Technique for Projective Workload Assessment. *Human Factors*, *33*(6), 677–691. doi:10.1177/001872089103300605

Village, J., Salustri, F. A., & Neumann, W. P. (2016). Cognitive mapping links human factors to corporate strategies. *European Journal of Industrial Engineering*, *10*(1), 1–20. doi:10.1504/EJIE.2016.075126

Vincent, C., Ward, J., & Langdon, P. (2012). Unraveling complex system. *The Ergonomist*, *506*, 12–13.

Voaklander, K. D., Kelly, D. C., Rowe, B. H., Schopflocher, D. P., Svenson, L., Yiannakoulias, N., & Pickett, W. (2006). Pain, medication, and injury in older farmers. *American Journal of Industrial Medicine*, *49*(5), 374–382. doi:10.1002/ajim.20292 PMID:16526061

Wachter, J. K., & Yorio, P. L. (2014). A system of safety management practices and worker engagement for reducing and preventing accidents: An empirical and theoretical investigation. *Accident; Analysis and Prevention*, *68*, 117–130. doi:10.1016/j.aap.2013.07.029 PMID:23993683

Wan, J. J., Qin, Z., Wang, P.-Y., Sun, Y., & Liu, X. (2017). Muscle fatigue: General understanding and treatment. *Experimental & Molecular Medicine*, *49*(10), e384. doi:10.1038/emm.2017.194 PMID:28983090

Waring, S. P. (2016). *Taylorism transformed: Scientific management theory since 1945*. UNC Press Books.

Waterson, P., Robertson, M. M., Carayon, P., Hoonakker, P., Holden, R., Hettinger, L., … Waterson, P. (2014). Macroergonomics and Sociotechnical Methods Current and Future Directions. *Proceedings of the Human Factors and Ergonomics Society Annual Meeting*, *58*(1), 1536–1540. 10.1177/1541931214581320

Waterson, P., Robertson, M. M., Cooke, N. J., Militello, L., Roth, E., & Stanton, N. A. (2015). Defining the methodological challenges and opportunities for an effective science of sociotechnical systems and safety. *Ergonomics*, *58*(4), 565–599. doi:10.1080/00140139.2015.1015622 PMID:25832121

WCED. (1987). *Our common future. World Commission on Environment and Development*. Oxford, UK: Oxford University Press.

Wei-Yuan, G., Shang-Ping, L., Chwei-Jen, F., & Chin-Fang, Y. (2013). Effects of organizational justice on organizational citizenship behaviors: Mediating effects of institutional trust and affective commitment. *Psychological Reports*, *112*(3), 818–834. doi:10.2466/01.21.PR0.112.3.818-834 PMID:24245075

Welborn, T. A., Dhaliwal, S. S., & Bennett, S. A. (2003). Waist-hip ratio is the dominant risk factor predicting cardiovascular death in Australia. *The Medical Journal of Australia*, *179*, 580–585. PMID:14636121

Welch, B. L. (1938). The significance of the difference between two means when the population variances are unequal. *Biometrika*, *29*(3-4), 350–362. doi:10.1093/biomet/29.3-4.350

Wickens, C. D. (1984). *The Multiple Resources Model of Human Performance: Implications for Display Design.* Urbana, IL: Illinois Univ. at Urbana. Retrieved from http://www.dtic.mil/docs/citations/ADP004516

Wickens, C. D. (1992). *Engineering Psychology and Human Performance* (2nd ed.). New York: Harper Collins.

Wierwille, W. W., & Casali, J. G. (1983). A Validated Rating Scale for Global Mental Workload Measurement Applications. *Proceedings of the Human Factors Society Annual Meeting*, *27*(2), 129–133. 10.1177/154193128302700203

Wilson, J. R. (2000). Fundamentals of ergonomics in theory and practice. *Applied Ergonomics*, *31*(6), 557–567. doi:10.1016/S0003-6870(00)00034-X PMID:11132041

Wilson, J. R. (2014). Fundamentals of systems ergonomics/human factors. *Applied Ergonomics*, *45*(1), 5–13. doi:10.1016/j.apergo.2013.03.021 PMID:23684119

Wilson, M. G., Dejoy, D. M., Vandenberg, R. J., Richardson, H. A., & McGrath, A. L. (2004). Work characteristics and employee health and wellbeing: Test of a model of healthy work organization. *Journal of Occupational and Organizational Psychology*, *77*(4), 565–588. doi:10.1348/0963179042596522

Womack, J. P., & Jones, D. T. (1996). *Lean thinking: banish waste and create wealth in your corporation.* New York: Simon and Schuster.

Wood, S. J., Stride, C. B., Wall, T. D., & Clegg, C. W. (2004). Revisiting the use and effectiveness of modern management practices. *Human Factors and Ergonomics in Manufacturing*, *14*(4), 415–432. doi:10.1002/hfm.20006

World Health Organization (WHO). (2006). *Prevención del suicidio. Un instrumento en el trabajo. Departamento de Salud Mental y Abuso de Sustancias.* Manejo de Trastornos Mentales y Cerebrales.

World Health Organization (WHO). (2012). *Estadísticas sanitarias mundiales.* Retrieved from http://www.who.int/gho/publications/world_health_statistics/2012/es/

World Health Organization (WHO). (2016). *Obesidad y Sobrepeso, datos y cifras.* Retrieved from http://www.who.int/mediacentre/factsheets/fs311/es/

World Health Organization. (2001). *International Classification of Functioning, Disability and Health: ICF.* Geneva: WHO.

Wreathall, J. (2011). Monitoring - A Critical Ability in Resilience Engineering. In E. Hollnagel, J. Paries, D. Woods, & J. Wreathall (Eds.), *Resilience Engineering in Practice: A Guidebook* (pp. 61–68). Surrey, UK: Ashgate Publishing Limited.

Xiang, H., Wang, Z., Stallones, L., Keefe, T. J., Huang, X., & Fu, X. (2000). Agricultural work-related injuries among farmers in Hubei, People's Republic of China. *American Journal of Public Health*, *90*(8), 1269–1276. Retrieved from http://www.ncbi.nlm.nih.gov/pubmed/10937008

Yang, Y., & Asan, O. (2016). Designing patient-facing health information technologies for the outpatient settings: A literature review. *Journal of Innovation in Health Informatics*, *23*(1), 441–449. doi:10.14236/jhi.v23i1.185 PMID:27348487

Yeung, A., Ulrich, D., Nason, S., & VonGlinow, M. (2000). *Las capacidades de aprendizaje en la organización.* Oxford University Press.

Yin, R. K. (1981). The Case study Crisis: Some Answers. *Administrative Science Quarterly*, *26*(1), 58–65. doi:10.2307/2392599

Yin, R. K. (1994). *Case Study Research: Design and Methods* (2nd ed.). Thousand Oaks, CA: Sage Publications.

Young, M. S., Brookhuis, K. A., Wickens, C. D., & Hancock, P. A. (2015). State of science: Mental workload in ergonomics. *Ergonomics*, *58*(1), 1–17. doi:10.1080/00140139.2014.956151 PMID:25442818

Young, S., Tu, Y. M., & Tseng, Y. T. (1999). Organizational Learning as a feedback system: a Conceptual Framework. *17th International Conference of The System Dynamics Society and the 5th Australian & New Zealand Systems Conference.*

Yukl, G. (2002). *The Nature of Leadership* (5th ed.). London: Prentice-Hall International Inc.

Zambrano, I. I. B., Véliz, V. M. B., & Barzola, W. J. F. (2017). Factores del clima laboral que influyen en el rendimiento de los trabajadores del sector público en el Ecuador. *Dominio de las Ciencias*, *3*(3), 917–937.

Zhao, D., McCoy, A. P., Kleiner, B. M., Smith-Jackson, T. L., & Liu, G. (2016). Sociotechnical systems of fatal electrical injuries in the construction industry. *Journal of Construction Engineering and Management*, *142*(1), 04015056. doi:10.1061/(ASCE)CO.1943-7862.0001036

Zhao, X., Hsu, C.-Y., Chang, P.-C., & Li, L. (2016). A genetic algorithm for the multi-objective optimization of mixed-model assembly line based on the mental workload. *Engineering Applications of Artificial Intelligence*, *47*, 140–146. doi:10.1016/j.engappai.2015.03.005

Zhou, C., & Roseman, J. M. (1994). Agricultural injuries among a population-based sample of farm operators in Alabama. *American Journal of Industrial Medicine*, *25*(3), 385–402. doi:10.1002/ajim.4700250307 PMID:8160657

Zink, K. J. (2000). Ergonomics in the past and the future: From a german perspective to an international one. *Ergonomics*, *43*(7), 920–930. doi:10.1080/001401300409116 PMID:10929827

Zink, K. J. (2014). Designing sustainable work systems: The need for a systems approach. *Applied Ergonomics*, *45*(1), 126–132. doi:10.1016/j.apergo.2013.03.023 PMID:23608710

About the Contributors

Arturo Realyvásquez has studied Ergonomics since 2008. He contributed to this science developing an expert system to measure ergonomic compatibility of advanced manufacturing technology. Most recently, he developed the Macroergonomic Compatibility Questionnaire (MCQ), he proved that macroergonomic compatibility has positive effects on work systems performance, and he proposed the macroergonomic compatibility index (MCI). Currently, he works as full-time professor in Instituto Tecnológico de Tijuana. He offers the classes of Ergonomics, Operations Research, and Research Fundamentals.

Aide Maldonado-Macías works as a professor-Investigator for the Autonomous University of Juarez Mexico. Dr. Maldonado has published several papers in various international journals and conference proceedings. She has been a member of the programme committee and scientific committee of several national and international conferences, in most cases also serving as referee. She is a certified professional in ergonomics in Mexico and she serves as Member of the Mexican Society on Ergonomics and the Society of Industrial Engineers. She also serves as Associate Editor of the AcademiaJournals.com journals and the Occupational Ergonomics Advances and Applications. She serves as referee for the International Journal of Advanced Manufacturing Technology and the World Conference of Industrial Engineering. She has also been evaluator and reviewer for research projects in Mexico. She has obtained the recognition with her research the best investigation in Industrial Engineering in Mexico for the Tecnological Institute of Juarez México and has participated in the most prestigious International Conferences and Congresses on Ergonomics and Computers and Industrial Engineering. She has publications on indexed journals in her country and the International Journal of Advanced Manufacturing Technology and the International Journal of Industrial Engineering. Also has contributions in book chapters for prestigious editorials.

Karina Cecilia Arredondo-Soto studied Industrial Engineer at the Tijuana Institute of Technology where she continued studying a Master Degree in Industrial Engineering too; with a minor in Manufacturing. She has a Ph.D. in Industrial Engineering at the Ciudad Juarez Institute of Technology. She is a full-time teacher at the Autonomous University of Baja California since 2012 and an invited Professor at the Tijuana Institute of Technology since 2008. Her interest area is work studies to improve productivity in organizations through the collaborative work with the industrial regional sector.

* * *

Liliana Avelar-Sosa has a Master Degree in Industrial Engineering by the Technological Institute of Ciudad Juárez and a Doctorate Degree in Engineering Sciences by the Autonomous University of Ciudad Juárez.

Daniel Azpeitia is a Business Administration PhD from UNAM.

Yolanda Báez López is a professor of industrial engineering at the School of Engineering, Architecture and Design. She is a member of the National System of Researchers of the National Council of Science and Technology in Mexico. Her research interests include topics related to lean manufacturing, modeling of production processes, statistical inference and human reliability. She is president of the Academy of Industrial Engineering in this faculty. She collaborates in research groups in Colombia, Spain and Brazil. Dr. Báez has an international patent. She has participated in several research projects related to Lean Manufacturing, Six Sigma and Design of Experiments and is the author / co-author of more than 25 works of jounal, book chapters and conference documents. She has been a reviewer of scientific articles in journals indexed in Journal Citation Reports. Dr. Báez has been an evaluator of Innovation Projects and Graduate Programs. She is a member of the Academic Quality and Academic Productivity Group, recognized in Mexico as Consolidated.

Cesar Balderrama Armendariz is a professor-Researcher in the Institute of Architecture, Design and Arts. Head of prototype laboratory in Industrial design Programs: Undergraduate Program of Industrial Design, Master in Design and Product Development, Master in Studies and Creative Processes in Art and Design

Manuel Barajas has a Master Degree in Industrial Engineering by the Technological Institute of Ciudad Juárez and is a PhD Student in the Autonomous University of Ciudad Juárez.

Jorge Alberto Cazarez is a graduate of the Software Engineering degree program at the Enrique Díaz de León University. He has five years of experience developing applications under the object-oriented paradigm. Expert in technologies such as Java, C #, UML, web and mobile development, databases and hybrid programming. He is a consultant for the firms Stratia consultants, active laboratories and for the Autonomous University of Guerrero. National workshop developer in development for Android and Java. Currently working in the Development Area of the University of Guadalajara, in financial management systems.

Edgar Cossio received his PhD in computer systems. Master in Software Engineering from the Universidad del Valle de Atemajac (UNIVA) Guadalajara in 2011. His research interest is the parallel computing, software engineering, bio-inspired algorithms. He is part time professor in the Enrique Díaz de León University.

Jose Flores Figueroa is a professor-researcher with a PhD in Literature and a master's degree in arts in visual arts. He has published 11 books and more than 15 papers.

Yolanda Frausto is a teacher in the industrial engineering program, Education master degree

Gabriela Jacobo Galicia, has a master's degree with emphasis in quality and production systems. She has professional experience in organizational production system improvements and working in sociothecnical systems focus in psychosocial risks

Jorge García-Alcaráz has a Master Degree in Industrial Engineering by the Technological Institute of Colima and a Doctorate Degree in Industrial Engineering by the Technological Institute of Ciudad Juárez, a Doctorate in Innovation in Engineering, Product and Industrial Process.

Murray Gibson is Founder & Ergonomics Consultant of Saturn Ergonomics Consulting in Auburn, AL. Murray has over 20 years experience applying ergonomics at the plant & corporate-level in industry, and as a consultant in over 150 companies. Murray holds a bachelor's and master's degree in Industrial & Systems Engineering from Auburn University. He is a licensed PE (Professional Engineer) and a CPE (Certified Professional Ergonomist). Saturn Ergonomics, founded in 2015, provides ergonomics consulting services and develops innovative ergonomics technology solutions.

Juan Hernández-Arellano is a researcher and a full-time teacher of the Design Department at the Autonomous University of Ciudad Juarez, Mexico. His research interests are about industrial ergonomics, ergonomic product design, statistics applied to research ergonomics, and biomechanics. Dr. Hernandez have published numerous articles, books and books chapters in indexed Journals and Internationals books. Currently, Dr. Hernandez is the coordinator of the Ergonomic Product Design Lab at the Autonomous University of Ciudad Juarez.

Guadalupe Hernandez-Escobedo received a PhD in Information Behaviour and Cognitive Engineering at the University of Leeds, UK and has more than 20 years of experience in teaching, research and application of Industrial Engineering.

Xhimi Hysa is the Head of Business Administration Department at Epoka University. He has a PhD in Organizational Behavior with a thesis on group dynamics and systems thinking from Sapienza University of Rome, where previously he has worked as course instructor. His research appears on journals such as Systems Research and Behavioral Science, Journal of Organisational Transformation & Social Change, The TQM Journal, etc. Dr. Hysa is also an organizational consultant, and among all his contribution appears in international organizations such as United Nations, where Dr. Hysa has pioneered on behalf of UNDP a Behavioral Change Model (BCM) for "Changing Attitudes and Behaviors of Public Servants with Regards to Roma and Egyptians' Access to Public Services". Dr. Hysa is also passionate about the emerging field of Social Business and Sustainable Development, and for this purpose he has founded inside the business department the "Epoka University Cluster on Social Business and Sustainable Systems".

Judith Lara Reyes has an M.C in industrial engineering and master's degree in instruction specialist in mathematics education.

Jorge Limón Romero is a professor of industrial engineering in the Faculty of Engineering, Architecture and Design at the Autonomous University of Baja California, Mexico. He has been President of the Industrial Engineering Academy at this faculty and he has also participated in several research projects.

Dr. Limón is a member of the National System of Researchers of the National Council of Science and Technology in Mexico. His research interests include topics related to Six Sigma Methodology, Lean Manufacturing and applied multivariate and univariate statistics for process improvement.

Marisela Lucero Gaytán is a teacher in the industrial engineering program.

Sharon Macías Velásquez is a PhD student at the Universidad Autónoma Baja California working on the ergonomic and occupational health research line (Actually). Master's degree in industrial engineering from Instituto Tecnológico of Saltillo (2013-2015) with line of research on distribution and supply of systems. Exhibitor at international conference semac 2018 with the presentation "Comparison of the techniques to measure the Burnout in the average and superior controls of the Manufacturing Industry."

Luis Maldonado has a Business Administration PhD from UACJ.

Ismael Mendoza Muñoz has a master's degree in mechanical engineering and design. His research interests are product design to manufacture, material choosing and process route is an important role in industry.

Debesh Mishra is a research scholar of KIIT. He has published many journals.

Francisco Molins-Correa is a Graduate in Psychology from Universitat de València.

Mildrend Ivett Montoya Reyes, has a master's degree in engineering methods and work study, also with master degree in education with emphasis in organizational development.

Beata Mrugalska received the M.Sc. Eng. in Corporate Management in 2001, the Ph.D. degree in Machines Building and Operations in 2008, all of them from the Poznan University of Technology, Poland. Since 2009, she has been working in the Chair of Ergonomics and Quality Engineering at the Faculty of Engineering Management in the Poznan University of Technology. She is a Centre for Registration of European Ergonomists (CREE) Council representative as a member of the Polish National Assessment Board. Her research interests include: robust design and control, parameter optimization techniques, system faults and failures, quality management, quality control, machinery safety, process safety and reliability. ORCID: 0000-0001-9827-9449.

Carlos Raul Navarro Gonzalez, part of the "Universidad Autónoma de Baja California" as a staff member of the Industrial Engineering area, as university professor. With many papers related to "Lean Manufacturing", "Six Sigma" and "process optimization"; also with research in "ergonomics and occupational health" as a novelty area in this university. He's grades are Industrial Engineering, with a Master Degree in Engineering with Industrial Process and PhD in Engineering since 2008. As 2014 has been acting as full-time research professor in this University.

Toivo Niskanen, PhD (Eng.), received Doctor of Technology in 1993 (September 21th) in the Technical University of Helsinki (nowadays Aalto University) in Finland. Niskanen has prepared the Finnish implementations of the EU directives in safety and health at work in the Finnish Ministry of Social Affairs and Health. Published articles are in the following journals: Safety Science, Applied Ergonomics, Journal of Loss Prevention in the Process Industries, International Journal of Occupational Safety and Ergonomics (JOSE), Business Ethics: A European Review. Niskanen acted as an editor and chapters' writer in the book 'Integrated Occupational Safety and Health Management: Solutions and Industrial Cases' published by Springer International Publishing.

Alberto Ochoa (BS 1994–Eng. Master 2000, PhD 2004 and Postdoctoral Researcher 2006 & Industrial Postdoctoral Research 2008) participated in the organization of different international congress and workshops like HAIS, HIS, ENC, and MICAI. His research interests include Evolutionary Computation, Natural Processing Language and Social Data mining, he is part time professor at the Social Science department at Juarez City University. He is member of the National System of Researchers Level 2 in Mexico (SNI).

Gilberto Rivera received the M.Sc degree in Computer Science (in 2010) and the PhD degree in Computer Science (in 2015) from the Tecnológico Nacional de México (Mexican National Institute of Technology). His doctoral research was recognised in 2015 as the best one of the PhD projects presented to the Mexican Society of Artificial Intelligence. He is currently a full-time professor at the Autonomous University of Juarez City. His primary interests are in the areas of multi-criteria optimisation, applied soft computing, logistics and swarm intelligence.

Manuel Rodriguez is a research Professor in the Division of Research and Graduate Studies at ITCJ.

Edgardo Rojas is an English teacher in the Language Center at ITCJ.

Suchismita Satapathy is currently working as an Associate Professor in the School of Mechanical Sciences, KIIT University. She has published more than 20 articles, many national and international journals and conferences. She has more than ten years of teaching and research experience.

Miguel Serrano is a full-time professor and researcher by the Universidad de Valencia, Spain.

Tetiana Shmelova is a professor of Air Navigation Systems Department in National Aviation University (CV in the appendix). Doctor of Science. Title of thesis "Scientific and Methodological Basis of Decision-Making in the Air Navigation System" Teaching courses: Theory of decision-making, Air Navigation, Operations Research, Mathematical Programming, Informatics of Decision Making, Effectiveness of Air Traffic Management Areas of Scientific Interests: Mathematical Models of Decision-Making by human operator, especially in Emergency Situation; Decision Support System; Research of Air-Navigation System as Socio-Technical System; Methodology of scientific research; Aviation Sociometry and Socionics. Author of more than 250 scientific articles and guides (about 190 articles is in Ukrainian and Russian, 30 articles in English, 14 methodical manuals, 12 copyright certificates for computer programs), 4 monographs in fields of aviation, economics, mathematics, theory of system. Has got two children, daughter and son, and a husband.

Yuliya Sikirda is an Associate Professor of Management, Economy and Law Department; Vice-Dean of Management Faculty at the Kirovograd Flight Academy of National Aviation University. She has graduated from the Kirovograd Flight Academy of National Aviation University (former State Flight Academy of Ukraine) in 2001, received a Master degree on speciality "Air Traffic Service" specialization "Air Traffic Controller". (2001). She has got a diploma of PhD in Technical Sciences on speciality "Automated Control Systems and Advanced Information Technologies" (2004). PhD thesis "Modelling of Decision Support System for an Air Traffic Controller in Flight Emergency Situations", thesis has defended in Kyiv Automation Institute. In 2006 she has received an attestat of Associate Professor of Management and Economy Department. Since 2014 she is the member of the specialized scientific council for the defence of dissertations for PhD in Technical Sciences on the speciality "Navigation and Traffic Control". Together with the scientific work she has actively engaged in the methodological provision of the educational process, during the years of labour activity 70 teaching and methodological works have issued. Areas of scientific interests. The increasing of the decision-making efficiency by the human-operator of the Air Navigation System in flight emergency situations; the assessment of the influence of aviation enterprises' internal and external management environment factors on the activity of the Socio-technical system operators'. Teaching courses. "Introduction to the Speciality", "Transport and Transport Infrastructure", "Information Systems in Management", "Methodology of Scientific Researches", "Mathematical Modelling of Professional Tasks". The author and co-author of 160 scientific works, of which two monographs, 41 articles in specialized scientific publications (including three foreign and 18 included in international science-computer databases), 117 abstracts to conferences, three copyright certificates on computer programs.

Ludovico Soto Nogueira is a research professor.

Luz Tarango received her Bachelor's degree in industrial engineering and her M.SC. degree in administrative sciences from the Ciudad Juarez Institute of Technology (ITCJ). She worked as an industrial processes engineer and production supervisor in several local industries. Was head of the Technologic Management and Linking department for six and a half years at the ITCJ, while formerly being the coordinator of the Engineering Graduate programs for eight years in the same institute. She coordinated the Gender Equality Model and achieved the certification in the same area for the ITCJ in 2013. She has been a Professor at the ITCJ for twenty-five years, currently a part of the graduate studies and research faculty, and head of the Administrative Engineering council.

Diego Tlapa is a professor of industrial engineering in the School of Engineering, Architecture and Design at the Autonomous University of Baja California, Mexico. He is member of the National System of Researchers of the National Council of Science and Technology in Mexico and member of the optimization of industrial processes network (ROPRIN); his research focuses on supply chain management, and processes improvement projects including six sigma and lean manufacturing. He has participated in several research projects related to process improvement and is author/coauthor of more than 30 journal papers, book chapters and conference papers.

Index

H

I

J

K

L

M

O

P

Q

R

S

T

V

W

www.ingramcontent.com/pod-product-compliance
Lightning Source LLC
LaVergne TN
LVHW060418300926
849162LV00006B/1465